STERILISATION OF TISSUES USING IONISING RADIATIONS

Editors

JOHN F. KENNEDY **BSc, PhD, DSc, CEng, CSci, EurChem CChem FRSC, EurProBiol CBiol FIBiol, FInstMC, CEnv CIWEM, CText FTI, FCMI, FIFST**

Director of Birmingham Carbohydrate and Protein Technology Group,
School of Chemical Sciences, The University of Birmingham & The University of Birmingham Research Park, Birmingham B15 2TT, England, UK,

Director of Chembiotech Ltd,
University of Birmingham Research Park, Birmingham B15 2SQ, England, UK,

Director of Inovamed Ltd,
Chembiotech Laboratories, University of Birmingham Research Park, Vincent Drive, Birmingham B15 2SQ, England, UK, and

Professor of Applied Chemistry,
The North East Wales Institute of Higher Education, Plas Coch, Mold Road, Wrexham, Clwyd, LL11 2AW, Wales, UK.

GLYN O. PHILLIPS **BSc, PhD, DSc, HonDSc, HonLlB, CChem FRSC**

Chairman of Research Transfer Ltd,
2 Plymouth Drive, Radyr, Cardiff, CF15 8 BL, Wales, UK,

Chairman of Phillips Hydrocolloid Research Ltd,
45 Old Bond Street, London, W1S 4AQ, England, UK, and

Professorial Fellow, and Chairman, Phillips Hydrocolloids Research Centre,
The North East Wales Institute of Higher Education, Plas Coch, Mold Road, Wrexham, Clwyd, LL11 2AW, Wales, UK.

PETER A. WILLIAMS **BSc, PhD, CChem FRSC**

Director of the Centre for Water Soluble Polymers,
The North East Wales Institute of Higher Education, Plas Coch, Mold Road, Wrexham, Clwyd, LL11 2AW, Wales, UK,

Director of the Centre for Advanced and Renewable Materials at the North East Wales Institute and University of Wales, Bangor,
The North East Wales Institute of Higher Education, Plas Coch, Mold Road, Wrexham, Clwyd, LL11 2AW, Wales, UK, and

Professor of Polymer and Colloid Chemistry,
The North East Wales Institute of Higher Education, Plas Coch, Mold Road, Wrexham, Clwyd, LL11 2AW, Wales, UK.

CRC Press
Boca Raton Boston New York Washington, DC

WOODHEAD PUBLISHING LIMITED
Cambridge England

Published by Woodhead Publishing Limited, Abington Hall, Abington
Cambridge CB1 6AH, England
www.woodheadpublishing.com

Published in North America by CRC Press LLC, 2000 Corporate Blvd, NW
Boca Raton FL 33431, USA

First published 2005, Woodhead Publishing Ltd and CRC Press LLC

British Library Cataloguing in Publication Data
A catalogue record for this book is available from the British Library.

Library of Congress Cataloging in Publication Data
A catalog record for this book is available from the Library of Congress.

Woodhead Publishing ISBN 1 85573 838 4
CRC Press ISBN 0-8493-3797-6
CRC Press order number: WP3797

Printed by Antony Rowe Limited, Chippenham, Wilts, England

CONTENTS

Preface

G. O. Phillips ix

PART 1: STANDARDS FOR TISSUE BANKS AND FOR RADIATION STERILISATION OF TISSUES . 1

1. **IAEA international standards for tissue banks**

   B. Loty 3

2. **The development of a code of practice for the radiation sterilisation of tissue allografts**

   B. J. Parsons, E. Kairiyama and G. O. Phillips . . 39

3. **Use of the IAEA code of practice for the radiation sterilisation of bone allografts**

   E. Castro Gamero and K. L. Palomino . . 65

PART 2: METHODOLOGY IN THE STERILISATION AND PRESERVATION OF TISSUES 77

4. **Radiation sources: types and suitability for dose delivery to tissues for sterilisation**

   J. T. M. Jansen, F. W. Schultz and J. Zoetelief . . 79

5. **Improved method for gamma irradiation of donor tissue**

   R. Garcia, A. Harris, M. Winters, B. Howard, P. Mellor, D. Patil and J. Meiner 105

6. **Rapid heat transfer dynamics and cold gamma sterilisation methods for soft tissue allografts**

   M. Hayzlett, S. Griffey and G. Greenleaf. . 117

7. Comparison of different thawing methods on cryopreserved rabbit aorta

S. B. Sim, Y. M. Oh and S. H. Lee . . . 123

PART 3: EFFECTS OF RADIATION ON BONE, TISSUES, AND THEIR COMPONENTS 131

8. Effects of gamma irradiation on bone – clinical experience

W. W. Tomford . . . 133

9. Effects of gamma irradiation on the mechanical properties of human cortical allograft bone

M. H. Zheng, R. A. Power, J. N. Openshaw, R. I. Price, R. E. Day, J. Winter, A. Cowie and D. J. Wood 141

10. The effect of cold gamma radiation sterilisation on the properties of demineralised bone matrix

A. A. Gertzman, M. H. Sunwoo, D. Raushi and M. Dunn . . 151

11. Complications of structural allografts for malignant bone tumours

Y.-K. Kang, J.-Y. Jeong, Y.-G. Chung, W.-J. Bahk and S.-K. Rhee . 157

12. Effects of radiation on the integrity and functionality of soft tissue. Current situation: cartilage, heart valves, tendons and other tissues. Changes with increasing dose/dose limits

D. M. Strong . . 163

13. The effect of preservation procedures and radiation sterilisation conditions on connective tissue grafts and their constituents

A. Dziedzic-Goclawska, A. Kaminski, I. Uhrynowska-Tyszkiewicz, J. Michalik and W. Stachowicz 173

14. Effects of radiation on the integrity and functionality of amnion and skin grafts

J. Koller . . 197

15. Structural effects of radiation sterilisation on sodium hyaluronate

J. F. Kennedy, M. P. C. da Silva, L. L. Lloyd and C. J. Knill . 221

PART 4: VIRAL ASPECTS OF TISSUES FOR TRANSPLANTATION 233

16. Viruses and their relevance for gamma irradiation sterilisation of allogenic tissue transplants
A. Pruss, R. von Versen and G. Pauli . 235

17. Viral infections transmitted through tissue transplantation
T. Eastlund . . . 255

PART 5: MICROBIOLOGICAL ASPECTS OF TISSUES FOR TRANSPLANTATION 279

18. Bacterial inactivation in tissues
M. Winters and J. Nelson . 281

19. Novel pathogen inactivation of soft tissue allografts using optimised gamma irradiation
T. A. Grieb, R.-Y. Forng, J. Lin, Ll. Wolfinbarger, J. Sosa-Melgarej, C. Sharp, W. N. Drohan and W. H. Burgess 287

20. Establishing an appropriate terminal sterilisation dose based upon post-processing bioburden levels on allograft tissue
C. J. Ronholdt, S. Bogdansky and T. F. Baker . . 303

21. Determination of microbial bioburden levels on pre-processed allograft tissues
C. J. Ronholdt and S. Bogdansky . . 311

22. Bioburden estimation in relation to tissue product quality and radiation dose validation
N. Yusof, A. R. Shamsudin, H. Mohamad, A. Hassan, A. C. Yong and M. N. F. A Rahman 319

23. Protective effects on microorganisms in radiation sterilised tissues
J. H. Hendry . . 331

Index . 339

THE CELLUCON TRUST

Incorporating

Cellucon Conferences

International Educational Scientific Meetings
on Carbohydrate Polymers and their Parent Matrices

Cellucon Conferences as an organisation was initiated in 1982, and Cellucon '84, which was the original conference, set out to establish the strength of British expertise in the international field of cellulose and its derivatives. This laid the foundation for subsequent conferences on carbohydrate etc. polymer topics in Wales (1986), Japan (1988), Wales (1989), Czechoslovakia (1990), USA (1991), Wales (1992), Sweden (1993), Wales (1994), Finland (1998), Japan (1999), and Wales (2000 & 2003). These conferences have had truly international audiences drawn from the major industries involved in the production and use of cellulose pulp and fibre derivatives of cellulose, plus representatives of academic institutions and government research centres. This diverse audience has allowed the cross-fertilisation of many ideas, which has done much to give the field of cellulose in its diverse forms the higher profile that it rightly deserves. More recently other carbohydrate polymers have been the centre of focus, particularly hyaluronan, with the conference in 2000 - *Hyaluronan 2000* - being the first major international conference on this majorly important carbohydrate polymer. Studies of hyaluronan, a major tissue component, lead into the stability and sterilisation of the human tissues in which this polymer is a key component and so with the help of the International Atomic Energy Authority

Cellucon Conferences are organised by ***The Cellucon Trust***, an official UK Charitable Trust with world-wide objectives in education in wood and cellulosics. The Cellucon Trust is continuing to extend the knowledge of all aspects of cellulose, lignin, hyaluronan and other national polymers world-wide. At least one book has been published from each Cellucon Conference as the proceedings thereof. This volume arises from the 2003 conference held in Wrexham, Wales. Further conferences are intended and in turn these will generate further useful books in this area.

The Cellucon Trust is a registered charity, UK Registration No: 328582 and a company limited by guarantee, UK Registration No: 2483804 with its registered offices at Chembiotech Laboratories, The University of Birmingham Research Park, Vincent Drive, Birmingham, B15 2SQ, UK.

THE CELLUCON CONFERENCES

1984 Cellucon '84 UK	CELLULOSE AND ITS DERIVATIVES Chemistry, Biochemistry and Applications
1986 Cellucon '86 UK	WOOD AND CELLULOSICS Industrial Technology, Biotechnology, Structure and Properties
1988 Cellucon '88 Japan	CELLULOSICS AND WOOD Fundamentals and Applications
1989 Cellucon '89 UK	CELLULOSE: SOURCES AND EXPLOITATION Industrial Utilisation, Biotechnology and Physico-Chemical Properties
1990 Cellucon '90 Czechoslovakia	CELLULOSE New Trends in the Complex Utilisation of Ligno-cellulosics (Phytomass)
1991 Cellucon '91 USA	CELLULOSE A Joint Meeting of: ACS Cellulose, Paper and Textile Division, The Cellucon Trust, and 11th Syracuse Cellulose Conference
1992 Cellucon '93 UK	SELECTIVE PURIFICATION AND SEPERATION PROCESSES
1993 Cellucon '93 Sweden	CELLULOSE AND CELLULOSE DERIVATIVES Physico-Chemical Aspects and Industrial Applications
1994 Cellucon '94 UK	CHEMISTRY AND PROCESSING OF WOOD AND PLANT FIBROUS MATERIALS The Chemistry and Processing of Wood and Plant Fibrous Materials
1998 Cellucon '98 Finland	PULP AND PAPER MAKING Fibre and Surface Properties and other Aspects of Cellulose Technology
1999 Cellucon '99 Japan	RECENT ADVANCES IN ENVIRONMENTALLY COMPATABLE POLYERS
2000 Hyaluronan 2000 UK	HYALURONAN 2000 An International Meeting Celebrating the 80th Birthday of Endre A Balazs
2003 Tissue Sterilisation 2003 UK	STERILISATION OF TISSUES USING IONISING RADIATIONS 2003

The proceedings of each conference were formerly published by Ellis Horwood, Simon and Schuster International Group, Prentice Hall, Campus 400, Maylands Avenue, Hemel Hempstead, Herts, HP2 7EZ, UK and from 1993 are published by Woodhead Publishing Limited, Abington Hall, Abington, Cambridge CB1 6AH, UK.

The 13th International Cellucon Conference

STERILISATION OF TISSUES USING IONISING RADIATIONS

ACKNOWLEDGEMENTS

This book arises from the International Conference – *Sterilisation of Tissues Using Ionising Radiations* – which was held at The North East Wales Institute of Higher Education, Wrexham, Wales, UK. The Conference was organised in association with the International Atomic Energy Agency and the funding from this organisation is gratefully acknowledged. This meeting also owes its success to the invaluable work of its Executive Committee. The editors gratefully acknowledge the input from Mr Haydn Hughes for conference co-ordination, and from Mrs Patricia A. Johnston and Dr Charles J. Knill in the production of this book.

EXECUTIVE COMMITTEE

PREFACE

Existing methods and processing for sterilising tissues are proving, in many instances, inadequate. Infections have been transmitted from the graft to the recipient and in the USA, the Centre for Disease Control and other regulatory bodies, have drawn attention to the need for a reliable end sterilisation method which does not damage the functionality of the final tissue. Safety of surgical allografts is, therefore, a major concern, due to microbial and viral contamination of tissues, which is now a problem, even in the most sophisticated centres.

The US Food and Development Agency has now reacted to the infections that have been transmitted via allograft tissues and stressed also their requirement for an end sterilisation method, which can provide greater safety for the graft in its final packaging. The proposed European Directive is also concerned mainly with tissue grafts safety, and their requirements will again focus on the degree of sterility assurance attained for the final product

To address this concern, the Presidents of the main Professional Associations of Tissue Banks: American, European, and Latin American met in Vienna to review the situation and concluded that the time was opportune to organise an international high level expert meeting, which would identify the best method of using radiation technology to assist in the production of safe tissue allografts. This is technically the most reliable method of delivering an end-sterilisation step when the graft is in its final packaging.

This volume provides the information on this subject presented at this international meeting in Wales, which was supported by the International Atomic Energy Agency. Experts throughout the world contributed. New methods of protecting the tissues were presented, which at the same time allow the use of sufficiently high doses of ionising radiations to inactivate the invading organisms. A Code of Practice for the Radiation Sterilisation of Tissues was evaluated and the outcome and the full Code are published in this volume, which also covers all the methodologies used in the field.

No such volume now exists and so tissue bankers, users of tissues for transplantation, and the regulators who oversee safely, will find this reference book invaluable for accessing information not readily available elsewhere.

Glyn O. Phillips

Conference Organiser & Technical Adviser
to the International Atomic Energy Agency

PART 1

STANDARDS FOR TISSUE BANKS AND FOR RADIATION STERILISATION OF TISSUES

IAEA INTERNATIONAL STANDARDS FOR TISSUE BANKS

Bernard Loty

Établissement Français des Greffes (French Transplantation Agency)
5 Rue Lacuée F 75012, Paris, France

ABSTRACT

The IAEA aimed at developing complete Standards for Tissue Banks, including all steps of the activity, and being applicable in all parts of the world. The IAEA draft International Standards for Tissue Banks are presented.

INTRODUCTION

Many standards for Tissue Banking have been published over the years, by national or international Scientific Associations (e.g. AATB, EATB...) and by national governments or intergovernmental organisations (e.g. Council of Europe, WHO...). The IAEA prepared a Code of Practice for the radiation sterilisation of biological tissues but aimed at developing complete Standards including all steps of the activity and applicability in all parts of the world. Radiation sterilisation of Tissues can only be successfully achieved when the entire tissue banking activity is carried out with respect of complete control of the activity. Experts from all continents developed IAEA International Standards. Workshops were organised in Vienna in 7/2001 and Seoul in 5/2002 and took into account the existing documents already published on the matter, trying not to reinvent the wheel. They described the ethical and legal Rules, the organisational aspects of a Tissue Bank and the processing methods. They also included a 'Guide for Legal and Regulatory Control' intended to assist Governmental Control Authorities and Tissue Banks in their joint task of improving the quality of human tissues for transplantation through Regulation and Legislation that interfaces with Standards. The IAEA will encourage all Tissue Banks participating in the IAEA Radiation and Tissue Banking Programme to apply these Standards, in accordance with their national conditions, with the purpose of ensuring the safe clinical use of the tissues produced. The goal of this work is to ensure that the Tissue Banks have validated their systems, demonstrating their compliance with the Standards.

INTRODUCTION

International Standards for Tissue Banks

Standards have been established by the IAEA that should be used as a starting point for Good Tissue Banking Practices. These Standards describe the safety and quality dimensions of human tissue for transplantation, Quality Management, processing methods, tissue sterilisation and validation. These Standards apply to all types of tissues including corneas and to cells (see Definitions).

Guide for Legal and Regulatory Control

In order for a Tissue Banking Programme to be successfully implemented, there is need for a variety of Laws and Regulations to be legislated and enforced. These Laws and Regulations should cover the safety of the tissue to recipients as well as ethical concerns such as maintaining the dignity of the donor and his/her family and respect the gratuity of the donation. Regulations should be based on Standards adopted by the country, individual Tissue Banks or associations representing Tissue Banks in the specific country/area. An International or Intergovernmental approach to the development of Laws and Regulations is suggested for those areas of the world that have common legal systems, eliminating redundant or conflicting Regulations. In the International Atomic Energy Agency's view, this Guide for Legal and Regulatory Control shall present requirements in a form that can be used for establishing Tissue Banks and determining whether a Tissue Bank complies with current Good Tissue Banking Practices. It shall also serve as an aid for interpreting and clarifying the Standards. It is also intended to support the harmonisation of inspection and internal audit procedures. The reasons for the justification of this Guide for Legal and Regulatory Controls are clear: there is need to protect the health and well-being of the citizens, encourage cost-effective and improved healthcare, promote social programmes that work for the well being of the community, prohibit unethical practices, avoid health hazards associated with the distribution and transplantation of tissues and to protect against tissue banks that refuse to adhere to acceptable practices.

The document is divided into two parts:

Part 1: International Standards for Tissue Banks

The Standards include two sections. Section A deals with general and organisational policies. Section B deals with the implementation of these policies.

Part 2: Guide for Legal and Regulatory Control

Part 2 includes a Guide which advises regulatory bodies about the aspects which must be considered in setting up a regulatory system and evaluating compliance with the system.

TABLE OF CONTENTS:

Part 1: INTERNATIONAL STANDARDS FOR TISSUE BANKS

SECTION A: General and Organisational Policies

A 1.000 **Introduction**
A 1.100 General
A 1.110 Scope
A 1.120 Purpose of the Document
A 1.130 Concerns
A 1.200 Definitions: see Annex 1
A 2.000 **Ethical and Legal Rules**
A 2.100 General
A 2.200 Permission for Tissue Retrieval
A 2.210 Living Donor Consent
A 2.212 Collection of Surgical Residues
A 2.220 Non-Living Donor Consent
A 2.230 Consent Documentation
A 2.300 Monetary Inducement for Donation
A 2.310 Prohibition of Payment to Donor
A 2.320 Compensation for Donation-Related Expenses
A 2.400 Anonymity
A 3.000 **Organisation of a Tissue Bank**
A 3.100 Institutional Identity

A 3.110 General
A 3.120 Authorisation, Licensing and Registration
A 3 130 Collaboration with other organisations
A 3.200 Personnel
A 3.210 Medical Director
A 3.220 Administrative Director
A 3.230 Staff
A 3.240 Training
A 3.300 Quality Management System
A 3.310 Quality Requirements
A 3.320 Quality Management
A 3.330 The Basic Elements of an Appropriate Quality Management System
A 3.340 Methods for Detecting, Correcting and Preventing Quality Failures from Recurring
A 3.350 Competency
A 3.400 Facilities and Equipment
A 3.410 General
A 3.420 Design
A 3.430 Security
A 3.440 Environmental Monitoring
A 3.450 Sanitation
A 3.460 Equipment
A 3.470 Environmental Safety

SECTION B: Implementation

B 1.0 00 Donor Selection
B 1.1 00 General
B 1.2 00 Medical and Behavioural History
B 1.2 10 Donor History Review
B 1.2 20 Exclusion Criteria
B 1.3 00 Physical Examination
B 1.4 00 Cadaveric Donor Autopsy Report
B 1.5 00 Transmissible Diseases Blood Tests
B 1.5 10 General
B 1.5 20 Blood Tests
B 1.5 30 Exclusion Criteria
B 1.6 00 Bacteriological Studies of Donor and Tissues
B 1.6 10 Bacteriological Testing Methods
B 1.6 20 Bacteriological Bioburden Limits
B 1.7 00 Non Microbiological Tests
B 1.8 00 Age Criteria
B 1.9 00 Cadaver Donor Retrieval Time Limits
B 2.0 00 Tissue Retrieval
B 2.1 00 Rationale
B 2.2 00 Non-Living Donor Tissue Retrieval
B 2.2 10 Determination of Death
B 2.2 20 Donor Identification
B 2.2 30 Retrieval Conditions
B 2.2 40 Body Reconstruction
B 2.3 00 Surgical Residues Collection

B 2.4 00	Living Donor Tissue Retrieval
B 2.5 00	Packaging and Transportation to the Tissue Bank
B 2.5 10	Procurement Container
B 2.5 20	Procurement Container Integrity
B 2.5 30	Procurement Container Label
B 2.6 00	Retrieval Documentation
B 3.0 00	**Tissue Banking General procedures**
B 3.1 00	General
B 3.1 10	Written Procedures
B 3.1 20	Process Validation
B 3.1 30	Quality Controls
B 3.1 40	Records Management
B 3.2 00	Unique Tissue Identification Number
B 3.3 00	Reagents, Container and Packaging
B 3.3 10	Reagents
B 3.3 20	Tissue Container
B 3.3 30	Tissue Outer Package
B 3.4 00	Pooling
B 3.5 00	Environmental Control
B 3.6 00	Storage Conditions
B 3.6 10	Temperature
B 3.6 20	Storage of Quarantined or Unprocessed Tissue
B 3.7 00	Documentation Reviewing and Tissue Inspection
B 3.7 10	Incoming Inspection
B 3.7 20	Review of Donor Eligibility
B 3.7 30	Sizing of Specimens
B 3.7 40	Inspection Prior to Release Into Finished Inventory
B 3.7 50	Final Inspection
B 3.8 00	Non-Conforming Tissues
B 3.9 00	Expiry Dates
B 4.0 00	**Specific Processing Procedures**
B 4.1 00	General
B 4.2 00	Disinfectant or Antibiotic Immersion
B 4.3 00	Fresh Tissue
B 4.4 00	Frozen Tissue
B 4.5 00	Cryopreserved Tissue
B 4.6 00	Freeze-Dried Tissue
B 4.7 00	Simply Dehydrated Tissue
B 4.8 00	Irradiated Tissue
B 4.9 00	Ethylene Oxide Sterilised Tissue
B 4.10 00	Other Processing Methods
B 4.10 10	Other Inactivation Methods
B 4.10 20	Bone Demineralisation
B 5.0 00	**Labelling**
B 5.1 00	General Requirements
B 5.1 10	Rationale
B 5.1 20	Nomenclature
B 5.1 30	Labelling Integrity
B 5.1 40	Visual Inspection
B 5.2 00	Tissue Containers Labelling

B 5.3 00 Package Insert
B 5.3 10 General
B 5.3 20 Accompanying Documentation Requirements
B 5.4 00 Tissue Outer Package Labelling
B 6.0 00 Distribution
B 6.1 00 General
B 6.2 00 Traceability
B 6.3 00 Transportation
B 6.4 00 Accompanying Documentation
B 6.5 00 Return into Inventory
B 6.6 00 Adverse Events
B 6.7 00 Recall
B 6.8 00 Distribution to Storage Facilities Outside the Tissue Bank (depot)
B 6.8 10 General
B 6.8 20 Labelling
B 6.8 30 Storage
B 6.8 40 Records
B 6.9 00 Distribution to Another Tissue Bank
B 6.10 0 Acquisition of Tissue from Another Tissue Bank
B 6.10 10 Medical Director Approval
B 6.10 20 Labelling
B 6.10 30 Distribution Record

ANNEXES

Annex 1: Glossary

Annex 2: Guidelines of Factors to be Considered for Determining Risk for Human Immunodeficiency Virus or B or C Hepatitis

Annex 3: Primary Tumours of the Central Nervous System: Evaluation of a Suitable Donor. Reference List

Annex 4: Example of Algorithm for Calculating the Hemodilution of a Donor Having Received Blood, Blood Components, or Plasma Volume Expanders Within 48 Hours Prior to Death

Annex 5: References and Contact Addresses

Part 1: GUIDE FOR LEGAL AND REGULATORY CONTROL

INTRODUCTION

HISTORICAL PROGRESSION

LAWS AND REGULATIONS

1 - Donation/Transplantation/Recovery/Waiting Lists
2 - Consent
3 - Organisation of the Tissue Bank
4 - Interrelationships with Organ Donation Programmes
5 - Registration/Licensing/Accreditation/Authorisation of the Tissue Bank
6 - Import/Export of Tissue
7 - Financial Aspects of Tissue Banking
8 - Enforcement and Compliance

CONCLUSIONS

Part 2: INTERNATIONAL STANDARDS FOR TISSUE BANKS

SECTION A: General and Organisational Policies

A 1.000 Introduction

A 1.100 General

A 1.110 Scope

These Standards apply to human tissues used for therapeutic purposes, excluding reproductive and genetically modified tissues. They do not apply to animal tissues.

A 1.120 Purpose of the Standards

These standards bring together current State of the Art practice on selection of donors, tissue retrieval, testing, processing, storage, labelling and distribution of finished tissue, in order to provide safe tissue of reliable quality while respecting the ethical rules.

A 1.130 Concerns

The therapeutic use of tissues raises ethical and safety concerns. Safety of tissues includes the following aspects:

- Avoiding transmission of communicable diseases including bacteria, parasites, viruses, prions and of tumours;
- Avoiding adverse events due to additives and residues from chemical or physical methods of processing;
- Preserving efficient biological qualities and assuming reproducibility and traceability.

Besides bacterial and parasitic infection, several cases of viral disease (*) and Creutzfeldt-Jakob (**) disease transmissions have been reported in the literature. These events should be compared with the thousands of patients that have received tissues successfully, but imply the need for preventive measures. Not only the risks, but also the risk-benefit balance has to be considered. Risks include known risks, which imply preventive measures, and unknown risks, which call for precautionary measures. The benefit and the existence or absence of alternative treatments should be appreciated. The factors of clinical safety are well known and include donor selection, retrieval conditions, processing protocol and controls, distribution protocol, traceability and record keeping, including proper indication, surgical technique and postoperative care.

* (2 cases of HIV, three of Hepatitis B and two of Hepatitis C)

** (3 cases through corneas, and > 60 through non-viable freeze-dried dura-mater)

A 1.200 Definitions (See Annex 1)

A 2.000 Ethical and Legal Rules

A 2.100 General

In each country, the applicable Inter-governmental, National, Regional and Local Law or Regulation governing consent and retrieval of tissues from living or cadaver donors should be followed. Recommendations about the Ethical aspects of the use of human tissues for therapeutic purpose have been published by the World Health Organisation (WHA 44.25 - May 1991) and Council of Europe (78-29 May 1978). Council of Europe also adopted a Convention on the Human Rights and Biomedicine (Oviedo, 4 April 1997) and is preparing an Additional Protocol to the Convention on Transplantation of Organs and Tissues of Human Origin. Recommendations about the safety aspects of Tissue Banking were also adopted by the Council of Europe (Recommendation No R (94) 1 on Human Tissue Banks) and by the European Group on Ethics in Science and New Technologies to the European Commission (Opinion on Ethical Aspects of Human Tissue Banking, adopted on 21 July 1998).

A 2.200 Permission for Tissue Retrieval

If there is no applicable Inter-governmental, National, Regional and Local Law or Regulation, the following principles shall be applied:

A 2.210 Living Donor Consent
A 2.211 Voluntary Donation of Tissue

Appropriate medical investigation shall be made to evaluate and reduce the risk to the health of donor and recipient. The donor must be given appropriate information before the removal about the possible consequences of this removal, in particular medical, social and psychological, as well as the importance of the donation for the recipient. An Informed Consent in writing shall be obtained from the living donor. Consent before an official body may be necessary according to applicable Inter-governmental, National, Regional and Local Law or Regulation. In case of a minor or otherwise legally incapacitated person, Informed Consent shall be obtained from his legal representative, if the donor does not object to it. The appropriate authority shall be consulted in accordance to applicable Inter-governmental, National, Regional and Local Law or Regulation. The donation of substances that cannot regenerate is usually confined to transplantation between family related persons and restricted to major and capable persons.

A 2.212 Collection of Surgical Residues

Surgical residues are collected during a surgical procedure where the material is collected for therapeutic purpose other than to obtain tissue (e.g. femoral head, skin and amnion). Informed Consent shall be obtained from the donor according to applicable Regulation.

A 2.220 Non-Living Donor Consent

No removal of tissue will take place when there was an open or presumed objection on the part of the deceased. Permission or confirmation of the absence of objection for tissue donation shall be obtained from the next of kin in case of a minor or legally incapacitated person; the consent of his legal representative is required. Removal of tissue can be effected if it does not interfere with a forensic examination or autopsy as required by Law.

A 2.230 Consent Documentation

Consent for tissue donation shall be documented. The consent form shall specify whether there is a general permission for organs and / or tissues or permission for specified organs and / or tissues only.

A 2.300 Monetary Inducement for Donation
A 2.310 Prohibition of Payment to Donor

Monetary payment or advantages for the donation shall not be made to living donors, cadaver donor's next of kin or any donor-related party.

A 2.320 Compensation for Donation-Related Expenses

Donors or their family shall not be financially responsible for expenses related to retrieval of tissues.

A 2.400 Anonymity

Anonymity between donor and unrelated recipient shall be strictly preserved.

Anonymity between donor and recipient shall allow tracking of tissues, through anonymous identification numbers.

A 3.000 Organisation of a Tissue Bank

A 3.100 Institutional Identity

A 3.110 General

The purpose of a Tissue Bank shall be clearly established and documented. The Tissue Bank shall state whether it is a freestanding entity or part of an Institution.

A 3.120 Authorisation, Licensing, or Registration

The Tissue Bank shall comply with all applicable Inter-governmental, National, Regional and Local Law or Regulation for authorisation, licensing or registration.

A 3.130 Collaboration with other Organisations

A 3.131 Written Agreement – Contract

Each Tissue Bank shall have written agreements or contracts with all other organisations that perform donor screening services, tissue retrieval, processing or distribution for the Tissue Bank. Tissue Banks which contract for laboratory services shall verify the laboratory licensing or accreditation, according to applicable Inter-governmental, National, Regional and Local Law or Regulation.

A 3.132 On-site Audit

Documentation, which is audit-specific for the services performed for the Tissue Bank, shall be maintained by the Tissue Bank. Such documentation shall itemise all operational systems that were audited to determine compliance with Standards or applicable Regulation.

A 3.200 Personnel

A 3.210 Medical Director

A 3.211 Qualification

The Medical Director shall be qualified by training and experiences for the scope of activities being pursued in accordance with applicable Inter-governmental, National, Regional and Local Law or Regulation.

A 3.212 Responsibilities

The Medical Director shall be responsible for medical operations, including compliance with these Standards. His/Her responsibilities include determining what tissues are to be collected, define donor screening policies and prescribe technically acceptable means for their processing, Quality Assurance, storage and distribution. The Medical Director shall be responsible for policies and procedures regarding donor suitability and adverse events.

A 3.213 Medical Advisory Board

It is recommended that a Tissue Bank sets up a Medical Advisory Board to provide medico-technical and scientific advice (external from the Tissue Bank).

A 3.220 Administrative Director

The Administrative Director, when applicable, shall be responsible for administration, management, and other general activities. The Administrative Director shall not be responsible for medical activities.

A 3.230 Staff

A 3.231 General

The Tissue Bank shall have sufficient personnel for pursuing the various tasks.

A 3.232 Qualification

The Tissue Bank staff must possess the educational background, experience and training, sufficient to assure assigned tasks are performed in accordance with the Tissue Banks established procedures.

A 3.233 Responsibilities

The technical staff shall be responsible for implementation of policies and procedures as established by the Medical Director. The duties of each staff member shall be described in a written job description. Staff must demonstrate competency in operations to which they are assigned.

A 3.240 Training

The scope of activities, specific staff responsibilities and reporting structure shall be established by the Medical Director. The Medical Director shall ensure that all staff members have adequate training to perform their duties safely and competently. The Medical Director shall be responsible for ensuring that technical staff maintain their competency by participation in training courses and technical meetings or other educational programmes. All staff shall review applicable institutional policies and procedures annually and when changes are made.

A 3.300 Quality Management System

In order to reduce the risk for patients by the transplantation of tissues to an acceptable level, it is necessary to operate an effective Quality Management System. The System may include extensive testing of donor blood and tissue samples, but this alone is not sufficient guarantee of safety and efficacy and the System should include other management and control measures. Those involved in procuring, processing and supplying tissues for transplantation. In addition, a risk analysis of procedures prone to error to disease transmission should be used to develop safe procedures to implement a Quality Management System based on clearly identified requirements for tissues.

A 3.310 Quality Requirements

The Quality Requirements form the basis of all Quality Assurance and Quality Control Programmes. It is necessary to define the Quality Requirements not only for the final product, but also for the starting material collected, reagents and equipment used, staff competencies, testing techniques, packaging materials, labels and process intermediates. These Quality Requirements are best prescribed and quantified in written specifications. These specifications determine the Quality Control testing or inspection performed on which release decisions are based. Quality Requirements will be based on characteristics that effect both patient safety and maintaining the clinical effectiveness of the product.

A 3.320 Quality Management

It is recognised that quality has to be managed in an organisation and that a systematic approach is the only way to ensure that the quality of products produced and services delivered consistently meets the Quality Requirements. The high level of Quality Assurance required for safety, critical therapeutic medical products and clinical services can only be achieved through the implementation of an effective Quality Management.

The International Standard for Quality Management is the ISO 9000 series. Specific principles to be incorporated into the Quality Systems covering the manufacture and Quality Control of medicines are known as Good Manufacturing Practice (GMP). The ISO Standards, GMP or other applicable Standards and other applicable Inter-governmental, National, Regional and Local Law or Regulation, should be consulted when developing a Quality Management for Tissue Banking organisations and other procurement organisations.

A 3.330 Basic Elements of an Appropriate Quality Management System
A 3.331 Organisational Structure and Accountability

This is necessary to achieve the Quality Requirements and for reviewing the effectiveness of the arrangements for Quality Assurance. There should be a suitably qualified and experienced member of staff appointed who verifies that the Quality Requirements are being met, and that there is compliance with the Quality Management System. The Quality Manager should be a designated individual who should be independent of production (not directly responsible for or involved in the procurement, processing and testing of tissue) and preferably of other responsibilities within the Tissue Bank. The Quality Manager should be generally familiar with the specific work being reviewed and be responsible for each Quality Assurance review. This individual should report, for his function, specifically to this Medical Director and/or his/her designee. Where a Tissue Bank is operated within a large organisation with its own Quality Department and possibly its own Quality Manager then strong working links should exist between the Tissue Bank's Quality Manager and the relevant Quality Department staff, as well as to the Medical Director.

A 3.332 Documentation Rationale

The objectives of thorough documentation are to define the system of information and control, to minimise the risk of misinterpretation and error inherent in oral or casually written communication and to provide unambiguous procedures to be followed. Documents should clearly state the Quality Requirements, organisational structures and responsibilities, the organisation's policies and standards, the management and technical procedures employed and the records required.

General: All procedures in the processing of tissue should be documented and the documents controlled. Documentation should be legible, readily identifiable and retrievable. Documentation should clearly identify the way in which it is to be used and by whom. Documentation should be available to staff to cover all procedures. Any correction should be hand-written clearly and legibly in permanent ink and signed and dated by an authorised person.

Control of Documentation: The system for document control should identify the current revision status of any document and the holder of the document. The system in place should demonstrate that all controlled documents meet the following criteria: They are current and authorized; they are reviewed at regular intervals; multiple copies are controlled with a distribution list; obsolete documents are removed and controlled to prevent further use. Changes to documents should be acted upon promptly. They should be reviewed, dated and signed by the authorised person and formally implemented

Storage and Retention of Documentation: Documented procedures should be established and maintained for identification, collection, filing, storage, retrieval and maintenance of all documents. Master copies of obsolete copies should be archived in a secure and safe environment for 10 years or in accordance with applicable Inter-governmental, National, Regional and Local Law or Regulation.

A 3.333 Control of Processes (SOPs)

Written instructions of Standard Operating Procedures (SOPs) shall be produced where it is essential that tasks must be performed in a consistent way. Equipment, processes and procedures shall be validated as effective before being implemented or changed. Equipment essential to the quality of the product shall be routinely serviced and calibrated, if appropriate. The processing environment and staff performing processes shall meet minimum, prescribed Standards of cleanliness and hygiene. The Tissue Bank shall maintain a SOPs Manual, which details in writing all aspects of these Standards. The SOPs shall be utilised to ensure that all material released for transplantation meet at least minimum requirements defined by professional Standards and applicable Inter-governmental, National, Regional and Local Law or Regulation. The SOPs Manuals should include, where relevant, but should not be limited to the following:

- Standard procedures for donor screening, consent, retrieval, processing, preservation, testing, storage and distribution
- Quality Assurance and Quality Control Policies
- Laboratory procedures for tests performed in-house and in contracted laboratories
- Specifications for materials used including supply, reagents, storage media and packaging materials
- Personnel and facility safety procedures
- Standard procedures for facility maintenance, cleaning and waste disposal procedures
- Methods for verification of the effectiveness of sterilisation procedures
- Equipment maintenance, calibration and validation procedures
- Environmental and microbiological conditions and the methods used for controlling, testing and verification
- Physiological and physical test specifications for materials
- Methods for determination of shelf life, storage temperature and assigning expiry dates of tissues
- Determination of insert and or label text
- Policies and procedures for exceptional release of material
- Procedures for adverse events reporting and corrective actions
- Donor/recipient tracking and product recall policies and procedures

All SOPs, their modification and associated process-validation studies shall be reviewed and approved by either the Medical or Administrative Director as dictated by content. All medically related SOPs shall be reviewed and approved by the Medical Director. Copies of the SOPs Manual shall be available to all staff, and to authorised individuals for inspections upon request. Upon implementation, all SOPs shall be followed as written. SOPs shall be updated at regular intervals to reflect modifications or changes. The authorised person, depending on the content shall approve each modification or change. Appropriate training shall be provided to pertinent staff. Obsolete SOPs Manuals shall be archived for 10 years (minimum) taking into account material shelf life.

A 3.334 Record Keeping

General: Records shall be confidential, accurate, complete, legible and indelible. All donor, processing, storage, and distribution records should be maintained for 30 years or in accordance with applicable Inter-governmental, National, Regional and Local Law or Regulation. Records shall hold all information that identifies the origins of the product and to demonstrate that the product meets all the Quality Requirements. Records shall show that all the required processing steps and all Quality Control tests have been performed correctly by trained staff and that the product has only been released for use after the correct authorisation. Records shall also demonstrate correct handling and storage of materials and track the final status of products, whether transplanted, discarded or used for research. The use and storage of records shall be controlled.

Contract Records: When two or more Tissue Banks participate in tissue procurement, processing, storage or distribution functions, the relationships and responsibilities of each shall be documented and ensure compliance with relevant scientific and quality professional Standards by all parties. Tissue Banks should perform on-site audits of contract laboratories to ensure their compliance with relevant scientific and professional Standards, Technical Manuals and the Tissue Bank's own requirements.

Donor Tracking: Each component shall be assigned one unique identifier that shall serve as a lot number to identify the material during all steps from collection to distribution and utilisation. This unique number shall link the final packaged material to the donor. This number shall be used to link the donor to all tests, records, organs and other material, and for tracking purposes to the recipient. Records shall include identification and evaluation of the donor, blood testing and micro-biological evaluation of the donor, conditions under which the material is procured, processed, tested and stored and its final destination. Records shall indicate the dates and identity of staff involved in each significant step of the operation.

Inventory: A record of unprocessed, processed, quarantined and distributed tissues shall be maintained.

Recipient Adverse Events and Non-compliances: An adverse events file shall be maintained including any non-compliance.

Electronic Records: If a computer record-keeping system is used, there shall be a system to ensure the authenticity, integrity and confidentiality of all records but retain the ability to generate true paper copies. A description of the system, its function and specified requirements must be documented. The system shall record the identity of persons entering or confirming critical data. Alteration to the system or programme shall only be made in accordance with defined procedures. When the release of finished batches is conducted by computerised systems it must identify and record the person (s) releasing the batches. Alternative management systems should be available to cope with failures in computerised systems.

A 3.340 Methods for Detecting, Correcting and Preventing Quality Failures from Recurring

A 3.341 Quality Failures

Quality failures include in-use product deficiencies (complaints, adverse events, etc.), failures to meet Quality Control specifications and non-compliance with procedures. Methods for detecting failures include Quality Control tests, inspections, Quality Audits, staff and end-user feedback. The ability to trace, locate, quarantine and recall materials, consumables and products at any stage, is essential to patient safety. Serious failures shall be thoroughly registered, investigated and appropriate changes to specifications, systems and procedures implemented to prevent further failures of a similar nature.

A 3.342 Audit

The Tissue Bank shall participate in an Audit Programme. Quality Assurance staff shall perform internal audits. Focused audits shall be conducted to monitor critical areas and when problems with quality have been identified. Regular audits shall be performed by qualified staff that does not have direct responsibility for the processes being audited.

A 3.350 Competency

The educational and training requirements for each member of staff shall be determined and specified. There shall be regular and formal appraisal of staff competency.

Training and education shall include the requirements for quality, Standards of Practice and Good Hygiene as well as appropriate continuing professional development. Records of training shall be maintained up to date.

A 3.400 Facilities and Equipment

A 3.410 General

The facilities of the Tissue Bank shall be of suitable size and location and shall be designed and equipped for the specialised purposes for which they are to be used.

A 3.420 Design

The design of the facilities shall prevent errors and cross-contamination. Critical procedures shall be performed in designated areas of adequate size.

A 3.430 Security

Access to the Tissue Bank shall be limited to authorised persons.

A 3.440 Environmental Monitoring

Environmental monitoring procedures shall be established, when appropriate, as part of the Quality Assurance Programme. The procedures shall include acceptable test parameters. The monitoring may include particulate air samplings and work surface cultures. Each monitoring activity shall be documented.

A 3.450 Sanitation

Facilities used for retrieval, processing or preservation, where there is potential for cross-contamination of material or exposure to blood-borne pathogens, shall be subjected to routine, scheduled and documented cleaning procedures.

A 3.460 Equipment and Instruments

Equipment and instruments shall be of appropriate quality for their intended function. Equipment and non-disposable supplies that come into contact with tissue shall be constructed so surfaces do not alter the safety or quality of the material. Equipment shall be designed, manufactured and qualified for appropriate cleaning and shall be sterilised or decontaminated after each use. Multiple use of disposable instruments for several donors shall be excluded.

There shall be SOPs for monitoring, inspection, maintenance, calibration, and cleaning procedures for each piece of equipment. Storage equipment shall be inspected on a regularly scheduled basis. Appropriate certification and maintenance records shall be maintained for equipment and instruments.

A 3.470 Environmental Safety

A 3.471 General

Each Tissue Bank shall provide and promote a safe work environment by developing, implementing and enforcing safety procedures. Safety precautions and procedures for maintaining a safe work environment shall be included in the SOPs Manual and shall conform to applicable Inter-governmental, National, Regional and Local Law or Regulation.

A 3.472 Safety Procedures

Safety procedures shall include, but are not limited to the following:

- Instructions for fire prevention and evacuation routes in case of fire or natural disaster
- Procedures for prevention of worker injury including possible exposure to biohazard materials
- Procedures for proper storage, handling and utilisation of hazardous materials, reagents and supplies
- Procedures outlining the steps to be followed in cleaning biohazard spills
- Hazardous material training including chemical, biological and radioactive hazards
- Immunisation appropriate vaccinations should be offered to all non-immune personnel whose job-related responsibilities involve potential exposure to blood-born pathogens. Personnel files should include documentation of receipt of vaccination or refusal of vaccination

Personnel: personnel engaged in the retrieval, processing, preservation and packaging of tissues shall be suitable attired to minimise the spread of transmissible pathogens among and between donors, tissue and staff. Any staff member with a serious infectious condition shall be excluded from Tissue Banking activities until the condition is resolved.

A 3.473 Waste Disposal

Human tissue and other hazardous waste items shall be disposed of in such a manner so as to prevent hazards to Tissue Bank personnel or the environment and shall conform to applicable Inter-governmental, National, Regional and Local Law or Regulation. Dignified and proper disposal procedures shall be applied to human remains.

SECTION B: Implementation

B 1.0 00 Donor Selection

B 1.1 00 General

The suitability of a specific donor for tissue allograft donation is based upon medical and behavioural history, medical records review, physical examination, cadaveric donor autopsy findings (if an autopsy is performed) and laboratory tests.

B 1.2 00 Medical and Behavioural History

B 1.2 10 Donor History Review

Donor evaluation includes an interview of the potential living donor or the cadaveric donor's next of kin, performed by suitably trained personnel, using a questionnaire. A qualified physician shall approve donor evaluation.

B 1.2 20 Exclusion Criteria

B 1.2 21 General Contraindications

The following conditions contraindicate the use of tissues for therapeutic purposes:

- History of chronic viral Hepatitis
- Presence of active viral Hepatitis, or jaundice of unknown etiology
- History of, or clinical evidence or suspicion or laboratory evidence of HIV infection. Risk factors for HIV, HBV and HCV have to be assessed by the Medical Director according to existing National Regulations taking into account national epidemiology. Annex 2 includes a generally agreed list of risk factors
- Presence or suspicion of central degenerative neurological diseases of possible infectious origin, including dementia (e.g. Alzheimer's Disease, Creutzfeldt-Jakob Disease or familial history of Creutzfeldt-Jakob Disease and Multiple Sclerosis)

- Use of all native human pituitary derived hormones (e.g. growth hormone), possible history of dura-mater allograft, including unspecified intracranial surgery
- Septicemia and systemic viral disease or mycosis or active tuberculosis at the time of procurement precludes procurement of tissues. In case of other active bacterial infection, tissue may be used only if processed using a validated method for bacterial inactivation and after approval by the Medical Director
- Presence or history of malignant disease. Exceptions may include primary basal cell carcinoma of the skin, histologically proven and unmetastatic brain tumour (see Annex 3)
- Significant history of connective tissue disease (e.g. systemic lupus erythematosus and rheumatoid arthritis) or any immunosuppressive treatment
- Significant exposure to a toxic substance that may be transferred in toxic doses or damage the tissue (e.g. cyanide, lead, mercury and gold)
- Presence or evidence of infection or prior irradiation at the site of donation
- Unknown cause of death: if at the time of death the cause of death is unknown, autopsy shall be performed to establish this cause.

B 1.2 22 Specific Tissue Selection Criteria

Cornea donors with solid extra-ocular malignancies are generally accepted.

B 1.3 00 Physical Examination

Prior to procurement of tissue, the donor body shall be examined for general exclusion signs and for signs of infection, trauma or medical intervention over donor sites that can affect the quality of the donated tissue.

B 1.4 00 Cadaveric Donor Autopsy Report

If an autopsy is performed, the results shall be reviewed by the Medical Director or designee before tissue is released for distribution.

B 1.5 00 Transmissible Diseases Blood Tests
B 1.5 10 General
B 1.5 11 Law and Practice

Tissues shall be tested for transmissible diseases in compliance with Law and practice in the country concerned. In the case of living donors, applicable consent procedure for blood testing shall be followed.

B 1.5 12 Tests

Tests shall be performed and found acceptable on properly identified blood samples from the donor, using recognised and if applicable, licensed tests and according to manufacturer's instructions. Tests shall be performed by a qualified, and if applicable, licensed laboratory and according to Good Laboratory Practice (GLP).

B 1.5 13 Timing of Blood Sampling

Blood for donor testing should be drawn at or within seven days of the donation and preferably within 24 hours after death.

B 1.5 14 Recent Blood Transfusion

For potential tissue donors who have received blood, blood components, or plasma volume expanders within 48 hours prior to death, if there is an expected hemodilution of more than 50%, based on calculation algorithm (see example of algorithm in Annex 4), a pretransfusion blood sample shall be tested.

B 1.5 15 Notification of Confirmed Positive Test Results

The living donor or cadaver donor's next of kin or physician shall be notified in accordance with State Laws of confirmed positive results having clinical significance. Confirmed positive donor infectious disease tests shall be reported to Local/National Health Authorities, when required.

B 1.5 16 Donor Serum Archive

A sample of donor serum shall be securely sealed and stored frozen in a proper manner until 5 years after the expiration date of the tissue or according to applicable Inter-governmental, National, Regional and Local Law or Regulation.

B 1.5 20 Blood Tests

B 1.5.21 Minimum Blood Tests

Minimum Blood Tests shall include:

- Human Immunodeficiency Virus Antibodies (HIV-1/2-Ab)
- Hepatitis B Virus Surface Antigen (HBs-Ag)
- Syphilis: nonspecific (e.g.. VDRL) or preferably specific (e.g.. TPHA)

B 1.522 Optional Blood Tests

Optional Blood Tests could be necessary for compliance with applicable Inter-governmental, National, Regional and Local Law or Regulation and/or to screen for endemic diseases:

- Hepatitis B core antibodies (HBc-Ab): HBc-Ab should be negative for tissue validation. Though, if the HBc-Ab test is positive and the HBs-Ag is negative, confirmation cascade should be entered. If the antibodies against the surface antigen are found (HBs-Ab), the donor can then be considered to have recovered from an infection and the tissue can be used for transplantation.
- Antigen test for HIV (p24 antigen) or HCV or validated Molecular Biology Test for HIV and HCV (e.g. PCR), if performed by an experienced laboratory.
- Antibody to HTLV 1: depending on the prevalence in some regions.
- Cytomegalovirus (CMV), Epstein-Barr Virus (EBV) and Toxoplasmosis Antibodies: for immunosuppressed patients
- Alanine Aminotransferase (ALT) for Living Donors:
- In addition to the general testing requirements, testing living donors of tissue for Alanine Aminotransferase (ALT) is recommended.

B 1.523 Living Donors Retesting

Retesting of living donors for HIV and HCV at 180 days is recommended. If another method of increasing safety, rather than retesting (antigen testing, Molecular Biology or viral inactivation method) is used (and allowed by applicable Regulation), it shall be documented and validated.

B 1.5 30 Exclusion Criteria

B 1.5.31 General Exclusion Criteria

Positive results for HIV, Hepatitis and HTLV-1 are reasons for exclusion.

B 1.5.32 Specific Exclusion Criteria

In life threatening situations for the recipient (e.g. related HPC donation), positive results for Hepatitis are no reason for exclusion, in accordance to applicable Regulations.

In these situations, tissues with a higher risk for recipient may be offered as long as full information is given to the recipient or, if it is not possible, to his relatives.

B 1.6 00 Bacteriological Studies of Donor and Tissues

B 1.6 10 Bacteriological Testing Methods

Representative samples of each retrieved tissue have to be cultured, if the tissues are to be aseptically processed without terminal sterilisation. Samples shall be taken prior to exposure of the tissue to antibiotic containing solution. The culture technique shall allow for the growth of both aerobic and anaerobic bacteria as well as fungi. Results shall be documented in the donor record. Blood culture, if procurement is performed on a cadaver donor, may be useful in evaluating the state of the cadaver and interpreting the cultures performed on the grafts themselves.

B 1.6 20 Bacteriological Bioburden Limits

If bacteriological testing of tissue samples obtained at the time of donation reveals growth of low virulence microorganisms, which are commonly considered nonpathogenic, the tissue may not be distributed without being further processed in a way that effectively decontaminates the tissue. Tissue from which high virulence microorganisms have been isolated are not acceptable for transplantation, unless the procedure has been validated to effectively inactivate the organisms without harmful potential effects, taking in account possible endotoxins.

B 1.7 00 Non Microbiological Tests

Non-microbiological tests depend upon the tissues and cells to be transplanted. Haematopoeitic Progenitor Cell donor selection requires as a minimum:

- ABO Blood Group and Rhesus Group
- Human Leukocyte Antigen Typing (HLA)
- Whole Blood Cell Count

B 1.8 00 Age Criteria

Donor age criteria for each type of tissue shall be established and recorded by the Tissue Bank.

B 1.9 00 Cadaver Donor Retrieval Time Limits

Tissues shall be retrieved as soon after death as is practically possible. Specific time limits vary with each tissue obtained, which shall be determined by the Medical Director. Usually, procurement of tissues should be completed within 12 hours after death (or circulatory arrest if also an organ donor). If the body has been refrigerated within 4 to 6 hours of death, procurement should start within 24 hours and no later than 48 hours.

B 2.0 00 Tissue Retrieval

B 2.1 00 Rationale

There shall be documented procedures that detail all requirements for retrieval to ensure that these processes are carried out under controlled conditions. Retrieval shall be performed using techniques appropriate to the specific tissue recovered, taking into consideration the eventual utilisation of the tissue.

B 2.2 00 Non-Living Donor Tissue Retrieval

B 2.2 10 Determination of Death

Tissue Bank physicians or physicians involved in removal or transplantation shall not pronounce death nor sign the death certificate of any individual from whom tissue will be collected. Inter-governmental, National, Regional and Local Law or Regulation concerning determination of death shall be respected.

B 2.2 20 Donor Identification

Precise identification of the cadaver donor shall be performed before procurement begins.

B 2.2 30 Retrieval Conditions

B 2.2 31 Facility for Retrieval

Procurement shall be accomplished in an operating room or adequate mortuary facility.

B 2.2 32 Procurement Equipment Sterility

All instruments and equipment used for procurement shall be sterilised between procurements.

B 2.2 33 Aseptic or Clean/Non-Sterile Procurement Techniques

Tissues may be removed using either aseptic or clean/non-sterile procurement techniques:

Aseptic technique: Aseptic technique shall be observed throughout the procurement procedure. Procurement sites shall be prepared using a standard surgical technique; all methods shall be consistent with standard operating room practice.

Clean/non-sterile technique: Allografts procured using clean/non-sterile techniques are suitable for transplantation, if efficient validated sterilising methods are used to eliminate pathogens after retrieval.

B 2.2 34 Samples for Microbiological Testing

Samples for microbiological testing shall be taken, where applicable.

B 2.2 40 Body Reconstruction

Following tissue procurement, the donor's body is to be reconstructed to approximate its original anatomical configuration and to make usual funeral proceedings possible.

B 2.3 00 Surgical Residues Collection

Surgical residues shall be collected under aseptic conditions during a surgical procedure in the operating theatre.

B 2.4 00 Living Donor Tissue Retrieval

Tissues must be removed under conditions representing the least possible risk to the donor, in properly equipped and staffed institutions.

B 2.5 00 Packaging and Transportation to the Tissue Bank

B 2.5 10 Procurement Container

Each tissue segment shall be packaged individually as soon as possible after retrieval, using sterile containers in a manner that will prevent contamination. Containers shall conform to Inter-governmental, National, Regional and Local Law or Regulation, as appropriate. Proper reagents or preservation solution shall be used, as specified in SOPs. Procedures shall be used for ensuring and documenting proper temperature storage during transit.

B 2.5 20 Procurement Container Integrity

After filling and closing the container, it shall not be re-opened nor the tissue removed until further processing by the Tissue Bank.

B 2.5 30 Procurement Container Label

At all times, the container shall be labelled with the donor and tissue identification; in such manner that traceability of tissues will be achieved. The container shall be labelled as containing human tissue, the name and address of the shipping facility and the name and address of the intended receiving facility. Containers shall comply with additional labelling requirements established by common carriers or by Inter-governmental, National, Regional and Local Law or Regulation.

B 2.6 00 Retrieval Documentation

Appropriate records of each donation procedure and all tissues retrieved shall be available and kept by the Tissue Bank. All retrieved tissue shall be provided with an accompanying retrieval form including, at a minimum:

- The donor identity
- The date, time and place of the procedure
- The identity of the person (s) performing the retrieval
- The tissue(s) retrieved donor and tissue selection information

B 3.0 00 Tissue Banking General Procedures

B 3.1 00 General

B 3.1 10 Written Procedures

Specific methods employed for processing may vary with type of tissue and with the manner in which it has been retrieved. Each type of tissue shall be prepared according to a written procedure, which shall conform to these Standards and other applicable Standards, resulting in processed tissues appropriate for safe and efficient clinical use.

B 3.1 20 Process Validation

All steps involved during the processing of tissues shall be validated, when appropriate, to demonstrate the effectiveness of procedures. When computers are used as part of a processing or Quality Management System, the computer software shall be validated. When validation cannot be adequately evidenced through testing, validation shall be evidenced through documentation demonstrating adequate design, development, verification and maintenance procedures.

B 3.1 30 Quality Controls

Tests and procedures shall be performed to measure, assay or monitor processing, preservation and storage methods, equipments and reagents to ensure compliance with established tolerance limits. Results of all such tests or procedures shall be recorded.

B 3.1 40 Records Management

Appropriate records of each tissue processed shall be kept by the Tissue Bank. Records shall allow traceability of tissues, including the different steps in the preparation, the date and time of the procedure, the identity of the person performing the procedure and the record of the materials used. Laboratory results (e.g. microbiology/processing cultures) and other test results used to determine final release shall be archived by the Tissue Bank distributing the tissue.

B 3.2 00 Unique Tissue Identification Number

Each individual tissue shall be marked with a unique identification number to relate each specimen to the individual donor.

B 3.3 00 Reagents, Container and Packaging
B 3.3 10 Reagents

The reagents used in preservation and processing shall be of appropriate grade for the intended use, be sterile, if applicable, and conform to existing Regulation. The origin, characteristics and expiration date of reagents shall be monitored and recorded.

B 3.3 20 Tissue Container

The type of tissue container may vary with the type of tissue and processing. They shall maintain the tissue sterility and integrity, withstand the sterilisation and storage methods utilised and avoid the production of toxic residues. They shall conform to applicable Inter-governmental, National, Regional and Local Law or Regulation. Each tissue container shall be examined visually for damage or evidence of contamination before and after processing and prior to its dispatch.

B 3.3 30 Tissue Outer Package

Packaging shall ensure integrity and effectively prevent contamination of the contents of the final container. It shall conform to applicable Regulation for Transportation.

B 3.4 00 Pooling

Tissue from each donor shall be processed and packaged in such a way as to prevent contact and cross-contamination with tissues from other donors. If tissues are subsequently treated in batches (e.g. sterilisation), a unique batch number shall be assigned and added to the records of the tissues.

Pooling of donors is not recommended and should only be accepted for specific tissues. The size of the pool should be limited to the minimum number of donors and traceability to each donor has to be ensured.

If pooling is used for specific tissues, a fully documented rationale and risk assessment shall be undertaken to document safety.

B 3.5 00 Environmental Control

Processing steps shall take place in an appropriately controlled environment. Tissue processing in an Open System shall have the environmental conditions and monitoring of the area clearly defined (such as for a "clean room" or laminar flow cabinet). Records shall be maintained to demonstrate that the area is monitored for microbiological contamination and air control.

B 3.6 00 Storage Conditions
B 3.6 10 Temperature

Acceptable temperature ranges for storage shall be established.

Temperature monitoring of storage: Low temperature (refrigerated or frozen) storage devices and incubators shall be connected to a central alarm system or each shall be equipped with an audible alarm system, that will sound when the temperature deviates from the acceptable storage range. The alarm system shall be connected to an emergency power source. Continuous recording and daily review of data are recommended.

B 3.6 20 Storage of Quarantined or Unprocessed Tissue

There shall be a system of Quarantine for all tissues to ensure that they cannot be released for clinical use until they have met the defined acceptable criteria for release.

Storage areas of quarantined or unprocessed tissue shall be separate from storage areas of tissue approved for processing or ready for distribution. The storage areas shall be clearly labeled as containing quarantined, released for processing or processed finished tissue.

B 3.7 00 Documentation Reviewing and Tissue Inspection
B 3.7 10 Incoming Inspection

Staff shall inspect the tissue container upon arrival from the procurement facility in order to ensure the integrity of the container(s), the presence of proper identification and documentation.

B 3.7 20 Review of Donor Eligibility

The donor's medical history, the physical examination, the results of tissue procurement microbiologic tests and donor blood testing, and if performed, the results of an autopsy, shall be reviewed by the Medical Director or designee. Quarantined tissues shall be reviewed prior to distribution after all testing has been satisfactorily completed.

B 3.7 30 Sizing of Specimens

Specimen sizing may be made by actual measurements or by imaging sizing techniques.

B 3.7 40 Inspection Prior to Release Into Finished Inventory

Prior to the release of tissue into the Finished Inventory, a final review shall be made of donor suitability, procurement, production, processing records, Quality Control tests, the finished tissue, containers, closures and labels shall be inspected and approved by the Medical Director or designee.

B 3.7 50 Final Inspection

Prior to distribution, final inspection of the container, label and documentation shall be performed to ensure accuracy and integrity.

B 3.8 00 Non-Conforming Tissues

Tissues failing any portion of the review process shall be maintained in quarantine pending disposal and shall not be released for clinical use. There shall be a documented policy for discard of tissue unsuitable for clinical use.

B 3.9 00 Expiry Dates

Expiry dates shall be established for all tissue released from a Tissue Bank. If the dating period is 72 hours or less, the hour of expiration shall be indicated on the label. Otherwise, the dating period ends at midnight of the expiration date.

B 4.0 00 Specific Processing Procedures
B 4.1 00 General

Section A relating to written procedures, process validation, quality control and record management always apply. All tissues rejected due to the ineligibility of the donor cannot be used for transplantation, even after processing including sterilisation or disinfection. Even if terminal sterilisation or disinfection using physical or chemical agents are used, the procurement and processing shall be adequate to minimise the microbial content of tissues to enable the subsequent sterilisation-disinfection process to be effective. Appropriate indicators for sterilisation must be included in each batch.

B 4.2 00 Disinfectant or Antibiotic Immersion

If disinfectants or antibiotics are used after retrieval, the tissues shall be immersed in a disinfectant or in an antibiotic solution following sterility testing and before final packaging. The type of solution used shall be specified on documentation.

B 4.3 00 Fresh Tissue

Fresh allografts (e.g. small fragments of articular cartilage and skin) are aseptically procured in an operating room. Fresh Tissue is usually stored refrigerated at 4°C or in accordance with written procedures. Fresh Tissue shall not be used in a patient until donor blood testing is completed according to these Standards, available bacteriologic results are acceptable and donor suitability has been approved by the Medical Director or designee.

B 4.4 00 Frozen Tissue

After aseptic procurement in the operating room, frozen tissue are placed in a -40°C or colder controlled environment within 24 hours of procurement. Subsequent manipulation of tissues (e.g. cleaning and cutting) shall be undertaken aseptically.

B 4.5 00 Cryopreserved Tissue

A cryopreservative solution (e.g. DMSO or Glycerol) is usually added to treat the tissue prior to freezing. Documentation of the concentration of cryoprotectants and nutrients or isotonic solutions in the cryopreservative solution shall be maintained. Properly packaged specimens are frozen by placing the specimens below -40°C, or may be subjected to control rate freezing using a computer assisted liquid nitrogen freezing device. If a programmed control-rate freezing method is employed, a record of the freezing profile shall be evaluated, approved and recorded.

B 4.6 00 Freeze-Dried Tissue
B 4.6 10 Freeze-Drying Methods

Various Protocols of freeze-drying tissues exist. Freeze-drying is a method for preservation, but is not a sterilisation method; sterility shall be assumed by Aseptic Protocol or additional sterilisation. After a standardised procedure for freeze-drying has been developed, a Quality Control Programme for monitoring the performance of the freeze-dryer shall be documented. Freeze-dried tissues shall be stored at room temperature or colder.

B 4.6 20 Freeze-Drying Controls

Each freeze-drying cycle must be clearly documented, including length, temperature and vacuum pressure at each step of the cycle. Representative samples shall be tested for residual water content.

B 4.7 00 Simply Dehydrated Tissue
B 4.7 10 Dehydration Method

The use of simple dehydration (evaporation) of tissues as a means of preservation shall be controlled in a manner similar of freeze-drying. Temperatures of simple dehydration shall be below 60°C.

B 4.7 20 Dehydration Controls

Each dehydration cycle shall be monitored during operation for temperature. Following dehydration, representative samples shall be tested for residual moisture.

B 4.8 00 Irradiated Tissue
B 4.8 10 Irradiation Methods

Commercial or hospital radiation facilities are available for ionising irradiation. The minimum recommended dose for bacterial decontamination is 15 kGy (kiloGray). The minimum recommended dose for bacterial sterilisation is 25 kGy (kiloGray). Viral inactivation would require higher doses and depends on numerous factors. For this reason no specific dose can be recommended, but shall be validated, when applicable. The used Protocol shall be validated taking in account the initial bioburden, and shall be performed by facilities following good irradiation practices (see IAEA Code of Practice for the Radiation Sterilisation of Biological Tissues).

B 4.8 20 Irradiation Sterilisation Controls

Sterilisation by ionising radiation shall be documented (see IAEA Code of Practice for the Radiation Sterilisation of Biological Tissues). The processing records include the name of the facility and the resultant dosimetry for each batch.

B 4.9 00 Ethylene Oxide Sterilised Tissue
B 4.9 10 Ethylene Oxide Sterilisation Method

Care should be taken when using ethylene oxide since the residues may have toxic effects already demonstrated for musculoskeletal allografts in the literature. Following appropriate processing procedures, the tissues are placed in ethylene oxide permeable containers and exposed to the ethylene oxide gas mixture following the manufacturer's suggested Guidelines. T he conditions of exposure may need to be individualised depending upon the nature of the specimens to be sterilised. A Quality Control Programme shall demonstrate that equipment meets requirements in temperature, humidity and gas concentration for the selected period. Following ethylene oxide sterilisation, an appropriate aeration procedure shall be followed, to allow removal of residual ethylene oxide and/or its breakdown products (Ethylene Chlorhydrin and Ethylene Glycol).

B 4.9 20 Ethylene Oxide Sterilisation Controls

Chemical indicator strips shall be included in each batch. A validated procedure shall be run with each lot of tissue to document that sterilisation has been achieved. Monitoring for residual levels of chemicals or their breakdown products shall be conducted from representative samples of the finished tissues of each batch.

B 4.10 00 Other Processing Methods
B 4.10 10 Other Inactivation Methods

Some chemical agents only have a decontamination role. Other agents may have an inactivation effect on specific pathogens. The efficiency of these agents towards the treated type of tissue shall be validated. The use of chemical and possible presence of trace residuals shall be included in the information accompanying the tissue. Under specific conditions, heat may be used to decontaminate or sterilise some type of tissues. The used Protocol shall be validated taking in account the initial bioburden and shall be performed by a recognised facility.

B 4.10 20 Bone Demineralisation

Several methods and procedures for the formation of demineralised bone are available and acceptable. Controlled quality reagents shall be used. Residual calcium obtained by the method shall be determined.

B 5.0 00 Labeling

B 5.1 00 General Requirements

B 5.1 10 Rationale

There shall be written procedures designed and followed to ensure that correct labels and labeling are used for tissue identification.

B 5.1 20 Nomenclature

Standard measurement nomenclature shall be used to describe tissues and the processing they have undergone.

B 5.1 30 Label Integrity

The tissue label applied by the Tissue Bank facility shall not be removed, altered or obscured.

B 5.1 40 Visual Inspection

When visual inspection through the container is possible, a sufficient area of the container shall remain uncovered to permit inspection of the contents.

B 5.2 00 Tissue Containers Labelling

Tissue containers shall be labelled so as to identify, as a minimum:

- The human nature of the contents
- Product description
- Name and address of Tissue Bank
- Tissue identification number
- Expiration date

The following information shall be included on the label, if possible, otherwise on the accompanying documentation:

- Amount of tissue in the container expressed as volume, weight or dimensions or such combination of the foregoing as needed, for an accurate description of the contents
- Sterilisation or inactivation procedure used, if applicable
- Batch number, if applicable
- Potential residuals of added preserving and processing agents/solution (e.g. antibiotics, ETOH, ETO, DMSO)
- Recommended storage conditions

B 5.3 00 Package Insert

B 5.3 10 General

All tissues shall be accompanied by a document describing the nature of tissue and processing methods and instructions for proper storage and reconstitution, when appropriate. Specific instructions shall be enclosed with tissue requiring special handling.

B 5.3 20 Accompanying Documentation Requirements

Accompanying documentation shall contain all the information described for container labeling and the following additional information:

- Origin of tissue (country of procurement)
- The nature and results of biological tests performed on the donor using appropriate and licensed tests
- Processing methods used and results of sterility tests or inactivation controls
- Special instructions indicated by the particular tissue for storage or implantation
- Tissue that is to be reconstituted at or prior to the time of use shall include information on the conditions, under which such tissue shall be stored and reconstituted prior to implantation
- Indications and contraindications for use of tissue, when necessary
- Statement that each tissue shall be used for a single patient only

B 5.4 00 Tissue Outer Package Labeling

Labeling of the tissue outer package shall conform to Transportation Regulations, when applicable.

B 6.0 00 Distribution

B 6.1 00 General

Tissues can be distributed for a specific patient to a physician, dentist and other qualified medical professional or a storage facility located in another institution for local use or distributed to another Tissue Bank. Distribution for therapeutic use shall be based on medical criteria on equitable bases, in accordance with Inter-governmental, National, Regional and Local Law or Regulation and practice. There shall be written procedures and documentation for all tissues distributed. The clinical team using the tissue shall have instructions for contacting the Tissue Bank for any question they have and shall be made aware of the following:

- Action to be taken in the event of loss of integrity of the package
- Action for reporting of adverse event
- Action for the return or the disposal of unsuitable or unused tissue

B 6.2 00 Traceability

There shall be an effective system that enables the traceability of tissues between the donor, the processed tissue and the recipient. It is the responsibility of the hospital tissue storage and distribution facility or clinician to implement recipient records and to inform the Tissue Bank of the destination of tissues (implantation date, surgeon and recipient identification). Tissue Banks shall maintain records which document the destination of distributed tissue: implantation (date, surgeon and recipient identification), destruction (date and place) and of any adverse event reports.

B 6.3 00 Transportation

Maintenance of (upper and/or lower parameters) environmental conditions during transit, as defined in the written procedure of the Tissue Bank, shall be ensured. Use of hazardous elements such as dry ice or liquid nitrogen shall comply with relevant Regulations.

B 6.4 00 Accompanying Documentation

The release of tissue from storage shall include all documentation originating from the Tissue Bank. Surgeons shall be aware that copies of this documentation shall be maintained in the recipient's medical records.

B 6.5 00 Return into Inventory

Issued tissues shall not be returned to the Tissue Bank without prior consultations with the Medical Director or designee. Tissue must be in its original unopened container and the storage conditions must have been maintained as required.

B 6.6 00 Adverse Events

Reports of adverse events shall be evaluated by the institution where the tissue was used and reported immediately to the Tissue Bank. All adverse events shall be reviewed by the Medical Director and appropriate action documented, in accordance with Inter-governmental, National, Regional and Local Law or Regulation. Identified transmission of disease shall be reported to the Public Health Authorities, processing Institutions, to the donor's personal physician, if clinically relevant and to physicians involved in implantation of the tissue, in accordance with Inter-governmental, National, Regional and Local Law or Regulation on Confidentiality. When donor to recipient disease transmission through tissue use is discovered, all facilities involved in the procurement and distribution of organs or tissues from the infected donor shall be notified without delay. A written report of the investigation of adverse events, including conclusions, follow up and corrective actions, shall be prepared and maintained by the Tissue Bank in an adverse event file.

B 6.7 00 Recall

A written procedure shall exist for recall of tissues.

B 6.8 00 Distribution to Storage Facilities Outside the Tissue Bank (depot)
B 6.8 10 General

When a storage facility is located outside the Tissue Bank, the institution where this facility is located is responsible for establishing acceptable storage and record keeping procedures to ensure the maintenance of the safety and efficacy of tissue from receipt to use and the trace ability of tissue and recipients. The relevant part of these Standards shall be made available to these institutions. These storage facilities (depot) shall be subjected to Quality Audit and Control from the Tissue Bank.

B 6.8 20 Labeling

Labels on tissue containers shall not be altered, made invisible or removed.

B 6.8 30 Storage

Tissue storage shall conform to Guidelines established by the distributing Tissue Bank.

B 6.8 40 Records

Records shall document, as a minimum, the receival date of tissue and the destination (transplant date, the recipient's identity and transplant surgeon). These destination records shall be transmitted to the Tissue Bank.

B 6.9 00 Distribution to Another Tissue Bank

The associated Tissue Bank should adhere to these Standards.

B 6.10 00 Acquisition of Tissue from Another Tissue Bank
B 6.10 10 Medical Director Approval

Prior to acquiring tissue from another Tissue Bank, the Medical Director shall ensure that the Tissue Bank works according to these Standards or according to comparable recognised Standards.

B 6.10 20 Labelling

Labels on processed tissue acquired from another Tissue Bank shall not be altered, made invisible or removed.

B 6.10 30 Distribution Record

Accompanying documentation from the original Tissue Bank shall be forwarded with the tissue to the clinical team. After implantation, the destination record (transplant date, the recipient's identity and transplant surgeon) shall be forwarded to the original Tissue Bank.

ANNEXES

Annex 1: Glossary

ADVERSE EVENTS [Adverse Outcome / Reaction]: undesirable effect or untoward complication in a recipient consequent to or related to tissue transplantation.

ALLOGRAFT: graft transplanted between two different individuals of the same species.

ASEPTIC RETRIEVAL: retrieval of tissue using methods that restrict or minimise contamination with microorganisms from the donor, environment, retrieval personnel and/or equipment.

BRAIN DEATH / BRAIN STEM DEATH: complete and irreversible cessation of brain stem and brain encephalic functions and certified according to National Laws.

CLEAN ROOM: room in which the concentration of airborne particles is monitored and controlled to defined specification limits.

COMPLIANCE: conforming to established Standards or Regulations.

CONTAINER: enclosure for one unit of transplantable tissue.

CONTROLLED ENVIRONMENT: environment which is controlled with respect to particulate contamination, both viable or non-viable particles are controlled. May also include temperature and humidity controls.

CORONER: (see Medical Examiner).

CORRECTIVE ACTION: steps taken to ameliorate non-compliance.

COST: actual costs for retrieval, processing, preservation, storage, distribution, education, research and development.

CROSS-CONTAMINATION: transfer of infectious agents from tissues to other tissue or from one donor's tissue to another donor's tissue.

DEATH: (see Brain Death).

DISINFECTION: process that reduces the number of viable cellular microorganisms, but does not necessarily destroy all microbial forms, such as spores and viruses.

DISTRIBUTION: transportation and delivery of tissues for storage or use in recipients.

DONOR MEDICAL HISTORY INTERVIEW: documented dialogue with an individual or individuals who would be knowledgeable of the donor's relevant medical history and social behaviour; such as the donor, if living, the next of kin, the nearest available relative, a member of the donor's household, other individual with an affinity relationship and/or the primary treating physician. The relevant social history includes questions to elicit whether or not the donor met certain descriptions or engaged in certain activities or behaviours considered to place such an individual at increased risk for HIV and Hepatitis or other diseases.

DONOR REGISTRY: formal compilation of individuals intent relating to donation that may be maintained by a Governmental agency or private establishment.

DONOR SELECTION / DONOR SCREENING: evaluation of information about a potential donor to determine whether the donor meets qualifications specified in the SOPs and Standards. This includes but is not limited to, medical social and sexual histories, physical examination and laboratory test results (and autopsy findings, if performed).

DONOR: living or deceased individual who is the source of tissue for transplantation in accordance with established medical criteria and procedures.

END-USER: healthcare practitioner who performs transplantation procedures.

FACILITY: any area used in the procurement, processing, sterilisation, testing, storage or distribution of tissue and tissue components.

FINISHED INVENTORY: Storage of finished tissue.

FINISHED TISSUE: tissue that has undergone all of the stages of processing, packaging and is approved for distribution.

GIFT DOCUMENT: legally recognised document in which an individual indicates his/her wishes as they relate to donation of organs and tissues.

GOOD TISSUE BANKING PRACTICES: practices that meet accepted Standards as defined by relevant Government or professional organisations.

HPC: Haematopoietic Progenitor Cells

INSPECTION: Tissue Bank examination to ascertain Good Tissue Banking Practices.

LABELLING MATERIAL: any printed or written material including labels, advertising, and/or containing information (for example package insert, brochures, pamphlets) related to the tissues.

LABELLING: includes steps taken to identify the material and to attach the appropriate labels on the container or package so that it is clearly visible. Includes the package insert which is the written material accompanying a tissue graft bearing information about the tissue, directions for use and any applicable warnings.

MEDICAL EXAMINER [Coroner]: Governmental official (usually a pathologist) charged with investigating deaths and determining cause of death.

NATIONAL REGULATORY AUTHORITY [NRA]: body appointed by the Government with the goal of controlling Tissue Banking practices.

NEXT OF KIN: person(s) most closely related to a deceased individual as designated by applicable law.

NON-COMPLIANCE: non-conformance to established standards or regulations.

OPEN SYSTEM: system which has been breached but where every effort is made to maintain sterility by the use of sterile material and aseptic handling techniques in a clean environment.

ORGAN: (see Vascular Organ).

PACKAGING: (see Container).

PROCESSING: any activity performed on tissue, other than tissue recovery, including preparation, preservation for storage and/or removal from storage, to assure the quality and/or sterility of human tissues.

QUALITY: totality of characteristics of a product, process or system that bare on its ability to satisfy customers or other interested parties.

QUALITY ASSURANCE (part of Quality Management): planned and systematic actions necessary to provide confidence in fulfilling Quality Requirements (see Quality Requirements).

QUALITY AUDIT: documented review of procedures, records, personnel functions, equipment, materials, facilities, and/or vendors in order to evaluate adherence to the written SOPs, Standards, or government laws and regulations.

QUALITY CONTROL (part of Quality Management): operational techniques and activities that are used to fulfil Requirements for Quality.

QUALITY MANAGEMENT: all activities of the overall management function that determine the Quality Policy, Objectives and Responsibilities, and their implementation by means of Quality Planning, Quality Control, Quality Assurance and Quality Improvement, within the Quality System.

QUALITY REQUIREMENTS: Requirements for the characteristics of a product, a process or a system.

QUALITY MANGEMENT SYSTEM: (see Quality Management).

QUARANTINE: status of retrieved tissue or packaging material, or tissue isolated physically or by other effective means, whilst awaiting a decision on release or rejection.

RECALL: requested return of finished tissue known or suspected to be non-compliant to the Tissue Bank, in accordance with the instructions contained in an advisory notice.

RECIPIENT: individual into whom organs, tissue is transplanted.

RETRIEVAL [Recovery, Procurement, Removal, Harvest]: removal of tissues from a donor for the purpose of transplantation.

SAFETY: Quality of tissue indicating handling according to standards and substantial from the potential for harmful effects from recipients.

STANDARD OPERATING PROCEDURES [SOPs]: group of Standard Operating Procedures detailing the specific policies of a Tissue Bank and the procedures used by the staff / personnel. This includes, but is not limited to procedures to: assess donor suitability and retrieve, process, sterilise, quarantine, release to inventory, label, store, distribute and recall tissue.

STERILISATION: validated process to destroy, inactivate, or reduce microorganisms to a sterility assurance level of 10-6.

STERILITY ASSURANCE LEVEL: probability of detecting an unsterile product, tissue.

STORAGE: maintenance of tissues in a state ready for distribution.

TERMINAL STERILISATION: sterilisation that takes place at the end of processing the tissue, in the final packaging.

TISSUE: human tissue includes all constituted parts of a human body, including surgical residues and amnion, but excluding organs, blood and blood products, as well as reproductive tissues such as sperm, eggs and embryos. New products engineered from human tissue are included. The word 'Tissue' in this text applies to all types of tissues including corneas and to cells.

TISSUE BANK: entity that provides or engages in one or more services involving tissue from living or cadaveric individuals for transplantation purposes. These services include assessing donor suitability, tissue recovery, tissue processing, sterilisation, storage, labelling and distribution.

TRACEABILITY: ability to locate tissue during any step of its donation, collection, processing, testing, storage and distribution. It implies the capacity to identify the donor and the medical facility receiving the cells and/or tissue or the recipient.

TRANSPLANTATION: removal of tissues and / or cells and grafting of these tissues whether immediately or after a period of preservation and / or storage. Transplantation may be from one person to another (allogeneic) or from a person to themselves (autologous).

VALIDATION: refers to establishing documented evidence that provides a high degree of assurance that a specific process will consistently produce a product meeting its predetermined specifications and quality attributes. A process is validated to evaluate the performance of a system with regard to its effectiveness based on intended use.

VASCULAR ORGANS: Any part of a human body consisting of vascularised, structured arrangement of cells, which removed, cannot be replicated by the body. Example: heart, liver, lung, kidney, pancreas, intestine.

Annex 2: Guidelines of Factors to be Considered for Determining Risk for Human Immunodeficiency Virus or B or C Hepatitis

- Men who have had sex with another man in the preceding 12 months. Persons who report non-medical intravenous, intramuscular or subcutaneous injection of drugs in the preceding 12 months.
- Men and women who have engaged in sex in exchange for money or drugs in the preceding 12 months.
- Persons with a history of chronic Hemodialysis.
- Persons with a history of Haemophilia or related clotting disorders who have received human-derived clotting factor concentrates.
- Persons who were sexual partners of persons having a HIV or B or C Hepatitis history, manifestations, or risk factors previously described, in the past 12 months.
- Percutaneous exposure or contact with an open wound, non-intact skin or mucous membrane to blood thought to be at high risk for carrying HIV or Hepatitis in the preceding 12 months.
- Inmates of correctional systems in past 12 months.
- Diagnosed or treated for Syphilis or Gonorrhoea in past 12 months.
- A potential tissue donor who has received a blood transfusion within 12 months prior to death may only be accepted as a tissue donor after individual approval from the Medical Director.
- The donor is not eligible if in a deferral status of any Blood Services Donor Deferral Register. The local blood centre(s) shall be checked each time possible (blood donor card available).
- Tattoo, ear piercing, body piercing, and/or acupuncture, unless by sterile, non-reused needle or equipment, in the preceding 12 months.

Annex 3: Primary Tumours of the Central Nervous System: Evaluation of a Suitable Donor. Reference List

No Contraindication
Pituitary Adenoma
Pinelcytoma
Hemangioblastoma
Schwannoma
Choroid Plexus Papilloma
Ependimoma
Oligodendroglioma differentiated
Craniopharyngioma
Benign Meningioma
Pilocytic astrocytoma
Epidermoid tumours
Contraindication
Medulloblastoma
Chordoma
Glioblastoma multiforme
Highly anaplastic Oligodendroglioma
Anaplastic Epidimoma
Anaplastic Meningioma

Primary CNS Lymphoma
Pineoblastoma
CNS Sarcomas
Astrocytoma grade II
Astrocytoma grade III

Annex 4: Example of Algorithm for Calculating the Hemodilution of a Donor Having Received Blood, Blood Components, or Plasma Volume Expanders Within 48 Hours Prior to Death

The following equation allows calculation of a potential donor 50% plasma volume:
50% plasma volume (mL) = 21 x donor's body weight (kg)
The equation as been calculated as follows:
Total blood volume / kg = 1kg x 70 mL = 70 mL
Total plasma volume / kg = 70 mL x (1.0 –0.40) (normal adult hematocrit) = 42 mL
50% plasma volume / kg = 42 mL (total plasma volume per kg) x 0.50 = 21 mL / kg

Annex 5: References and Contact Addresses

REFERENCES

American Association of Tissue Banks (AATB): Standards for Tissue Banking (1984, 1985, 1987, 1989, 1993, 1996, 1998, 2001).
Australian Code of Good Manufacturing Practice - Human Blood and Tissues. Therapeutic Goods Administration, 2000.
Council of Europe: Guide on Safety and Quality Assurance for Organs, Tissues and Cells (Version 11. CDSP, Released for Consultation 1/2001)
European Association of Tissue Banks (EATB): EATB General Standards for Tissue Banking (1995), EATB and EAMST Standards for Musculoskeletal Tissue Banking (1997, revised1999), EATB Standards for Skin Banking and Banking of Skin Substitutes (1997).
IAEA Code of Practice for the Radiation Sterilisation of Biological Tissues. (IAEA Vienna)
Radiation and Tissue Banking, G.O. Phillips (ed.), World Scientific, Singapore, 2000.
UK Code Of Practice for Tissue Banks. Department of Health. United Kingdom. 2001.

CONTACT ADDRESSES

American Association of Tissue Banks (AATB)
1350 Beverly Road, Suite 220A, McLean, VA 22101, USA
www.aatb.org

Asia-Pacific Surgical Tissue Banking Association
Dr Norimah Yusof, Malaysian Institute of Nuclear Technology Research
Bangi, 43000 Kayang, Malaysia
Email: norimah@mint.gov.my

Council of Europe
Karl Friedrich Bopp, Health Department, Council of Europe
67075 Strasbourg, France
www.coe.int

European Association of Tissue Banks (EATB)
Dr Heinz Winkler, Vienna, Austria
www.eatb.de

Latin American Association of Tissue Banking
Dr Eulogia Kairiyama, Comision National de Energia Atomica, Centro Ezeiza
Presbitero Juan Gonzales y Aragon 15 (B1802 AYA)
Ezeiza, Pcia, Buenos Aires, Argentina
Email: kairiyam@cae.cnea.gov.ar

Part 2: GUIDE FOR LEGAL AND REGULATORY CONTROL

INTRODUCTION

This Section is intended to assist Governmental Control Authorities (GCA) and Tissue Banks in their joint task of improving the quality of human tissues for transplantation through Regulation and Legislation that interface with Standards. Each member of the IAEA and their Regulatory/Legislative bodies must necessarily determine the appropriate path for such Regulation/Law to follow, based on the technical capabilities of their region, religious beliefs and practices and healthcare systems. Within these key topics, many options are available for consideration.

HISTORICAL PROGRESSION

The first Tissue Banks were started in the 1950's, primarily in response to needs for bone, corneas and skin. Through the 1960's and 1970's, Tissue Banks began to proliferate, although they were usually small programmes that primarily served the hospital at which the Tissue Bank was located. Laws relating to organ and tissue donation, declaration of death and donor consent were passed in many countries in the 1970's through the 1980's. Starting in the mid-1980's, Standards for Tissue Banking were developed, and often were accompanied by accreditation programmes organised by Tissue Banking Associations. Even in countries with well-established Tissue Banks, the development and enforcement of Regulations and Laws did not occur until the 1990's, when concerns regarding safety of donated tissues increased. In the new century, Tissue Banking Regulations and Laws in developing countries have been passed, as have expanded Laws in other countries with well-developed tissue donation systems.

LAWS AND REGULATIONS

Laws and Regulations concerning a wide range of topics are necessary, including:

1 - Donation / Transplantation / Recovery / Waiting Lists
2 - Consent
3 - Organisation of the Tissue Bank
4 - Interrelationships with Organ Donation Programmes
5 - Registration / Licensing / Accreditation / Authorisation of the Tissue Bank
6 - Import / Export of Tissue
7 - Financial Aspects of Tissue Banking
8 - Enforcement and Compliance

1 – Donation / Transplantation / Recovery / Waiting Lists

a) A Law defining death (including Brain Stem Death or Brain Death) is mandatory in order for cadaveric donation of vascular organs and tissues. Because tissues may also be retrieved from a brain dead organ donor, the Standards should reference Brain Death Laws, if the Tissue Bank is prepared to accept tissues from donors meeting Brain Death criteria. Ideally, this Law will also address how death must be declared, and by whom (e.g., brain death may be determined by a registered medical practitioner not involved with the recovery or transplantation of organs / tissues and using clinical criteria).

b) A Law or Regulation covering the mechanisms for organ and tissue donation is mandatory. These Laws must outline how an individual may become or refuse to be a donor, the definition of organs and tissues that may be donated, the existence of a Referral System, to whom and how organs and tissues may be donated (allocation rules) and allow compensation for donation-related expenses.

c) Regulations that address at a minimum the donation, recovery, processing, storage and distribution of tissues are key to insuring the safety of the recipient. These Regulations must include a list of tissue that are applicable to the Regulations, Guidelines or Rules for donor screening criteria, donor approval systems, documentation, systems to guard against cross-contamination of tissue, labelling, quality systems, processing, validation of systems, storage, distribution and traceability of tissues. These Regulations should also be based upon Standards established for / by the Tissue Bank.

d) Regulations regarding the donation of tissue from living donors (including amnion and surgical residues) are necessary and should be based upon the Tissue Banking Standards.

2 - Consent

a) Laws addressing consent for donation are generally in place throughout the world, and vary widely, not only in content but also in practice. At a minimum, a Consent Law must include who may donate (e.g., the individual prior to his/her death, the individual's next of kin following death, or a patient prior to the donation of living tissues, etc.), whether the consent is presumed or informed ('opting out' or 'opting in') and whether the individual's wishes may be countermanded by his/her next of kin. In addition, a mechanism for an individual to change his/her mind about donation prior to death must be included as part of the Law/Regulation. Finally, Laws covering donor registries may also be considered as a way of insuring an individual's choice is carried out, and as a way to increase donation rates in the region.

b) *Presumed Consent*: Many countries have adopted Presumed Consent Laws, in which an individual is assumed to be a donor unless he/she has specifically indicated his/her wish not to be a donor. This decision may be made officially through a non-donor registry (e.g., Belgium, France, Portugal) or informally (e.g., family discussion). This system is also known as 'opting out'. The Presumed Consent Laws in several countries imply the family confirmation of Presumed Consent. Despite the fact that Presumed Consent Laws are in place in many countries, few tissue, eye or organ recovery agencies will proceed with the retrieval process without first discussing donation with the patient's family. They either obtain the next of kin's informed consent for donation or verify the patient's desire to be a donor. In other countries, however, consent confirmation for tissue donation may not be routinely obtained from family members. In some countries, the Medical Examiner / Coroner may allow the recovery of corneas and other tissues without family consent. However, this practice is under increasing scrutiny, due to the need for a family interview in order to determine medical suitability of the donor, and due to the perception that it may violate a donor family's rights.

c) *Informed Consent*: Informed Consent generally involves a discussion with the family of a recently deceased person regarding his / her desire or intent to be a donor, or in the absence of such knowledge or executed gift document, the family's desire to donate organs, eyes or other tissues for transplantation or research. In general, the consent conversation provides the potential donor family information about the recovery process and the uses of tissue for transplantation or research, what a 'reasonable person' would want to know in order to make an informed decision.

d) *Living Donor Consent*: Regulations for the donation of tissues from living donors should require, as a minimum, that Informed Consent be obtained from the donor or his / her legal guardian if he / she is not of majority age. Surgical residue collection (e.g. femoral, head, skin and amnion) imply information and consent from the patient before collection.

3 - Organisation of the Tissue Banks

Regulations addressing the organisation of the Tissue Bank should reference International Standards for Tissue Banks and may include:

- Personnel
- Training
- Building Design & Facilities
- Quality Management
- Equipment Requirements

4 - Interrelationships with Organ Donation Programmes

a) Collaboration between Tissue Banks and Organ Donation / Transplantation Programmes is necessary to minimise confusion within the general public and donor hospitals. It may be advisable to include language in Laws or Regulations that encourages such collaboration. Collaboration between recovery agencies can benefit all. It eliminates duplication of efforts (personnel, organisation, donor promotion / enlightenment programmes), minimises unnecessary expenditures and maximises recovery of organs and tissues when consent for all types are obtained at once. In addition, it reduces the possibility that a bereaved family will be approached with multiple requests to donate

b) Because Laws exist regarding organ donation in many areas, Tissue Banking Laws and Regulations should be written so as to coincide with them wherever possible.

5 - Registration / Licensing / Accreditation / Authorisation of the Tissue Bank

a) At a minimum, Regulations should provide some mechanism for Tissue Banks to be identified through Registration with the National Regulatory Authority (NRA) in order for the NRA to review the Tissue Banks' practices and to ensure compliance with established Regulations.

b) Licensing or official authorisation to operate may be preferred. Regulations requiring licensing must take into account the resources required (financial, personnel, technical) to perform in-depth inspections or evaluations of Tissue Banks. If the NRA does not have the requisite resources, registration can be a reasonable alternative.

c) In some cases, the NRA is unable to adequately inspect or license Tissue Banks. It may, however, choose to contract such activities to another agency or private accrediting body, such as one that accredits laboratories, hospitals or Tissue Banks.

6 - Import/Export of Tissue

a) With the global economy now extending into tissue donation and transplantation, it is critical for Laws and Regulations that address the import and export of donated human tissues. For instance, export of donated human tissues might be allowed only if all needs in the country have been met; or export of tissues might be allowed outright, depending on the Laws and Regulations of the other country.

b) Import of tissue requires specific rules in order to protect tissue recipients and compliance with these Standards or equivalent Standards, including ethical aspects, donor consent and safety issues.

7 - Financial Aspects of Tissue Banking

a) Tissue Banks may be funded in a variety of fashions, including: Governmental agency funding, private funding, funding through investors or through public or private hospitals or universities. Laws and Regulations outlining how Tissue Banks receive compensation or reimbursement for their costs, whether they may charge patients or hospitals for tissue are all necessary.

b) The required financial structure of a Tissue Bank should also be established (non-profit or for-profit or public).

c) Monetary payment or advantages for the donation may not be made to living donors, cadaver donor's next of kin or any donor-related party, excluding compensation for donation-related expenses. However, there are some locations that are considering pilot Programmes that would allow for some moderate financial compensation or reimbursement for travel or funeral expenses to donor families.

d) Commercial sale of tissues is of ethical and safety concern. However, many Laws allow for the cost recovering of all tissue transplantation operations, including research / development and educational costs. Several tissue-processing technologies are covered by patent rights that should be respected. Sale of tissues is a very vague statement and regulatory and legislative bodies would be well advised to clearly define and regulate what is allowable and what is not acceptable.

8 – Enforcement / Compliance

a) Laws and Regulations must include enforcement and compliance of the Regulations, for without such enforcement the Regulations will be far less effective.

b) Enforcement and compliance should include inspections of Tissue Banks and systems for addressing non-compliance or violation of Laws and Regulations. These could include requirements that the Tissue Bank destroy tissue, quarantine or retain tissue until corrective action is completed, notify hospitals, surgeons or patients of non-compliance, or issue a recall for all non-compliant tissue distributed. Penalties for non-compliance (e.g., closure of the Tissue Bank, financial penalties, civil and / or criminal prosecution) should also be considered and fully outlined.

c) *Adverse Events / Self-reporting of Non-compliance*: The Regulations should include requirements that the Tissue Bank have a system for receiving reports of adverse events, and for addressing those reports. In addition, the Regulations should require that the Tissue Bank notify the NRA in the event of serious instances of non-compliance with Standards, SOPs and/or Regulations.

d) *Internal Audits*: The Regulations should include the requirement that the Tissue Bank perform periodic internal audits in order to assure compliance with Standards, SOPs and/or Regulations.

e) If multiple organisations or Tissue Banks are involved in the same Tissue Banking process, the Regulations should address which organisation is ultimately responsible for the tissue. However, a Tissue Bank that engages another organisation or Tissue Bank under a contract, agreement or other arrangement, to perform any step in the process, should be responsible for ensuring that the work is performed in compliance with the requirements established in the Laws and Regulations.

CONCLUSIONS

The development and implementation of appropriate Laws and Regulations is a complex, time-consuming and difficult undertaking. In order for such a system to become a functional reality, it is vital for Tissue Banks to enlist the support of key stakeholders such as end-users (surgeons, dentists, physicians, etc.), tissue recipients and tissue donor families. It may also be possible to enlist the support of the general public and charitable organisations that support programmes intended to better the well being of their fellow citizens.

The importance of donor families should be emphasised, as they can be a powerful advocate for donation, if they are respected and included in the process of improving donation and transplantation. If they are ignored, disrespected or marginalised, they can become an even more powerful group, raising ethical questions about the donation and transplantation system that can result in an overall decrease in donation rates.

The IAEA will encourage all Tissue Banks participating in the IAEA Radiation and Tissue Banking Programme to apply these Standards, in accordance with their national conditions, with the purpose of ensuring the safe clinical use of the tissues produced.

THE DEVELOPMENT OF A CODE OF PRACTICE FOR THE RADIATION STERILISATION OF TISSUE ALLOGRAFTS

Barry J. Parsons [1*], Eulogia Kairiyama [2] and Glyn O. Phillips [3]

[1] *Department of Science, The North East Wales Institute, Mold Road, Wrexham, LL11 2AW, UK.*

[2] *Comision Nacional de Energia Atomica, Avenida Libertador 8250, 1429 Buenos Aires, Argentina.*

[3] *Phillips Hydrocolloids Research Ltd, 45 Old Bond Street, London, W15 4AQ, UK*

ABSTRACT

International Standards have already been developed concerning the use of ionising radiation for the sterilisation of health care products (*ISOs 11137:95, ISO/TR 13409:1996, AAMI/TIR 27:2001*). The International Atomic Energy Agency (IAEA) has been engaged in the development of a similar Code for the sterilisation of tissue allografts by ionising radiation. The process has been informed by the current best practice throughout the world and involved consultation with tissue bank practitioners, particularly in Central and South America, as well as in parts of Asia and Europe. This paper gives some of the essential features of the draft Code and in particular addresses the problems associated with non-uniformity of tissue allograft samples and also with the possibility of viral contamination. These and other issues crucial to the development of an international Code of Practice are described here.

KEYWORDS

Radiation; sterilisation; tissues; allografts; bacteria; viruses

INTRODUCTION

The sterilisation of health care products

International standards have been established for the radiation sterilisation of health care products which include medical devices, medicinal products (pharmaceuticals and biologics) and *in vitro* diagnostic products *(ISO 11137:1995 (E); ISO 11737-1:1995; ISO 11737-2:1998; ISO/TR 13409:1996, ISO/TR 15844:1998 and AAMI/TIR 27:2001).*

Following intensive studies of the effects of ionising radiation on chemical, physical and biological properties of tissue allografts and their components, these are now radiation sterilised using a variety of methods and practices.

Through its Radiation and Tissue Banking Programme, the International Atomic Energy Agency has sought during the period 2001-2002 to establish a Code of Practice for the Radiation Sterilisation of Tissue Allografts and its requirement for validation and routine control of the sterilisation of tissues.

The details set out in this paper describe the essential elements of such a Code and, in particular, the requirements of a process which would ensure that the radiation sterilisation of tissues produces standardised sterile tissue allografts suitable for safe clinical use. Although the principles adopted here are similar to those used for the sterilisation of health care products, there are substantial differences in practice arising from the physical and biological characteristics of tissues.

For health care products, the items for sterilisation come usually from large production batches. For example, syringes are uniform in size and have bacterial contamination arising from the production process, usually at low levels. It is the reduction of the microbial bioburden to acceptable low levels, which is the purpose of the sterilisation process, where such levels are defined by the Sterility Assurance Level (SAL). The inactivation of microorganisms by physical and chemical means follows an exponential law and so the probability of a surviving microorganism can be calculated if the number and type of microorganisms are known and if the lethality of the sterilisation process is also known. Two methods are used in ISO 11137:1995 to establish the radiation doses required to achieve low SAL values.

Method 1 of ISO 11137:1995 relies on knowing the bioburden (assuming a Standard Distribution of Resistances) before irradiation and uses this data to establish a verification dose, which will indicate the dose needed for a SAL of 10^{-2}. The method involves a statistical approach to setting the dose based on three batches and hence relatively large numbers of samples are required for both establishing the initial bioburden and the verification dose, both per product batch. A further adaptation of method 1 for a single production batch has also been developed (*ISO/TR 15844 – 1998*).

In Method 2 of ISO 11137:1995, the bioburden levels are measured after giving a series of incremental doses to the samples, these doses being well below the dose required for a SAL of 10^{-6}. In this method, 280 samples are required to determine the dose to produce a SAL value of 10^{-2}, from which the dose needed to yield a SAL value of 10^{-6} may be extrapolated. No assumptions are made in Method 2 about the distribution of microorganisms and their resistances.

In a later ISO/TR 13409:1996, Method 1 was adapted to allow the use of as few as 10 samples to determine the verification dose. In this modification, the dose needed for a SAL value of 10^{-1} is used to establish the dose required for a SAL value of 10^{-6}. The sole purpose, however, of this modification is to substantiate whether 25 kGy is an appropriate dose to achieve a SAL value of 10^{-6}. In AAMI/TIR 27:2001, another method to substantiate the sterilisation dose of 25 kGy has been developed which may replace the method in ISO/TR 13409 as the internationally accepted method of choice.

The sterilisation of tissue allografts

Tissues used as allografts comprise a wide range of materials and bioburden levels such that the above quality assurance methods developed for health care products cannot be applied without careful and due consideration given to the differences between health care products and tissue allografts.

Tissues that are sterilised currently include: bone, cartilage, ligaments, tendons, fascias, dura mater, heart valves, vessels, skin and amnion. Unlike health care products, the variability in types and levels of bioburden in tissues is much greater than that found for health care products where the levels of microbial contamination are usually low and relatively uniform in type and level.

In addition, tissue allografts are not products of commercial production processes involving large numbers of samples. These differences mean that extra attention must be given to the following:

a) Uniformity of sample physical characteristics (shape and density),

b) Uniformity of bioburden in sample,

c) Donor screening for viral contamination,

d) Whether low numbers of samples can be used for sterilisation dose setting purposes.

OBJECTIVE AND SCOPE

The objective of the Code is to provide the necessary guidance in the use of ionising radiation to sterilise tissue allografts in order to ensure their safe clinical use.

The Code specifies requirements for validation, process control and routine monitoring of the selection of donors, tissue processing, preservation, storage and the radiation sterilisation of tissue allografts. They apply to continuous and batch type gamma irradiators using the radioisotopes ^{60}Co and ^{137}Cs, electron beam accelerators and X-rays.

The principles adopted here are similar to those elucidated in ISO 11137:1995 in that statistical approaches to establishing doses to assure sterility of the tissue products are proposed.

VALIDATION OF PRE-STERILISATION PROCESSES

General

An essential step in the overall radiation sterilisation of tissues is rigorous donor selection to eliminate specific contaminants. Full details about donor selection, tissue retrieval, tissue banking general procedures, specific processing procedures, labelling and distribution are given in IAEA International Standards for Tissue Banks. Such tissue donor selection, retrieval, processing and preservation are processes which determine the characteristics of the tissue allograft prior to the radiation sterilisation process. The most important characteristics are those relating to use of the tissues as allografts, namely, their physical, chemical and biological properties, the latter including the levels and types of microbial contamination. Validation of these processes shall include the following:

a) Qualification of the Tissue Bank facilities,
b) Qualification of the tissue donors,
c) Qualification of the tissue processing and preservation,
d) Certification procedure to review and approve documentation of a), b) and c),
e) Maintenance of validation,
f) Process specification.

Qualification of the Tissue Bank facilities

Tissue Banks should have facilities to receive procured tissues and to prepare tissue allograft material for sterilisation. Such facilities are expected to include laboratories for the processing, preservation and storage of tissues prior to sterilisation. These laboratories and the equipment contained therein should meet international standards enunciated by the various Tissue Bank Professional Associations and now combined in the IAEA International Standards for Tissue Banks. A regular documented system should be established which demonstrates that these standards are maintained, with special emphasis on the minimisation of contamination by microorganisms throughout the tissue retrieval, transportation, processing, preservation and storage stages to bioburden levels which comply with the IAEA International Standards for Tissue Banks.

Tissue Banks should also have access to qualified microbiological laboratories to measure the levels of microorganisms on the tissue allografts at various stages in their preparation for the purposes of assessing both the levels of contamination at each stage and also typical bioburden levels of the pre-irradiated tissue allografts. The standards expected of such laboratories are specified in: *ISO 11737-1:1995 and ISO 11737-2:1998*. The overall purpose of the above facilities contained within Tissue Banks is to demonstrate that they are capable of producing preserved tissue allografts which have acceptably low levels of microorganisms in the preserved product prior to their sterilisation by radiation.

Qualification of tissue donors

The main aim of the tissue donor selection process carried out prior to processing, preservation, storage and sterilisation is to produce tissue allografts that are free from transmissible infectious diseases. Such a selection process in order to produce acceptable tissues should include the following minimal information:

a) Time of retrieval of tissue after death of donor, conditions of body storage,
b) Age of donor,
c) Medical, social and sexual history of donor,
d) Physical examination of the body,
e) Serological (including molecular biology) tests,
f) Analysis of autopsy as required by law.

Such information should be used to screen donors to minimise the risk of infectious disease transmission from tissue donors to the recipients of the allografts. The information so collected should be comprehensive, verifiable and auditable following Good Practice on Tissue Banking, as specified in the IAEA International Standards for Tissue Banks.

The following serological tests should be carried out as a minimum on each donor:

a) Antibodies to Human Immunodeficiency Virus 1 and 2 (HIV 1,2),
b) Antibodies to Hepatitis C virus (HCV),
c) Hepatitis B surface antigen (HBs-Ag),
d) Syphilis: non-specific (e.g. VDRL) or preferably specific (e.g. TPHA).

Other tests may be required by statutory regulations or when specific infections are indicated as specified in the IAEA International Standard for Tissue Banks. In using such laboratory-based tests to provide additional assurance that allografts are free of transmissible disease, due consideration should be given to the detection limits of such tests. It should therefore be verified that the combination of processing, preservation and irradiation is capable of reducing low levels of viral contamination, which might be implied by an otherwise negative test, to a SAL of 10^{-6}.

When addressing the problem of viral contamination, the same basic principles already advanced for elimination of bacterial contamination need to be applied with regard to donor screening, serology, processing, preservation and sterilisation by ionising radiation. It should be noted that the D_{10} values for viruses are, in general, higher than those for bacteria.

Qualification of tissue processing and preservation

The processing of tissue allograft materials such as bone, cartilage, ligaments, fascias, tendons, dura mater, heart valves and vessels, skin and amnion comprises various stages such as removal of bone marrow, defatting, pasteurisation, antibiotic treatment, percolation and treatment with disinfectants such as hypochlorite, ethyl alcohol and glycerol. The inclusion of any or all of these stages will depend on a number of factors including:

a) The preferred practice of the Tissue Bank,
b) The nature of the tissue (and its anticipated use in the clinic), and
c) The degree of contamination of the procured tissue.

The preservation of the processed tissue allografts may include:

a) Freeze drying,
b) Deep freezing,
c) Air drying,
d) Heat drying and
e) Chemical treatment.

An important function of the above processes is to produce tissue allografts that have low levels of microbial contamination and in particular less than 1000 cfu per allograft product when it is desired to substantiate a sterilisation dose of 25 kGy. In the latter case, for a bioburden of 1000 cfu per allograft product, a 25 kGy dose is sufficient to achieve a SAL of 10^{-6} for a Standard Distribution of Resistances. The capacity of all of the tissue processing and preservation procedures to remove microorganisms should be checked periodically and documented.

Maintenance of validation

For each of the qualifications detailed above, a validation process should be specified, which will demonstrate that the standards expected will be maintained. As a minimum, these validation processes should include:

a) An audit of the origin and history of the procured tissues,
b) A random, statistically significant sampling of procured tissues (that is, prior to processing and preservation) followed by a laboratory-based screening for viruses and infectious agents,
c) Measures of particle count and microbial contamination in the environment of each of the separate facilities of the Tissue Bank,
d) Random, statistically-significant sampling of tissue allografts prior to and after tissue processing and preservation for measurements of bioburden levels,
e) Determination of the ability of the tissue processing and preservation procedures to both reduce the levels of microorganisms and to produce the levels of bioburden required for the radiation sterilisation process. This should ensure a microbial contamination level of 1000 cfu per allograft product or less when it is required to substantiate a sterilisation dose of 25 kGy.

Process specification

A process specification should be established for each tissue allograft type. The specification should include:

a) The tissue allograft type covered by the specification,
b) The parameters covering the selection of tissue for processing,
c) Details of the tissue processing and preservation carried out prior to irradiation as appropriate to each tissue type,
d) Details of the equipment, laboratory and storage facilities required for each of the processing and preservation stages, particularly with regard to acceptable contamination levels,
e) Details of the routine preventative maintenance programme,
f) Process documentation identifying every processed tissue, including details of its origin, its processing and preservation, dates of performing all processes, details of process interruptions, details of any deviations from the adopted processing and preservation procedures.

VALIDATION OF THE STERILISATION PROCESS

General

The guidance given here is based on the procedures specified in previous documents (ISO 11137:1995, ISO/TR 13409:1996, ISO/TR 15844:1998 and AAMI TIR 27:2001) for the sterilisation of health care products. More emphasis is given here, however, on the factors which affect the ability of the sterilisation process to demonstrate that an appropriate Sterility Assurance Level (SAL) can be achieved with low numbers of tissue allografts, which may have more variability in the types and levels of microbial contamination than is found in health care products and which may also be more variable in size and shape.

More specifically, several approaches to establishing a sterilisation dose are proposed for the small numbers of tissue allografts typically processed. Emphasis is placed on the need to take into account both the variability of bioburden from one tissue donor to another, as well as the variability of size and shape of tissue allografts, which can affect both the accuracy of product dose mapping (and hence the sterilisation dose itself) and also the applicability of using Sample Item Portions (SIP) of a tissue allograft product.

Validation of the sterilisation process should include the following elements:

a) Qualification of the tissue allografts and their packaging for sterilisation,
b) Qualification of the irradiation facility,
c) Process qualification using a specified tissue allografts or simulated products in qualified equipment,
d) A certification procedure to review and approve documentation of a), b) and c),
e) Activities performed to support maintenance of validation.

Qualification of the tissue allografts for sterilisation

Evaluation of the tissue allograft and packaging

Prior to using radiation sterilisation for a tissue allograft, the effect that radiation will have on the tissue allograft and its components should be considered. The key references given in the Bibliography contain information on this aspect. Similarly, the effect of radiation on the packaging should also be considered. Guidance on the latter is given in Annex A of ISO 11137:1995. Using such information, a maximum acceptable dose shall be established for each tissue allograft and its packaging.

Sterilisation dose selection

Knowledge of the number and resistance to radiation of the microorganism population as it occurs on the tissue allografts should be obtained and used for determination of the sterilisation dose. For the sterilisation of health care products, a reference microbial resistance distribution was adopted in ISO 11137-1:1995 for microorganisms found typically on medical devices. Studies should be carried out to establish the types of microorganisms that are normally found on the tissue types to be sterilised as well as their numbers and resistance to radiation. Such studies should take account of the distribution of the microorganisms within the tissue allograft itself since this may not be uniform. This should be determined by taking Sample Item Portions (SIP) of the tissue and demonstrating that there are no significant statistical variations in distribution from SIP to SIP.

If such studies show a consistent distribution of microorganisms from one tissue allograft to another, and one which is less resistant than the Standard Distribution of Resistances (SDR) (See Table 1), then a Table similar to B24 in ISO 11137:1995 giving a distribution of resistances appropriate to the allografts may be constructed for the purpose of sterilisation dose setting. This would allow the use of appropriate and perhaps lower sterilisation doses than would be the case if Method 1 in ISO 11137:1995, based on the SDR in Table 1, were used. In the absence of such studies, the SDR may be used to establish sterilisation doses.

Table 1. Microbial Standard Distribution of Resistance (SDR).

D_{10} (kGy)	1.0	1.5	2.0	2.5	2.8	3.1	3.4	3.7	4.0	4.2
%	65.487	22.493	6.302	3.179	1.213	0.786	0.350	0.111	0.072	0.007

K. W. Davis, W. E. Strawderman and J. L. Whitby, 'The rationale and computer evaluation of a gamma sterilization dose determination method for medical devices using a substerilization incremental dose sterility test protocol', J. Appl. Bact., (1984) 57, 31-50.

To establish a sterilisation dose which will give a Sterility Assurance Level (SAL) of 10^{-6}, the methods based on those in ISO 11137:1995, ISO/TR 15844:1998, ISO/TR 13409:1996 and AAMI TIR 27:2001 should be used. A summary of these approaches as they apply to tissue allografts is given below.

Establishing a sterilisation dose

This section describes the practices and procedures for determining the bioburden levels of the tissue allografts and the application of this information to establish the radiation sterilisation dose. It must to be emphasised that such samples must be the end result of the series of validated donor screening and subsequent procedures as are described in the IAEA International Standards for Tissue Banks.

Selection of tissue allograft products

Tissue allografts can be prepared from a wide range of tissues such as skin, amnion, bone, cartilage tendons and ligaments. If samples can be prepared from these tissues, which are reasonably reproducible in shape, size and composition and also in sufficient numbers for statistical purposes, then the usual sampling procedures apply, as given, for example, in ISO 11137 and ISO/TR 13409. However, if allograft products are both few in number (less than 10) and cannot be considered as identical products then it may be necessary to take multiple Sample Item Portions of a single tissue allograft product for both bioburden analysis prior to sterilisation and also for the purpose of establishing a sterilisation dose. In such instances, it is important to have confidence in microorganism distribution throughout the sample, obtained e.g. by periodic monitoring of such products.

Sample Item Portion (SIP)

The SIP should validly represent the microbial challenge presented to the sterilisation process. SIPs may be used both to verify that microorganisms are distributed evenly, bioburden estimation and for establishing a sterilisation dose. It is important to ascertain that the SIPs are representative, not only in shape size and composition but also in bioburden. Statistical tests should be applied to establish this. At least 20 SIPs should be used (10 for bioburden testing and 10 for the verification dose experiments).

Bioburden determination

Bioburden determination could include the count of aerobic bacteria, spores, yeasts, molds and anaerobic bacteria. Many factors determine the choice of the tests most appropriate for the tissue allograft. At a minimum, the aerobic bacteria and fungi should be counted.

The objective of the bioburden determination is to:

a) Determine the total number of viable microorganisms within or on a tissue allograft and the packaging after completion of all processing steps before sterilisation,
b) Act as an early warning system for possible production problems,
c) Calculate the dose necessary for effective radiation sterilisation.

The validation of the bioburden estimation requires the determination of the effectiveness and reproducibility of the test method. The steps to estimate bioburden can be found in ISO 11737-1.1995.

Determination of the verification dose

In ISO 11137, the concept of establishing a verification dose for a SAL value which is much higher than 10^{-6}, for example, for a SAL value of 10^{-2} was proposed as an experimental method of establishing the sterilisation dose corresponding to a SAL of 10^{-6}. For such verification dose experiments, samples of tissue allografts should be taken from production batches and irradiated at the calculated verification dose. In these experiments it is assumed (and should be demonstrated statistically) that the tissue allograft products are reasonably uniform in shape, size, composition and bioburden distribution. For single batch sizes up to 999, the numbers of sample required may be obtained from Table 1 of ISO/TR 13409. For minimum batch sizes of 20-79, for example, 10 samples are required for the bioburden determination and 10 for the verification dose experiment. In general, the number of samples required for the bioburden determination and verification dose experiments will depend on the number of batches and the number of samples in each batch. For each circumstance, the number of positive sterility tests allowed in the verification dose experiment should be calculated statistically using an acceptable range of values of probability for 0, 1, 2, 3 etc positive tests of sterility. For the 100 samples used in Method 1 of ISO 11137, for example, there is a 92% chance of there being 1% positives when up to 2 positives are detected and a 10% chance of accepting a batch with 5.23% positives (*W. A. Taylor and J. M. Hansen, 'Alternative Sample Sizes for Verification Dose Experiments and Dose Audits', Radiation Physics and Chemistry, 1999, 54, 65-75).*

For the 10 samples taken in ISO/TR 13409:1996 from a batch of 20, up to 1 positive test of sterility is proposed. For 30 or more, up to 2 positive tests of sterility are proposed (ISO/TR 13409:1996). It should be noted here that these latter statistical tests do not offer the same degree of protection as obtained when accepting up to two positive tests of sterility for a sample size of 100. For example, when accepting up to one positive test of sterility in a sample size of ten, there is a 95% chance of accepting a batch with 3.68% positives and a 10% chance of accepting a batch with 33.6% positives. Alternative sampling strategies are now available (see *Taylor and Hansen 1999* above) which include for example; double sampling plans that can minimise sample sizes and yet offer similar protection. For single batches of low sample sizes, protection levels similar to those of the 100-sample approach in ISO 11137 can only be obtained by accepting a small number (possibly even zero) of positive sterility tests. For example, accepting up to one positive for a sample size of 50 offers similar protection.

Hence, in ISO/TR 13409:1996 the verification dose for 10 samples taken from a batch of 20 is that which is required to produce a SAL of 10^{-1} (the reciprocal of the number of SIPs used) and is that dose which will yield not more than one positive test of sterility from the ten irradiated SIPs.

In order to calculate the verification doses as well as the doses required to produce a SAL value of 10^{-6}, one of several approaches may be taken to establish an appropriate verification dose for low sample numbers (up to 100 but typically much less). The methods proposed here for the establishment of a sterilisation dose are based on statistical approaches used previously for the sterilisation of health care products (ISO 11137:1995, ISO 13409:1996, ISO 15844:1998, AAMI/TIR 27:2001) and modified appropriately for the typical low numbers of tissue allografts samples available. For a Standard Distribution of Resistance (SDR), the Tissue Bank may elect to substantiate a sterilisation dose of 25 kGy for microbial levels up to 1000 cfu per unit. Alternatively, for the SDR and other microbial distribution, specific sterilisation doses may be validated depending on the bioburden levels and radiation resistances (D_{10} values) of the constituent microorganisms.

a) *For establishing specific sterilisation doses for Standard Distribution of Resistance and other microbial distribution for samples sizes between 10 and 100* an adaptation of Method 1 of ISO 11137:1995 may be used. Method 1 of ISO 11137 is normally used for multiple batches containing a large number of samples per batch. For batches of 100 samples for example, verification dose experiments are carried out for a SAL of 10^{-2}. A successful experiment (up to 2 positive tests of sterility) will then enable the dose required to achieve a SAL value of 10^{-6} to be calculated from the survival curve of a Standard Distribution of Resistances (SDR). In this Code, an extension of Table 1 of ISO 11137 is given so that verification doses for SAL values between 10^{-2} and 10^{-1} may be found for bioburden levels up to 1000 cfu per allograft product. These SAL values correspond to relatively low sample sizes of 10-100. This allows Method 1 to be used for typical tissue allografts where relatively low numbers of samples are available and also where the distribution of microbial radiation resistances is known and different to the SDR. The worked example given later uses this approach and, in addition, applies it (with appropriate statistical sampling, see above) to a microbial population that has a different distribution of radiation resistances than the SDR. However, for low bioburden levels combined with low sample numbers, it may be anticipated that there is an increased probability using this adaptation of Method 1 that the verification dose experiment may fail. In the case of failure, the methods outlined in b) and/or c) may be used.

b) *For substantiation of a 25 kGy sterilisation dose, the Method in ISO/TR 13409:1996* may be used to calculate the verification dose. This is an accredited method and is essentially a modification of the method in a) above and applies only to a Standard Distribution of Resistances. In this method, the verification dose for a given SAL is approximated to the initial bioburden by a series of linear relationships. Each linear equation is valid for a particular ten-fold domain of bioburden level e.g., 1-10 cfu. The method in ISO/TR 13409:1996 can only be used to substantiate a dose of 25 kGy. It should be noted that the statistical approach allowing up to 1 positive test for sample sizes up to 30 and up to 2 positive tests for sample sizes above 30 does not offer the same level of protection as for the 100 samples in ISO 11137 until the sample size reaches 100. Alternative sampling strategies may be employed (*Taylor and Hansen 1999*) for all the verification dose methods proposed here.

c) ***For substantiation of a 25 kGy sterilisation dose,*** an alternative and more recent method in AAMI/TIR 27 may be used. The modification takes into account how the verification dose varies with bioburden level for a given SAL (and sample size) on the assumption that an SAL of 10^{-6} is to be achieved at 25kGy. Depending on the actual bioburden levels to be used (1-50 or 51-1000 cfu per allograft product), a linear extrapolation of the appropriate SDR survival curve is made from either (log N_0, 0 kGy) or (log 10^{-2}) to (log 10^{-6}, 25 kGy) for 1-50 cfu and 51-1000 cfu respectively. For bioburden levels less than 1000 cfu per allograft unit, these constructed survival curves represent a more radiation resistant bioburden than would otherwise be the case. The validity of this approach arises from the purpose of the method, which is to validate a sterilisation dose of 25 kGy. For all bioburden levels below 1000 cfu per allograft product, this means that for the reference microbial resistance distribution given in Table B24 of ISO 11137:1995 for medical devices, a more conservative approach to the calculation of a verification dose is taken. Hence, this modification allows the use of greater verification doses than would be allowed using the formula given in either Method 1 of ISO 11137 or in ISO/TR 13409:1996. The result is that there are fewer unexpected and unwarranted failures relative to verification doses experiments carried out using the method in ISO/TR 13409:1996. At a bioburden level of exactly 1000 cfu per allograft product (the maximum in both methods), there is no difference in the outcomes of the methods, i.e. the calculated verification doses are identical.

Procedures

a) Establish test sample sizes

Select at least 10 allograft products or SIPs, as appropriate, for the determination of the initial bioburden. The number of allograft products or SIPs should be sufficient to represent validly the bioburden on the allograft product(s) to be sterilised.

Select between 10 and 100 allograft products (or SIPs) for the verification dose experiments and record the corresponding verification dose SAL (= 1/n, where n is the number of allograft products or SIPs used). For 20-79 allograft products in a single batch, 10 allograft products may be used for both the bioburden determination and the verification dose experiment.

b) Determine the average bioburden

Using methods such as those in ISO 11737-1:1995 and as described above (Bioburden estimation), determine the average bioburden of at least 10 allograft products or SIPs (the number will depend on the number of batches and the number of samples in the batches).

For SIP values less than unity, the bioburden level for the whole product should be calculated and should be less than 1000 cfu per allograft product for verification dose experiments carried out to substantiate a 25 kGy sterilisation dose.

c) Establish the verification dose

The appropriate verification dose depends on the number of samples (allograft products or SIPs) to be used in the experiment (= 1/ number of samples). The verification dose calculation depends on which of the three methods above is being used, as follows:

i) For establishing specific sterilisation doses for Standard Distribution of Resistance and other microbial distribution for samples sizes between 10 and 100:
(an adaptation of Method 1 of *ISO 11137:1995)*

Calculate the dose required to achieve the required SAL from knowledge of the initial bioburden level and from the microbial distribution and associated radiation resistances. This may be calculated from the equation:

$$N_{tot} = N_{0(1)} . 10^{-(D/D1)} + N_{0(2)} . 10^{-(D/D2)} + \cdots\cdots + N_{0(n)} . 10^{-(D/Dn)}$$

Where N_{tot} , represents the number of survivors; where $N_{0(i)}$ represents the initial numbers of the various microbial strains i (where i = 1-n), D1, D2 ……D(n) represent the D_{10} values of the various microbial strains. D represents the radiation dose and n the number of terms in the equation for a standard distribution of resistances (n = 10). For the reference Standard Distribution of Resistances (K. W. Davis, W. E. Strawderman and J. L. Whitby, J. Applied Bacteriology, 1984, 57, 31-50) used in ISO 11137:1995 for medical devices (see Table 1), this equation will produce data similar to Table B1 of ISO 11137:1995 but for SAL values between 10^{-2} and 10^{-1} instead. By equating N_{tot} to the selected SAL value and by using the appropriate D_{10} values for each microbial type together with their numbers prior to irradiation, the verification dose, D, for SAL values between 10^{-2} and 10^{-1} can be calculated. The same calculation can be used to find the sterilisation dose for the desired SAL of 10^{-6} or reference can be made to Table B1 of ISO 11137:1995. In this method, the sterilisation dose is calculated using the bioburden level of the whole product. Alternatively, approximate values of the verification doses to achieve the same SAL values may be calculated using the equation given in ISO/TR 13409:1996 (see next paragraph).

***ii) For substantiation of a 25 kGy sterilisation dose* (Method ISO/TR 13409:1996):**

From knowledge of the average bioburden and the number of samples or SIPs to be used in the verification experiment, the verification dose for a Standard Distribution of Resistances is approximated by the equation:

Verification dose at a the selected SAL = I + [S x log (bioburden)]

Where appropriate values of I and S are given in *ISO/TR 13409:1996*

iii) For substantiation of a 25 kGy sterilisation dose (Method AAMI/TIR 27:2001):

The calculation of the verification dose follows the procedures by Kowalski and Tallentire (*Radiat. Phys. Chem., 1999, 54, 55-64.*) where the bioburden levels refer to either the SIP or whole product whichever is being used in the verification dose experiment:

For bioburden levels of 1 to 50 cfu per allograft product or SIPs:

Step 1 $D_{lin} = 25kGy/(6 + \log N_0)$

Step 2 Verification dose = $D_{lin} (\log N_0 - \log SAL_{VD})$

Where D_{lin} represents the D_{10} dose for a hypothetical survival curve that is linear between the coordinates ($\log N_0$, 0 kGy) and ($\log 10^{-6}$, 25 kGy) for initial bioburden levels, N_0, up to 1000 cfu per allograft product. This linear plot therefore represents a constructed survival curve in which there is 1 out of 10^6 probability of a survivor at 25 kGy. The method is valid therefore only for the substantiation of a 25 kGy sterilisation dose regardless of whether a lower dose could in fact be validated.

For bioburden levels of 51 to 1000 cfu per allograft product or SIPs:

Step 1 For a particular value of bioburden, use Table B1 of ISO 11137:1995 to identify doses (kGy) corresponding to SAL values of 10^{-2} [$D(10^{-2})$] and 10^{-6} [$D(10^{-6})$]. From these values, calculate TD_{10} from the following equation:

$TD_{10} = (Dose_{-6}\ kGy - Dose_{-2}\ kGy) / 4$

Where TD_{10} represents the hypothetical D_{10} value for a survival curve for a Standard Distribution of Resistances which has been modified such that it is linear between $\log 10^{-2}$ and $\log 10^{-6}$ (log SAL values) when plotted against dose, with the $\log 10^{-6}$ value being set at 25 kGy. Essentially, this produces a survival curve that is more resistant to radiation than the SDR (for bioburden levels less than 1000 cfu per allograft product) and one that is appropriate to substantiation of a 25 kGy sterilisation dose only.

Step 2: Verification dose $= 25kGy - [TD_{10} (\log SAL_{VD} + 6)]$

Where SAL_{VD} is the sterility assurance level at which the verification dose experiment is to be performed

d) Perform verification dose experiment

Irradiate the tissue allografts or SIPs thereof at the verification dose. Irradiation conditions of the samples for verification of the sub-sterilisation dose should be the same as the whole batch that is to be sterilised. For example, if the produced tissue batch is irradiated in frozen condition, the samples for the sub-sterilisation dose verification studies should be irradiated in the same condition and the frozen condition should be kept during the whole irradiation process.

The defined test sample size (SIP ≤ 1), according to the SAL and batch size, is exposed to radiation at the verification dose. The dose delivered should not be less than 90% of the calculated verification dose. Test the tissue allografts for sterility using the methods in ISO 11737- 2:1998 and record the number of positive tests of sterility. The irradiated SIPs, of all types of tissue allografts, are transferred to a growth medium and incubated for at least 14 days at an appropriate temperature. Positive and negative sterility tests results should be registered. For bone and skin allografts, an additional test is recommended to detect anaerobic bacteria.

e) Interpretation of results

For a verification dose experiment performed with up to 30 allograft products or SIPs, statistical verification is accepted if there is no more than one positive test of sterility observed. For 30 to 100 products or SIPs, statistical verification is accepted if there are no more than two positive tests of sterility observed (ISO/TR 13409:1996).

Where the verification dose experiment is successful, the dose required to produce a SAL of 10^{-6} for the whole allograft product should be calculated for procedure c (i) as indicated above. For procedures c (ii) and c (iii), a successful verification dose experiment substantiates the use of 25 kGy as a sterilisation dose.

Routine use of sterilisation doses

The routine use of a sterilisation dose calculated in procedure c (i) or of 25 kGy as substantiated by either procedure c (ii) or c (iii) should only be valid if the tissue selection and tissue processing procedures have been demonstrated to produce tissues allografts with consistent bioburden levels. It should be demonstrated that the level of variation in bioburden, is consistent with the sterilisation dose to be used routinely. In such cases, sterilisation dose audits should be carried out at regular intervals, at least every three months.

Technical requirements

The technical requirements to generate the information required for selection of the sterilisation dose shall be:

a) Access to qualified microbiological and dosimetric laboratory services,
b) Microbiological testing performed in accordance with ISO 11737-1:1995 and ISO 11737-2:1998,
c) Access to a ^{60}Co or ^{137}Cs radiation source, or electron beam or X-ray irradiators.

Transfer of sterilisation dose

The conditions for transferring the sterilisation dose between two irradiation facilities are the same as those given in ISO 11137:1995 (Section 6.2.3) and apply equally to tissue allografts.

QUALIFICATION OF THE IRRADIATION FACILITY

The principles covering the documentation of the irradiation system, its testing, calibration and dose mapping are covered in ISO 11137:1995 (Section 6.3) and apply equally to tissue allografts.

QUALIFICATION OF THE IRRADIATION PROCESS

Determination of the product-loading pattern

The principles given in ISO 11137:1995 (Section 6.4.1) covering this should also apply for the sterilisation of tissue allografts.

Product dose mapping

In general, the guidelines given in ISO 11137:1995 (Section 6.4.2) apply also to tissue allografts. However, it should be recognised that the product dose mapping of relatively uniform (i.e. in shape, size, composition and density) health care products is a more straightforward task than the product dose mapping of tissue allografts, which by their nature are more variable in their physical characteristics. In particular, the density of tissue allografts may vary depending on their water content. In addition, some tissue allografts may be heterogeneous in their distribution of density within the product, requiring an appropriate number of dosimeters for the dose mapping exercise. A consideration of these factors affecting the actual absorbed dose in tissue allografts must be undertaken so that the level of accuracy in delivering a dose to a particular tissue can be determined.

The acceptability of the accuracy of delivering a dose to tissue allografts will depend on the dose delivered in the verification dose experiments. If, for example, the actual dose delivered at its lowest possible accuracy limit is less than 90% of the verification dose, then the verification test must be repeated at a higher dose. Similarly, the minimum absorbed dose administered for sterilisation should take into account the likely variation in dose delivered so that sterilisation can be assured. As a guideline, uncertainties in the delivered dose should be within +/- 10%.

MAINTENANCE OF VALIDATION

The guidelines covering calibration of equipment and dosimetric systems, irradiator re-qualification and sterilisation dose auditing are the same as given in ISO 11137:1995 (Section 6.6) and apply equally to tissue allografts.

ROUTINE STERILISATION PROCESS CONTROL

The guidelines covering process specification, tissue allograft handling and packing in the irradiation container, sterilisation process documentation are similar to those given in ISO 11137:1995 (Section 7) and apply equally to tissue allografts.

QUALITY, SAFETY AND CLINICAL APPLICATION OF THE GRAFTS

A programme to demonstrate the quality, safety and clinical application of the tissue allograft throughout its shelf life shall be performed. Sampling procedures appropriate to the tissue type should be devised for this purpose.

DOCUMENTATION AND CERTIFICATION

Information gathered or produced while conducting the qualification and validation of the tissue allografts, Tissue Bank facilities and tissue processing, preservation and radiation sterilisation procedures shall be documented and reviewed for acceptability by a designated individual or group and retained in accordance with ISO 9001:2000 and the IAEA International Standard for Tissue Banks or revision thereof, whichever is applicable.

MANAGEMENT AND CONTROL

Control of the procedures involved in the selection of tissue donors, tissue processing and preservation prior to sterilisation by radiation and the radiation sterilisation process itself, shall be fully documented and managed in accordance with ISO 9001:2000 and IAEA International Standard for Tissue Banks, whichever is applicable.

CONCLUSIONS

This paper sets out the principles that the International Standards Organization (ISO) has applied to the radiation sterilisation of health care products. The same approach has been adapted to take into account the special features associated with human tissues, and the features which distinguish them from industrially produced sterile health care products.

The approach is described in detail and addresses directly the problem of viral as well as bacterial contamination here. Thus, it is emphasised that the human donors of the tissues must be medically and serologically screened, the latter taking into account the limits of detection of viruses. To further support this screening, it is recommended that autopsy reports are also reviewed if available. This adaptation of established ISO methods can thus only be applied for sterilisation of tissue allografts if the radiation sterilisation described here is the terminal stage of a careful detailed, documented sequence of procedures, involving:

- Donor selection,
- Tissue retrieval,
- Tissue banking general procedures,
- Specific processing procedures,
- Labelling and
- Distribution.

All of which are conducted according to the IAEA International Standards for Tissue Banks. It shall not be used outside this context.

The methods proposed here for the establishment of a sterilisation dose are based on statistical approaches used for the sterilisation of health care products (ISO 11137:1995, ISO 13409:1996, ISO 15844:1998, AAMI TIR 27:2001) and modified appropriately for the low numbers of tissue allograft samples typically available.

For a Standard Distribution of Resistances (SDR), the Tissue Bank may elect to substantiate a sterilisation dose of 25 kGy for microbial levels up to 1000 colony forming units (cfu) per allograft product.

Alternatively, for the SDR and other microbial distribution, specific sterilisation doses may be validated depending on the bioburden levels and radiation resistances (D_{10} values) of the constituent microorganisms.

BIBLIOGRAPHY

General

Advances in Tissue Banking, Volumes 1-4, (1997-2000), (Phillips, GO, Editor-in-Chief) World Scientific, Singapore, (ISBNs 981-02-3190-3; 981-02-3524-8; 981-02-3872-X; and 981-4287-5)

Radiation and Tissue Banking, (2000) (Phillips, Go, Ed.) World Scientific, Singapore, (ISBN 981-4287-7)

Biological Principles of Tissue Banking, (1982) (Klen, R, Ed., with Phillips, GO, English Ed.) Pergamon Press

Dziedzic-Goclawska A
Effect of radiation sterilization on biostatic tissue grafts and their constituents
In: *Sterilization by Ionizing Radiation* (Gughran WRL and Goudie AJ, Eds.) Multiscience, Montreal, 1978, Vol. 2, pp. 156-187

Bone

Akkus O and Rimnac CM
Fracture resistance of gamma radiation sterilized cortical bone allografts
J Orthop Res (2001) 19, 927-34

Cornu O, Banse X, Docquier PL, Luyckx S and Delloye C
Effect of freeze-drying and gamma irradiation on the mechanical properties of human cancellous bone
J Orthop Res (2000) 18, 426-31

Moreau MF, Gallois Y, Basle MF and Chappard D
Gamma irradiation of human bone allografts alters medullary lipids and releases toxic compounds for osteoblast-like cells
Biomaterials (2000) 21, 369-76

Silberman F and Kairiyama E
Radiation sterilisation and the surgical use of bone allografts in Argentina
Advances in Tissue Banking (2000), 4, 27-38

Hilmy N, Febrida A and Basril A
Validation of radiation sterilization dose for lyophilized amnion and bone grafts
J Cell Tissue Banking (2000) 1, 143-147

Araki N, Myoui A, Kuratsu S, Hashimoto N, Inoue T, Kudawara I, Ueda T, Yoshikawa H, Masaki N and Uchida A
Intraoperative extracorporeal autogenous irradiated bone grafts in tumour surgery
Clin Orthop (1999) 368, 196-206

Russell JL and Block JE
Clinical utility of demineralized bone matrix for osseous defects, arthrodesis and reconstruction: impact of processing techniques and stud methodology
Orthopedics (1999) 22, 524-31

Marczynski W, Tylman D, Komender J
Long-term follow up after transplantation of frozen and radiation sterilize bone grafts
Ann Transplant (1997) 2, 64-6

Russell J, Scarborough N and Chesmel K
Re: Ability of commercial demineralized freeze-dried bone allograft to induce new bone formation
J Peridontol (1997) 68, 804-6.

Zhang Q, Cornu O, Delloye C
Ethylene oxide does not extinguish the osteoinductive capacity of demineralized bone. A reappraisal in rats
Acta Orthop Scand (1997) 68, 104-8

Fideler BM, Vangsness CT Jr, Lu B, Orlando C and Moore T
Gamma irradiation: effects on biomechanical properties of human bone-patellar tendon-bone allografts
Am J Sports Med (1995) 23, 643-6

Goertzen MJ, Clahsen H, Burrig KF and Schulitz KP
Sterilisation of canine anterior cruciate allografts by gamma irradiation in argon. Mechanical and neurohistological properties retained one year after transplantation.
J Bone Joint Surg Br (1995) 77, 205-12 (Retracted publication)

White JM, Goodis HE, Marshall SJ and Marshall GW
Sterilisation of teeth by gamma radiation
J Dent Res (1994) 73, 1560-7

Loty B, Tomeno B, Evrard J and Postel M
Infection in massive bone allografts sterilised by radiation.
Int Orthop (1994) 18, 164-71

Yusof N
The use of gamma irradiation for sterilisation of bones and amnion
JSNM (1994) 12, 243-251

Yahia LH, Drouiin G and Zukor D
The irradiation effect on the initial mechanical properties of meniscal grafts
Biomed Mater Eng (1993) 3, 211-21

Zasacki W
The efficacy of application of lyophilized, radiation-sterilised bone graft in orthopedic surgery
Clin Orthop (1991) 272, 82-7

Komender J. Malczewska H and Komender A
Therapeutic effects of transplantation of lyophilized and radiation-sterilised, allogeneic bone
Clin Orthop (1991) 272 38-49

Dziedzic-Goclawska A, Ostrowski K, Stachowicz W, Michalik J and Grzesik W
Effect of radiation sterilisation on the osteoinductive properties and the rate of remodeling of bone implants preserved by lyophilization and deep-freezing
Clin Orthop (1991) 272 30-7

Ostrowski K, Dziedzic-Goclawska A, Stachowicz W and Michalik J
Radiation-induced paramagnetic centers in research in bone physiopathology
Clin Orthop (1991) 272 21-29

Angermann P and Jepsen OB
Procurement, banking and decontamination of bone and collagenous tissue allografts: guidelines for infection control
J Hosp Infect (1991) 17, 159-69

Loty B, Courpied JP, Tomeno B, Postel M, Forest M and Abelanet R
Bone allografts sterilised by irradiation. Biological properties, procurement and results of 150 massive allografts
Inst Orthop (1990) 14, 237-42

Weintroub S and Reddi AH
Influence of irradiation on the osteoinductive potential of demineralized bone matrix
Calcif Tissue Int (1988) 42, 255-60

MacDowell S
Irradiated cartilage
Plast Surg Nurs (1988) 8, 14-5

Wangerin K, Ewers R and Bumann A
Behaviour of differently sterilized allogenic lyophilized cartilage implants in dogs
J Oral Maxillofac Surg (1987) 45, 236-42

Linberg JV, Anderson RL, Edwards JJ, Panje WR and Bardach J
Preserved irradiated homologous cartilage for orbital reconstruction
Ophthalmic Surg (1980) 11, 457-62

Ostrowski K, Dziedzic-Goclawska A and Stachowicz W
Stable radiation-induced paramagnetic entities in tissue mineral and their use in calcified tissue research
In: *Free Radicals in Biology* (Pryor W, Ed.) Academic Press, New York, 1980, pp. 321-344

Horowitz M
Sterilisation of homograft ossicles by gamma radiation
J Laryngol Otol (1979) 93, 1087-9

Komender J, Malczewska H and Lesiak-Cyganowska E
Preserved bone in clinical transplantation
Arch Immunol Ther Exp (Warz) (1978) 26, 1071-3

Komender J
Evaluation of radiation-sterilized bone and clinical use
Acta Med Pol (1978) 19, 277-81

Burwell RG
The fate of freeze-dried bone allograft
Transplant Proc (1976) 8, 95-111

Dexter F
Tissue banking in England
Transplant Proc (1976) 8, 43-8

Komender J, Komender A, Dziedzic-Goclawska A and Ostrowski K
Radiation-sterilized bone grafts evaluated by electron spin resonance technique and mechanical tests
Transplant Proc (1976) 8, 25-37

Urist MR and Hernandez A
Excitation transfer in bone. Deleterious effects of cobalt 60 radiation-sterilization of bank bone
Arch Surg (1974) 109, 586-93

Imamaliev AS and Gasimov RR
Biological properties of bone tissue conserved in plastic material and sterilized with gamma rays (clinico-experimental study)
Acta Chir Plast (1974) 16, 129-35

Ostrowski K, Dziedzic-Goclawska A, Stachowicz W and Michalik J
Accuracy, sensitivity and specificity of electron spin resonance analysis of mineral constituents of irradiated tissues
Ann NY Acad Sci (1974) 238, 186-201

Ostrowski K, Dziedzic-Goclawska A, Stachowicz W, Michalik J, Tarsoly E & Komender A
Application of the electron spin resonance technique for quantitative evaluation of the resorption rate of irradiated bone grafts
Calcif Tissue Res (1971) 7, 58-66

Tarsoly E, Ostrowski K, Moskalewski S, Lojek T, Kurnatowski W and Krompecher S
Incorporation of lyophilized and radiosterilized perforated and unperforated bone grafts in dogs
Acta Chir Acad Sci Hung (1969) 10, 55-63

Ostrowski K, Kecki Z, Dziedzic-Goclawska A, Stachowicz W and Komender A
Free radicals in bone grafts sterilized by ionizing radiation
Sb Ved Pr Lek Fak Karlovy Univerzity Hradci Kralove (1969), Suppl: 561-3

Marquit B
Radiated homogenous cartilage in rhinoplasty
Arch Otolaryngol (1967) 85, 78-80

HIV

Smith RA, Ingels J, Lochemes JJ, Dutkowsky JP and Pifer LL
Gamma irradiation of HIV-1
J Orthop Res (2001) 19, 815-9

Hernigou P, Gras G, Marinello G and Dormont D
Inactivation of HIV by application of heat and radiation: implication in bone banking with irradiated allograft bone
Acta Orthop Scand (2000) 71, 508-12

Yusof N
Gamma irradiation for sterilising tissue grafts for viral inactivation
JSNM (2000) 18, 23-36

Campbell DG and Li P
Sterilization of HIV with irradiation: relevance to infected bone allografts
Aust N Z J Surg (1999) Jul 69, 517-21

Salai M, Vonsover A, Pritch M, von Versen R and Horoszowski H
Human immunodeficiency virus (HIV) inactivation of banked bone by gamma irradiation
Ann Transplant (1997) 2, 55-6

Fideler BM, Vangness CT Jr, Moore T, Li Z and Rasheed S
Effects of gamma irradiation on the human immunodeficiency virus. A study in frozen human bone-patellar ligament-bone grafts obtained from infected cadavera
J Bone Joint Surg Am (1994) 76, 1032-5

Campbell DG, Li P, Stephenson AJ and Oakeshott RD
Sterilization of HIV by gamma irradiation. A bone allograft model
Int Orthop (1994) 18, 172-6

Bedrossian EH Jr
HIV and banked fascia lata
Ophthal Plast Reconstr Surg (1991) 7, 284-8

Biomaterials

Holy CE, Cheng C, Davies JE and Shoichet MS
Optimizing the sterilization of PLGA scaffolds for use in tissue engineering
Biomaterials (2001) 22, 25-31

Andriano KP, Chandrashekar B, McEnery K, Dunn RL, Moyer K, Balliu CM, Holland KM, Garrett S and Huffer WE
Preliminary in vivo studies on the osteogenic potential of bone morphogenetic proteins delivered from an absorbable puttylike polymer matrix
J Biomed Mater Res (2000) 53, 36-43

Al-Assaf S, Meadows J, Phillips GO, Williams PA and Parsons BJ
The effect of hydroxyl radicals on the rheological performance of hylan and hyaluronan
Int J Biol Macromol (2000) 27, 337-348

Al-Assaf S, Hawkins CL, Parsons BJ, Davies MJ and Phillips GO
Identification of radicals from hyaluronan (hyaluronic acid) and cross-linked derivatives using electron paramagnetic resonance spectroscopy
Carbohydrate Polymers (1999) 38, 17-22

Deeble DJ, Phillips GO, Bothe E, Schuchmann H-P and von Sonntag C
The radiation induced degradation of hyaluronic acid
Radiat Phys Chem (1991) 37, 115-118

Cheung DT, Perelman N, Tong D and Nimni ME
The effect of gamma-irradiation on collagen molecules, isolated alpha-chains and crosslinked native fibers
J Biomed Mater Res (1990) 24, 581-9

Deeble DJ, Bothe E, Schuchmann H-P, Parsons BJ, Phillips GO and von Sonntag C
The kinetics of hydroxyl radical induced strand breakage of hyaluronic acid. A pulse radiolysis study using conductometry and laser light scattering
Z Naturforsch (1990) 45c, 1031-1043

Bruck SD and Mueller EP
Radiation sterilization of polymeric implant materials
J Biomed Mater Res (1988) 22, 133-44

Schwarz N, Redl H, Schiesser A, Schlag G, Thurnher M, Lintner F and Dinges HP
Irradiation-sterilization of rat bone matrix gelatin
Acta Orthop Scand (1988) 59, 165-7

Myint P, Deeble DJ, Beaumont PC, Blake SM and Phillips GO
The reactivity of various free radicals with hyaluronic acid: steady-state and pulse radiolysis studies
Biochim Biophys Acta (1987) 925, 194-202

Phillips GO
Chemical processes induced during radiation sterilisation of cellulose
Presented at: *Anselme Payen Award Symposium, American Chemical Society 188th National Meeting*, Philadelphia, USA (1984) August 26-31.

Nakamura Y, Ogiwara Y and Phillips GO
Free radical formation and degradation of cellulose by ionising radiations
Polymer Photochemistry (1985) 6, 135-159

Phillips GO
Radiation degradation of cellulosic systems
Proceedings of an International Symposium on *Fiber Science and Technology* (1985) August 20-24, Hakone, Japan, 88-90.

Edwards HE, Menzies AR and Phillips GO
Radiation effects on the biological activity and molecular weight parameters of heparin
Carbohydrate Polymers (1985) 5 , 473-478

Wozniak-Parnowska W and Najer A
Studies on the sterilization of pharmaceutical base materials with ionizing radiation and ethylene oxide
Acta Microbiol Pol (1978) 27, 161-8

Edwards HE, Moore JS and Phillips GO
Effects of Co-60 irradiation on chondromucoprotein
Int J Radiat Biol (1977) 32, 351-359

Soft tissues

Tyszkiewicz JT, Uhrynowska-Tyszkiewicz IA, Kaminski A and Dziedzic-Goclawska A
Amnion allografts prepared in the Central Tissue Bank in Warsaw
Ann Transplant (1999) 4, 85-90

Martinez Pardo ME, Reyes Frias ML, Ramos Duron LE, Gutierrez Salgado E, Gomez JC, Marin MA and Luna Zaragoza D
Clinical application of amniotic membranes on a patient with epidermolysis bullosa
Ann Transplant (1999) 4, 69-73

Johnson KA, Rogers GJ, Roe SC, Howlett CR, Clayton MK, Milthorpe BK and Schindhelm K
Nitrous acid pretreatment of tendon xenografts cross-linked with glutaraldehyde and sterilized with gamma irradiation
Biomaterials (1999) 20, 1003-15

Maeda A, Inoue M, Shino K, Nakata K, Nakamura H, Tanaka M, Seguchi Y and Ono K
Effects of solvent preservation with or without gamma irradiation on the material properties of canine tendon allografts
J Orthop Res (1993) 11, 181-9

Hinton R, Jinnah RH, Johnson C, Warden K and Clarke HJ
A biomechanical analysis of solvent-dehydrated and freeze-dried human fascia lata allografts. A preliminary report
Am J Sports Med (1992) 20, 607-12

Bumann A, Kopp S, Eickbohm JE and Ewers R
Rehydration of lyophilised cartilage grafts sterilized by different methods
Int J Oral Maxillofacial Surg (1989) 18, 370-2

Cantore G, Guidetti B and Delfini R
Neurosurgical use of human dura mater sterilized by gamma rays and stored in alcohol: long term results
J Neurosurg (1987) 66, 93-5

Edwards HE and Phillips GO
Radiation effects on human tissues and their use in tissue banking
Radiat Phys Chem (1984) 22, 889-900

Armand G, Baugh PJ, Balazs EA and Phillips GO
Radiation protection of hyaluronic acid in the solid state
Radiation Research (1975) 64, 573-580

Hall AN, Phillips GO and Rassol S
Action of ionizing radiations on a hyaluronate tetrasaccharide
Carbohydrate Research (1978) 62, 373-376

Edwards HE, Moore JS and Phillips GO
Effects of ionising radiations on human costal cartilage and exploration of the procedures to protect the tissue from radiation damage
Histochemical J (1978) 10, 389-398

Moore JS, Phillips GO and Rhys D
Chemical effects of γ-irradiation of aqueous solutions of chondroitin-4-sulphate
Int J Radiat Biol (1973) 23 (2), 113-119

Litwin SB, Cohen J and Fine S
Effects of sterilization and preservation on the rupture force and tensile strength of canine aortic tissue
J Surg Res (1973) 15, 198-206

Donnelly RJ, Aparicio SR, Dexter F, Deverall PB and Watson DA
Gamma-radiation of heart valves at 4 degrees C; a comparative study using techniques of histochemistry and electron and light microscopy
Thorax (1973) 28, 95-101

Mandelcorn MS and Crawford JS
Feasibility of a bank for storage of human fascia lata sutures
Arch Opthalmol (1972) 87, 535-7

Korlof B, Simoni E, Baryd I, Lamke LO and Eriksson G
Radiation-sterilization split skin: a new type of biological wound dressing. Preliminary report
Scand J Plast Reconstr Surg (1972) 6, 126-31

Rittenhouse EA, Sands MP, Mohri H and Meerendino KA
Sterilization of aortic valve grafts for transplantation
Arch Surg (1970) 101, 1-5

Welch W
A comparative study of different methods of processing aortic homograft
Thorax (1969) 24, 746-9

Malm JR, Bowman FO Jr, Harris PD, Kaiser GA and Kovalik AT
Results of aortic valve replacement utilizing irradiated valve homografts
Ann N Y Acad Sci (1969) 30, 740-7

Balazs EA, Davies JV, Phillips GO and Young M
Transient intermediates in the radiolysis of hyaluronic acid
Radiation Research (1967) 31, 243-255

USE OF THE IAEA CODE OF PRACTICE FOR THE RADIATION STERILISATION OF BONE ALLOGRAFTS

Emma Castro Gamero and Kety León Palomino

Laboratorio de Irradiación de Productos Médicos, Instituto Peruano de Energía Nuclear (IPEN)
Av. Canadá 1470, San Borja, Lima, Perú

ABSTRACT

Since 1998, tissue allografts are routinely gamma sterilised in Peru. Vital for the developing of tissue banking activity in the country has been the IAEA Programme on Radiation and Tissue Banking. As international standards for radiation sterilisation of health care products were not applicable for tissue allografts, IAEA elaborated the 'Code of Practice for the Radiation Sterilisation of Tissue Allografts - Requirements for Validation and Routine Control'. The purpose of this paper is to apply the IAEA Code of Practice for the sterilisation of bone allografts produced in tissue banks of Peru. For this, a thorough review of the document was made and each element was described according to our data. Results produced quality enhancement of bone allografts, as various procedures involved in tissue banking and radiation have been revised, optimised and systematised. It was concluded that products of improved quality will be available for users of bone allografts and that application of the IAEA Code of Practice supports and strengthens country regulations regarding radiation and tissue banking. Finally, the experience gathered in this work will boost the implementation of the Code of Practice for other tissue allografts.

KEYWORDS

Code of Practice; tissue banking; validation; bone allograft; radiation sterilisation

INTRODUCTION

Radiation sterilised tissue allografts are being routinely produced in Peru since 1998. The IAEA Programme for Radiation and Tissue Banking was crucial for the developing of tissue banking activity in the country. The Rosa Guerzoni Chambergo Tissue Bank (RGTB), established with the International Atomic Energy Agency (IAEA) in cooperation with the Peruvian Institute of Nuclear Energy (IPEN) and located at the Instituto de Salud del Niño (ISN); the Social Security (ESSALUD) Tissue Bank and the National Ophthalmologic Institute (INO) Corneal Bank are the institutions involved in tissue banking activities. Tissue banks are responsible for performing donor selection, procurement, processing and distribution as well as the follow-up of the medical application of tissue allografts. Tissues are sent to IPEN, where they are treated with gamma rays; the staff of the Irradiation of Medical Products Laboratory (IMPL) responsible for the sterilisation process, quality control of tissues and dose validation. Currently produced and radiation sterilised tissues for implants are human bone allografts, fresh and lyophilised, wedges chips and spheres; air-dried amnion dressings. Also, xenografts, such as pigskin dressings are produced. From 1998 to 2003, 23,586 tissues for implants have been produced and sterilised by gamma irradiation.

During the developing of the IAEA Programme in Radiation and Tissue Banking the need of standards for tissue irradiation was detected. Available international standards for radiation processes that involve health care products could not be applied to tissue allografts as they stand for medical and pharmaceutical products produced in large numbers with a uniform size and bioburden prior to sterilisation. Thus, IAEA elaborated the 'Code of Practice for the Radiation Sterilisation of Tissue Allografts: Requirements for Validation and Routine Control', document that was launched in 2002. It is intended that irradiation laboratories and facilities of countries involved in the IAEA Programme adopt the document and apply it routinely for the irradiation process for sterilisation of tissue allografts. The Irradiation of Medical Products Laboratory (IMPL) of IPEN has applied the Code to the sterilisation of bone allografts produced in RGTB and the procedure and its results are presented in this work.

BRIEF DESCRIPTION OF THE IAEA CODE OF PRACTICE FOR THE RADIATION STERILISATION OF TISSUE ALLOGRAFTS – REQUIREMENTS FOR VALIDATION AND ROUTINE CONTROL

This IAEA document takes into account the fact that tissue allograft bioburden can vary widely from one donor to another; the size of tissues is not standard, and that tissues are not products that can be produced in large numbers on a commercial basis. The objective of the Code (Table 1) is to provide guidance in the use of ionising radiation to sterilise tissue allografts in order to ensure their safe clinical use. It specifies requirements for validation, process control and routine monitoring for the selection of donors, tissue processing, preservation, storage and the radiation sterilisation of tissue allografts. It applies to continuous and batch type gamma irradiators using Co-60 or Cs-137. The principles for establishing the doses to assure sterility of the tissue products are similar to those adopted in ISO 11137:1995 and are related to statistical approaches. Human tissue donors should be serologically screened to avoid viral contamination; otherwise this Code cannot be applied. On the other hand, it can only be applied if radiation sterilisation is the terminal stage of a careful detailed documented sequence of procedures involving: donor selection, tissue retrieval, tissue banking general procedures, specific processing procedures, labelling and distribution.

Normative references relevant to this Code are listed below:

- ISO 9001: 2002 *Quality management systems - Requirements*
- ISO 11137:1995 *Sterilisation of health care products - Requirements for validation and routine control – Radiation sterilisation*
- ISO 11737-1: 1995 *Sterilisation of medical devices - Microbiological methods Part I*
- ISO 11737-2: 1998 *Sterilisation of medical devices - Microbiological methods Part 2*
- ISO TR 13409:1996 *Sterilisation of health care products – Radiation sterilisation – Substantiation of 25 kGy as sterilisation dose for small or infrequent production batches*
- ISO TR 15844:1998 *Sterilisation of health care products – Radiation sterilisation - Selection of sterilisation dose for single production batch*
- AAMI Technical Information Report (TIR 27): 2001 - *Sterilisation of health care products - Radiation sterilisation – Substantiation of 25 kGy as sterilisation dose – Method* VD_{max}
- ISO/ASTM 51261 (2002) *Guide for selection and calibration of dosimetry systems for radiation processing*
- IAEA (May, 2002) *International standards for tissue banking*

The IAEA Code of Practice contents are displayed below:

Code of Practice for the Radiation Sterilisation of Tissue Allografts: Requirements for Validation and Routine Control

Contents

1. Introduction
2. Objective
3. Scope
4. Normative References
5. Definitions
6. Personnel
7. Validation of the pre-sterilisation process
   7.1 General
   7.2 Qualification of the Tissue Bank Facilities
   7.3 Qualification of the tissue donors
   7.4 Qualification of the tissue processing and preservation
   7.5 Process specification
8. Validation of the sterilisation process
   8.1 General
   8.2 Qualification of the tissue allografts for sterilisation
   8.3 Qualification of the irradiation facility
   8.4 Qualification of the irradiation process
   8.5 Maintenance of validation
   8.6 Routine sterilisation process control
9. Quality, safety and clinical application of the tissue allograft
10. Documentation and certification procedures
11. Management and control

Annexes
A Establishing a sterilization dose
B Worked examples
C Tables 1, 2 and 3
D Key references for the sterilisation of tissues by ionizing radiation

Annex A describes the methods for selecting a sterilisation dose; Annex B provides three worked examples applying these methods; Annex C gives Tables which contain microbial survival data relating to Standard Distribution of Resistances (SDR), and gives Tables which contain microbial survival data relating to SDR, and Annex D gives a very complete list of references for the sterilisation of tissues by ionising radiation.

For a better understanding and following of the document the definitions below are emphasised:

- *Qualification:* Obtaining and documenting evidence concerning the processes and products involved in tissue donor selection, tissue retrieval, processing, preservation and radiation sterilisation that will produce acceptable tissue allografts.
- *Validation:* Refers to establishing documented evidence that provides a high degree of assurance that a specific process will consistently produce a product meeting its predetermined specifications and quality attributes. A process is validated to evaluate the performance of a system with regard to its effectiveness based on intended use.

APPLICATION OF THE CODE OF PRACTICE FOR RADIATION STERILISATION OF BONE ALLOGRAFTS PRODUCED IN PERU

In order to avoid confusion, and to follow the document closely and easily it was considered to maintain the numerals stated in the IAEA Code. Numerals 1-5 are not applicable for our intended purpose, thus the following document begins in heading Personnel (6).

6. Personnel

The RGTB staff as well as the IMPL staff were trained specifically for the tasks they perform. Tissue bank operators were trained on the Tissue Banking Operator Courses held at the Regional Centres of Buenos Aires and Singapore. Tissue Banks are in charge of donor selection, procurement, processing, distribution and the follow-up of the medical application of tissue allografts. The staff of IMPL has degrees on radiation sciences and were also trained through IAEA courses and workshops, they are in charge of the irradiation process as well as the quality control and dose validation. Functions and responsibilities of each staff are clearly specified in the Tissue Banking Quality Manual, according ISO 9001: 2002 *Quality management systems – Requirements* and IPEN quality documentation. In Figure 1, the organisational chart and interaction of both institutions are shown.

7. Validation of pre-sterilisation processes

7.1 General

Processes that determine the characteristics of tissue allografts prior to the irradiation process are very important for the radiation sterilisation of tissues. These procedures, specified in the IAEA International Standards for Tissue Banks are: donor selection; tissue retrieval; tissue banking general procedures and specific processing procedures; labelling and distribution. Characteristics given by these procedures are involved with physical, chemical and biological properties (levels and types of microbial contamination) of tissues. The elements of the validation of pre-sterilisation processes are described as follows:

7.2 Qualification of the Tissue Bank facilities

RGTB facility has been designed for processing, preservation and storage of tissues for implants in order to minimise the contamination of tissues with microorganisms during tissue banking operation. It meets the standards specified on the IAEA International Standards for Tissue Banking and the design is shown in Figure 2.

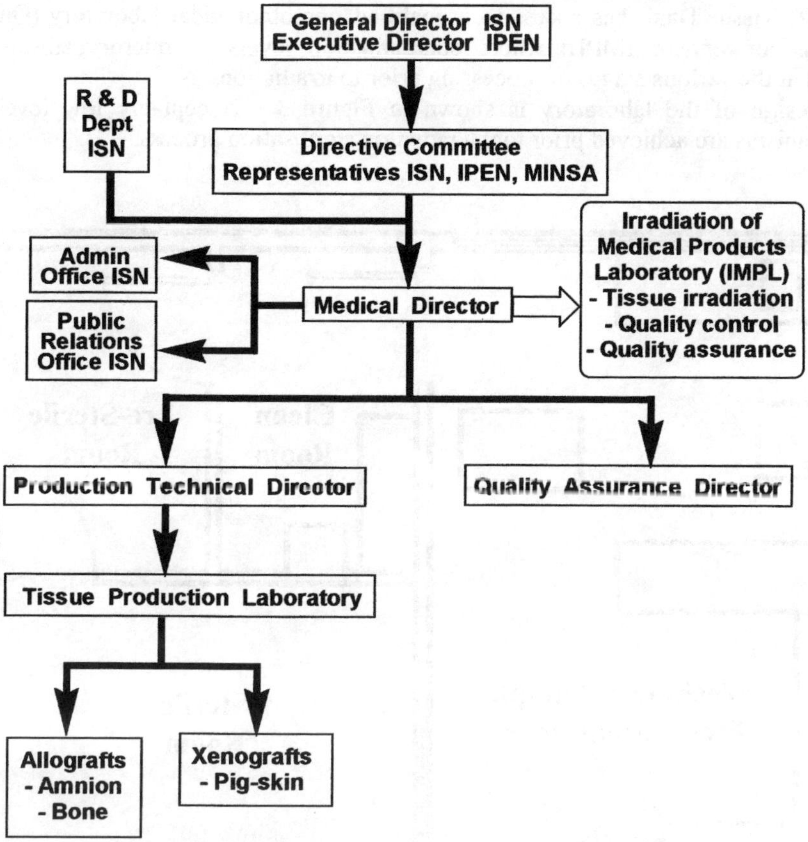

Figure 1. Organisational chart.

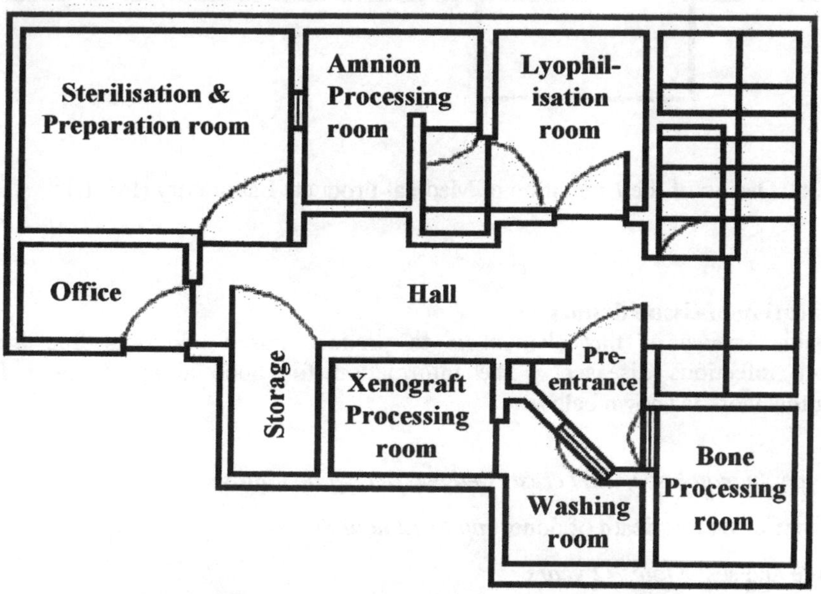

Figure 2. Design of the RGTB Tissue Bank.

RGCTB Tissue Bank has access to a qualified microbiological laboratory (Quality Control Laboratory of IMPL) where contamination levels of microorganisms are measured at the various stages of processing prior to irradiation.

The design of the laboratory is shown in Figure 3. Acceptably low levels of microorganisms are achieved prior to the radiation sterilisation process.

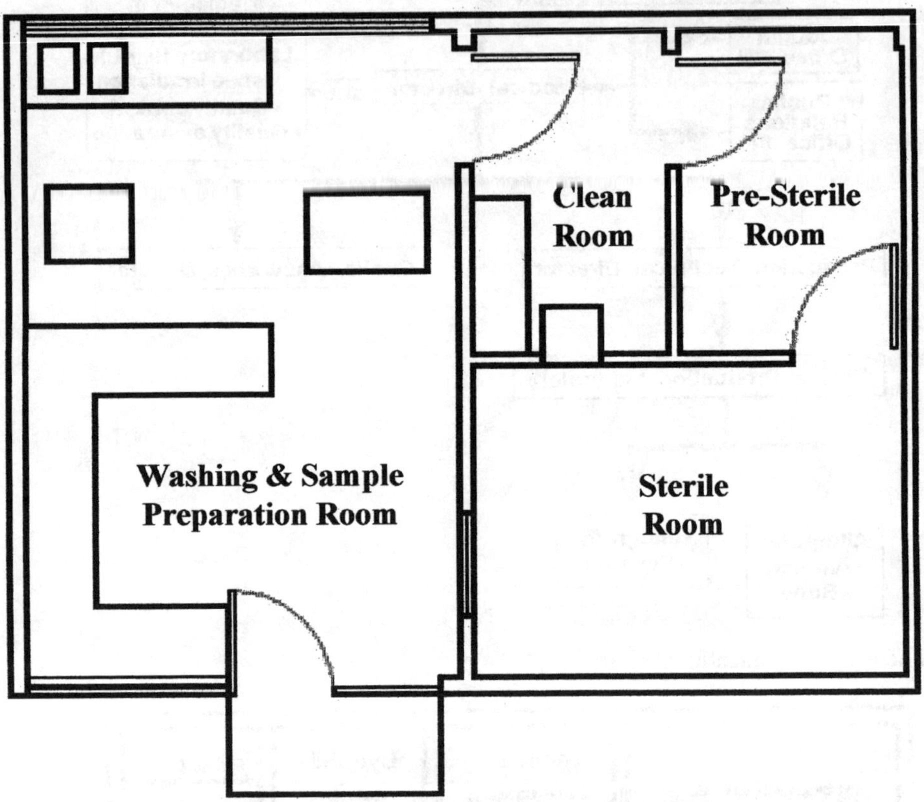

Figure 3. Design of the Irradiation of Medical Products Laboratory (IMPL).

7.3 Qualification of tissue donors

The main purpose of this element is to produce tissue allografts free from transmissible infectious diseases. The information for bone allografts used for performing this work is shown below:

- Tissue type: *human bone, iliac crest (wedges) and spine (chips)*
- Time of retrieval after death of donor: *up to 12 hours*
- Donor age and sex: *Male, 22 years*
- Medical social and sexual history of donor: *Yes*

- Physical examination for the body: *Yes*
- Serological tests performed: *Antibodies to HIV 1,2; Antibodies to Hepatitis C; Hepatitis B surface antigen; VDRL*
- Analysis of autopsy as required by law: *Always*
- Storage Conditions: *-80°C*

7.4 Qualification of tissue processing and preservation

Stages comprised in bone allograft processing and preservation were:

- *Pasteurisation*
- *Trimming*
- *Cutting*
- *Defatted*
- *Deproteinisation*
- *Deep-freezing*
- *Freeze-drying*

Maintenance of validation

For qualification of tissue bank facilities; tissue donors; tissue processing and preservation, a validation process should be specified, which will demonstrate that the standards expected will be maintained. Therefore, the following processes were performed:

- *Internal audit of the origin and history of procured tissues*
- *Sampling of procured tissues and laboratory based screening of bone allografts for viruses and infectious diseases (at ISN laboratories)*
- *Measurements of particle count and microbial contamination in the environment of each of the separate premises of the RGTB Tissue Bank*
- *IMPL at IPEN performs random, statistically-significant sampling of tissue allografts prior to and after tissue processing and preservation for measurements of bioburden levelsIML at IPEN determines the ability of the tissue processing and preservation procedures to both reduce the levels of microorganisms and to produce the levels of bioburden required for the radiation sterilisation process (< 1000 cfu / allograft product) for substantiating a sterilisation dose of 25 kGy. For this, bioburden measurements of washings during processing of bone allografts were performed. Samples afforded only 4 washes otherwise bioburden increased.*
- Number of pieces of bone obtained after completion of stages: *35 (8 wedges and 27 chips of 1cm x 1cm.*

7.5 Process specification

The established process specification for bone allografts is described as follows:

- Tissue allograft type covered by the specification: *lyophilised bone chips*
- Parameters covering the selection of tissue for processing: *Pasteurisation at 57°C for 3 hours, cut in chips of 1 cm x 1 cm, defatted and deproteinised, deep freezing at -80°C, lyophilised for 26 hours until RH < 6%*
- Details of equipment, laboratory and storage facilities required for each of the processing and preservation stages: *RGTB and IMPL quality documentation*
- Details on the routine preventative maintenance programme: *Maintenance Equipment Units of ISN and IPEN*
- Process Documentation: *Traceability - forms for donor selection, tissue retrieval, serological tests for cadaveric donors, transport of tissues to the tissue bank, processing and lyophilisation of bone allografts, bioburden and moisture tests, packaging of allografts operation, cleaning and maintenance of equipment, cleaning of premises*

8. Validation of the sterilisation process

8.1 General

Guidance given here is based on international standards for the sterilisation of health care products as ISO 1137:1995, ISO/TR 13409:1996, ISO/TR 15844:1998 and AAMI TIR 27:2001. Attention is given on the variability of bioburden from one tissue donor to another; on the variability of size and shape of tissue allografts, which can affect the accuracy of product dose mapping and therefore sterilisation itself and on the applicability of using Sample Item Portions (SIP) of a tissue allograft product.

The elements included in the validation of the sterilisation process are described below.

8.2 Qualification of the tissue allografts for sterilisation

This element takes into account the effect of radiation on the components of bone allografts and its packaging; the procedures for the selection of the sterilisation dose; the technical requirements to generate the information for the selection of the sterilisation dose and the condition for transferring the sterilisation dose between irradiation facilities.

Thus:

- Effect on packaging material (Annex A of ISO 11137:1995: *PE and Nylon, stable to radiation when treated at 25 kGy*)
- Sterilisation dose selection: In order to obtain a SAL of 10^{-6} the following methods can be used: ISO 11137:1995, ISO TR 15844:1998, ISO TR 13409:1996 and AAMI TIR 27:2001

Our experience:

Microorganisms present on bone allografts: Gram - positive cocci. Low bioburden (Item bioburden < 1000 cfu/g) and SDR (SIP <1). Limited number of bone samples, then method for selection of dose: ISO TR 13409:1996
Number of items produced: 35 chips (1 cm x 1 cm)

Stage I

- Test sample size for bioburden determination: 10
- Test sample size for verification dose experiment: 10

Stage II

- Samples obtained: 20

Stage III

- SIP: 1/ 35 = 0.028
- SIP bioburden (10 item samples): 15 cfu/item
- Average bioburden: 1/0,028 = 36 cfu/item <1000 cfu/item, method is valid

Stage IV

- Verification dose calculation: I + [S x log (average SIP bioburden)] kGy = 1.25 + (1.65xlog15) = 3.19 kGy

Stage V

- Verification Dose Experiment: 3.19 kGy dose delivered to 10 samples, the test sterility yielded 0 positive from the 10 SIPs tested

Stage VI

- Interpretation of results: The sterility test result was acceptable, 25kGy, as sterilisation dose was confirmed

- Technical requirements:
  - Access to qualified microbiological and dosimetric laboratory services: *IMPL at IPEN*
  - Microbiological testing: *according ISO 11737-1: 1995; ISO 11737-2: 1998*
  - Access to Co-60 radiation source: *IMPL at IPEN, Gammacell 220*

- Transfer of Sterilisation Dose: *ISO 11137:1995 (Section 6.2.3)*
  - Bone allograft production batches are of low numbers therefore, tissues are always sterilised in the Gammacell 220

8.3 Qualification of the Irradiation Facility

Principles involved in this element are covered in ISO 11137: 1995 (section 6.3), thus:

- Equipment Documentation: *Procedure Manual, Emergency Plan and Maintenance Programme of Gammacell 220. Work instructions, forms. Licensed equipment*
- Irradiation source: *Type I, lead shielded irradiator. Gammacell 220. Manufactured by Atomic Energy of Canada. Co-60 annular source, 12 stainless steel pencils, double encapsulated. Current total activity: 24 000 Ci. Drawer, shaft, up and down vertical movement.*
- Dimensions of irradiation chamber: *15.2 cm x 20.6 cm*
- Location: *IPEN*Equipment Testing: *Radiation level measurements performed at various distances from the source. Leak test performed*
- Equipment Calibration: *IPEN participates in IDAS IAEA programme for dosimetry system intercomparison*
- Irradiator dose mapping: *Performed specifically for bone allografts*

8.4 Qualification of the irradiation process

Items involved in this element are specified in ISO 11137:1995 (Section 6.4), thus:

- Determination of the product loading pattern (Section 6.4.1):
  Irradiation batch: Package of 35 units; distributed in a cylindrical way;
  weight: 500 g; height: 16 cm; width: 12 cm
- Product dose mapping (Section 6.4.2)
  Dosimetry system: Fricke (35 dosimeters, air and product both),
  Ethanol chlorobenzene (verification at minimum dose)
  Product Density: 0.141 g/cc
  Dose rate air: 18.5069 kGy/hr
  Minimum dose rate product: 14.3089 kGy/hr
  Maximum dose rate product: 23.0538 kGy/hr
- Certification: *SOPs for dosimetry and irradiation processes, forms and work instructions. The Head of the IMPL is responsible for the acceptability and reviewing of documentation.*

8.5 Maintenance of validation

Items involved in this element are specified in ISO 11137:1995 (section 6.6) and are described as follows:

- Calibration Programme: *Dosimetry system: irradiation times are adjusted according source decay*
- Irradiator requalification: *no equipment modification will be made*
- Sterilisation dose auditing: *every 6 months if no significant changes during allograft processing stages that could affect bioburden levels is reported*

8.6 Routine sterilisation process control

Items involved in this element are based on ISO 11137:1995 (section 7) and are:

- Documentation and registration of:
  Control, monitoring routine and preventive maintenance of the irradiator Gammacell 220; handling of product, prior during and after irradiation; packaging of bone allografts inside the irradiation chamber; product loading pattern; dose monitoring position (location of routinely dosimeter); relation between product density, dose rate and activity source: *IMPL quality documentation*
- Issue of a certificate with maximum dose received and sterilisation dose:
  Always issued and delivered to the Tissue Bank with irradiated bone allografts batches

9. Quality, safety and clinical application of the tissue allograft

Produced bone allografts are principally used in case of: Cyst bone tumours, traumatical and congenital spine diseases, benign bone tumours, congenital tibial pseudoarthrosis, post traumatic pseudoarthrosis, open fractures in long bones with bone loss, congenital hip displasia, maxillae bone defect, periodontal bone cyst.

A program to demonstrate the quality, safety and clinical application of the tissue allografts is being envisaged by RGTB. Tests to determine shelf – life of bone allografts must be designed and performed.

10. Documentation and certification procedures

Documentation produced while conducting validation of pre sterilisation processes is maintained by the RGTB Medical Director according with ISO 9001:2002 and IAEA International Standards for Tissue Banking.

Documentation produced while conducting validation of the sterilisation process is maintained by IMPL at IPEN. Both RGCTB and IMPL are responsible respectively for its documentation

11. Management and control

Control of procedures involved in selection of tissue donors, tissue processing and preservation prior to sterilisation by radiation and the radiation sterilisation process itself are documented and managed in accordance with ISO 9001, 2000 and IAEA International Standards for Tissue Banks.

RESULTS

The application of the IAEA Code of Practice has resulted in quality enhancement of produced bone allografts, as various procedures involved in tissue banking and radiation have been revised, optimised and systematised. Quality documentation of both RGTB and IMPL has been also revised or completed according to the Code of Practice requirements. Interaction between RGTB and IMPL is more efficient and organised. The importance of pre-sterilisation procedures has been demonstrated for achieving successfully the radiation sterilization process of bone allografts.

CONCLUSIONS

- **Products of improved quality will be available for users of bone allografts**
- **The application of the IAEA Code of Practice for Tissue Allografts supports and strengthens country regulations regarding radiation and tissue banking**
- **After this work, the Code of Practice will be easily applied to other tissues allografts**
- **Need to study in depth, properties of produced bone allograft as well as its shelf-life**
- **Need to interact with other Irradiation Laboratories for exchanging experiences on the application of the IAEA Code of Practice**

ACKNOWLEDGEMENTS

To the personnel of the Tissue Bank 'Rosa Guerzoni Chambergo', Dr. Moises Palti, Eng. Nancy Pérez and Biologist Renán Peña for providing samples and data related to RGTB.

REFERENCES

1. IAEA, Code of practice for the radiation sterilisation of tissue allografts: Requirements for validation of routine control, **2002**.
2. ISO 9001:**2002**, Quality management systems - Requirements
3. ISO 11137:**1995**, Sterilisation of health care products - Requirements for validation and routine control – Radiation sterilisation.
4. ISO TR 13409:**1996**, Sterilisation of health care products – Radiation sterilisation – Substantiation of 25 kGy as sterilisation dose for small or infrequent production batches.
5. IAEA (May, **2002**), International standards for tissue banking - ISO 9001:**2002** Quality management systems – Requirements.
6. ISO/ASTM 51261 (**2002**), Guide for Selection and Calibration of Dosimetry Systems for Radiation Processing.
7. Dziedzic-Goclawska, The application of ionising radiation to sterilize connective tissue allografts, In: *Radiation and Tissue Banking*, G. O. Phillips (ed.), World Scientific, Singapore, **2000**, pp. 57-99.

PART 2

METHODOLOGY IN THE STERILISATION AND PRESERVATION OF TISSUES

RADIATION SOURCES: TYPES AND SUITABILITY FOR DOSE DELIVERY TO TISSUES FOR STERILISATION

Jan Th. M. Jansen *, Frank W. Schultz and Johannes Zoetelief

Medical Physics, Radiation Technology, Interfaculty Reactor Institute
Delft University of Technology Mekelweg 15, 2629 JB Delft, The Netherlands
{ E-mail: JTJansen@iri.tudelft.nl}*

ABSTRACT

Different radiation sources for sterilisation of tissues and the resulting depth-dose distributions in several tissues are described. Dosimetric quantities and units are introduced and dosimetry equipment is briefly discussed. The influence of the irradiation geometry on the target dose is shown including the effects of backscatter, field size and density. The construction of a filter to improve field homogeneity is presented. The importance of a quality control system is stressed, including items that need special attention. Special considerations for the variability of the dose-effect relation of tissues due to changes in, for instance, temperature and oxygen supply are discussed.

KEYWORDS

Dosimetry; radiation sources; quality assurance (QA); tissues; sterilisation

INTRODUCTION

With the transplantation of tissues, it is also possible to transmit diseases of bacterial and viral origin from the donor to the recipient. To decrease the probability of transplanting contaminated tissues several actions can be taken such as donor screening, aseptic handling, disinfecting with antibiotics and thermal, chemical and/or radiation sterilisation. These latter treatments can be performed when the transplanted tissues do not contain living cells. However, the efficacy of the transplanted tissues in the patient can be decreased by these sterilisation treatments. The objective of a tissue bank is to deliver safe and effective tissues for transplantation. To reach this goal the tissue processing needs to be optimised and radiation sterilisation can be a part of this process.

Microorganisms have different sensitivity to radiation and this is expressed by the dose required to achieve 1 log microorganism kill, D_{10}. The bioburden level of microorganisms on a tissue sample is heterogeneous of composition and a conservative Standard Distribution of Resistances (SDR) [1] has been developed to estimate the required dose for a contamination probability of 10^{-6} (see Figure 1). Typical dose levels for sterilisation of tissues range from 25 to 50 kGy.

In this paper first the radiation sources used for sterilisation of tissues are described. The radiation quantities and units are presented, following the International Commission on Radiation Units and Measurements (ICRU). This includes quantities and units for radiometry, interaction coefficients, dosimetry and radioactivity. Radiation detectors are introduced including calorimeter, chemical dosimeters, gas-filled ionisation chambers, radiographic dosimeters and solid-state dosimeters.

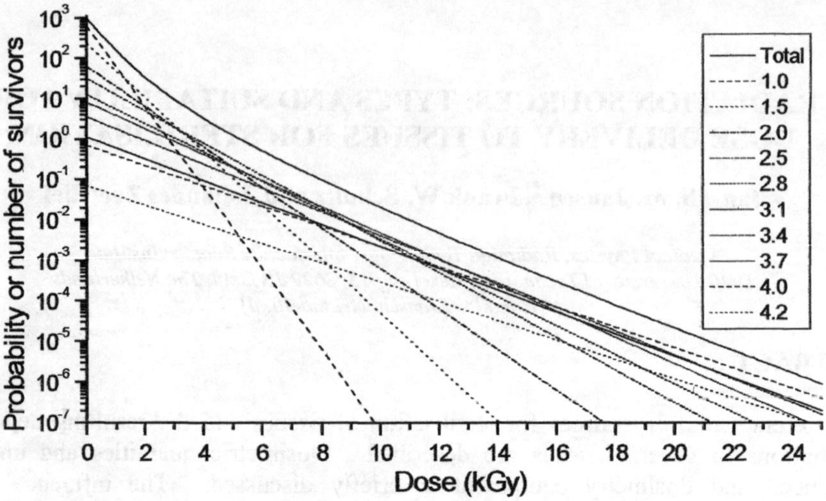

Figure 1. **The microorganisms survival curve (dashed line) composed of a conservative Standard Distribution of Resistances (SDR) with various 1 log cell kill doses, D_{10}, (solid lines) and bioburden levels (Kowalski et al. [1]).**

For sterilisation of tissues, low temperatures as for use of dry ice (CO_2) of 195 K and liquid nitrogen of 77 K are common. This results in extreme conditions for a dosimeter that need special attention. The section on dosimetry at low temperatures gives some indications on how to solve these problems and gives an overview of ongoing work to develop or calibrate dosimeters at these conditions. A section is devoted to quality control with special attention to X-ray tube or accelerator output, field homogeneity and monitoring. Finally, attention is paid to in-phantom measurements, factors affecting the dose homogeneity and biological response.

RADIATION SOURCES

Radiation sources can be roughly subdivided into two types; one group contains radionuclides and the other group is based on acceleration of charged particles. The differences between these groups of sources are large. Radionuclides emit radiation all the time and need shielding if they are not used, whereas X-ray equipment and accelerators can be switched off electrically. Radionuclides decay and in the course of time their activity decreases. To deliver the same dose, the irradiation time will increase with time. For X-ray equipment the photon energy spectrum depends on the high voltage applied between cathode and anode and the filtration, whereas for a single radionuclide the spectrum does not change. In the next paragraphs different radionuclides are discussed.

Radionuclides

One type of decay of radionuclides is called α-decay. In this process the nuclide disintegrates with the emission of a monoenergetic α-particle. The transition energy is distributed over the α-particle and the daughter nuclide according to the conservation of energy and momentum. The radionuclides emitting α-particles are heavy (large Z values).

The emission of an electron and an antineutrino by a radionuclide is called β^--decay and the electron is called a β^--particle. The transition energy is distributed over the electron and antineutrino. Each type of particle has an energy within a continuous spectrum between 0 and the transition energy. The β^--radionuclides are below the stability line on a chart of the nuclides. The emission of a positron (electron with a positive charge) and a neutrino by a radionuclide is called β^+-decay. The transition energy must be more then 1.022 MeV, the rest energy of a positron and an electron, where an orbit electron is emitted due to the decrease of the nuclide charge. Both positron and neutrino have a continuous energy distribution with a maximum energy of the transition energy minus 1.022 MeV. In matter the positron will annihilate with an electron and two photons will be created, each with 0.511 MeV energy. The radionuclides emitting positrons are located above the stability line.

In the electron capture process, a (most likely) K-shell electron is captured by the nucleus. If the transition energy is larger then 1.022 MeV, electron capture and β^+-disintegration are alternative processes. The transition energy is largely supplied to the neutrino. Other electrons, with emission of characteristic X-rays and/or Auger electrons, will fill the vacancy in the K-shell. The electron capture radionuclides are located above the stability line. A metastable (relatively long lived) radionuclide can lose its energy by emitting a monoenergetic photon (γ-radiation) and return to the ground state of the nuclide. This nuclear process is called isomeric transition. An alternative process to the emission of γ-radiation is internal conversion. In internal conversion the nucleus loses energy by emitting an orbital electron with a kinetic energy of the transition energy minus the binding energy of the orbital electron.

In spontaneous fission a nucleus splits into two daughter nuclei, a number of neutrons and maybe other particles. The transition energy is distributed over the neutrons and the daughter nuclei. These radionuclides are heavy ($Z \geq 92$) and above the stability line. Radioactive mother nuclides can decay, via excited stages to daughter nuclides by different combinations of disintegration processes. This can combine α- and γ-radiation. It is also possible that a radionuclide emits α- or β^--radiation.

Accelerators

The other group of sources of radiation is based upon acceleration of charged particles. In principle, all charged particles can be accelerated. The accelerator field can be electrostatic or dynamic. Examples of electrostatic field accelerators are X-ray tubes (voltages up to 400 kV, X-ray radiation perpendicular to electron beam) or Van de Graaff accelerators (accelerating potentials up to 12 MV, X-rays in electron beam direction). Linear accelerators and cyclotrons are dynamic field accelerators. For the types of linear accelerator one can distinguish travelling wave and standing wave fields. If electrons are needed for the irradiation then accelerators can deliver nearly monoenergetic beams whereas β^--radionuclides produce spectra. If the electrons collide with a target material to produce photons in the accelerator, a bremsstrahlungs spectrum is produced, whereas γ-radiation emitting radionuclides produce monoenergetic photon line(s).

RADIATION QUANTITIES AND UNITS

Radiometry

A radiation field can be described in terms of type, direction and energy of the particles at all relevant locations. Radiation fields are characterised by radiometric

quantities. The ICRU [2,3] defines the quantity particle number, N, as the number of particles emitted, transferred, or received. The radiant energy [2,3], R, is the energy of particles (excluding rest energy) emitted, transferred, or received; unit: J. One of the most useful radiometric quantities is the fluence. The (particle) fluence, Φ, is defined by the ICRU [3,4] as the quotient of dN by da, where dN is the number of particles incident on a sphere of cross-sectional area da, thus:

$$\Phi = \frac{dN}{da}$$

The unit of the quantity fluence is m^{-2}.

The energy fluence [2,3], Ψ, is the quotient of dR by da, where dR is the radiant energy incident on a sphere of cross-section area da, thus:

$$\Psi = \frac{dR}{da}$$

The unit of energy fluence is J m^{-2}.

Interaction coefficients

Interaction coefficients characterise the interactions between radiation and matter in non-stochastic quantities. They apply to specified radiation, both in type and in energy, specified materials and, when applicable, also for specified types of interactions. Such specifications are, however, not required for the definition.

The cross section [2,3], σ, of a target entity, for an interaction produced by incident charged or uncharged particles is the quotient of P by Φ, where P is the probability of the interaction for one target entity when subjected to the particle fluence Φ, thus:

$$\sigma = \frac{P}{\Phi}$$

The unit of cross section is m^2 and a special unit is introduced called the barn (b), where 1 b = 10^{-28} m^2.

The mass attenuation coefficient [2,3], μ/ρ, of a material for uncharged ionising particles is the quotient of dN/N by ρ dl, where dN/N is the fraction of particles that experience interactions in traversing a distance dl in a material of density ρ, thus:

$$\frac{\mu}{\rho} = \frac{1}{\rho N}\frac{dN}{dl}$$

The unit is m^2 kg^{-1}. The quantity μ is the total linear attenuation coefficient. The total linear attenuation coefficient can be split for photons into photoelectric, Compton, coherent, pair production and nuclear interaction parts. If interactions between target entities contained in a target of a given atomic species can be disregarded, the mass attenuation coefficient can be expressed in terms of the total cross section, σ. The mass attenuation coefficient is the product of σ and N_A/M, where N_A is the Avogadro constant, and M is the molar mass of the target element, thus:

$$\frac{\mu}{\rho} = \frac{N_A}{M}\sigma$$

The mass energy transfer coefficient [2,3], μ_{tr}/ρ, of a material for uncharged ionising radiation particles, is the quotient of dE_{tr}/EN by ρdl, where E is the energy of each particle (excluding rest energy), N is the number of particles, and dE_{tr}/EN is the fraction of incident particle energy that is transferred to kinetic energy of charged particles by interactions in traversing a distance dl in the material of density ρ, thus:

$$\frac{\mu_{tr}}{\rho} = \frac{1}{\rho EN}\frac{dE_{tr}}{dl}$$

The unit is $m^2\ kg^{-1}$.

The mass energy absorption coefficient [2,3], μ_{en}/ρ, of a material for uncharged ionising particles is the product of the mass energy transfer coefficient, μ_{tr}/ρ, and (1-g), where g is the fraction of the energy of secondary charged particles that is lost to bremsstrahlung in the material, thus:

$$\frac{\mu_{en}}{\rho} = \frac{\mu_{tr}}{\rho}(1-g)$$

The unit is $m^2\ kg^{-1}$.

The total mass stopping power [2,3], S/ρ, of a material for charged particles is the quotient of dE by ρdl, where dE is the energy lost by a charged particle in traversing a distance dl in the material of density ρ, thus:

$$\frac{S}{\rho} = \frac{1}{\rho}\frac{dE}{dl}$$

The unit is $J\ m^2\ kg^{-1}$ but also $eV\ m^2\ kg^{-1}$ is used. The quantity S is the total linear stopping power. The total linear stopping power can be split into electronic, radiative and nuclear stopping power parts.

Dosimetric quantities

Dosimetric quantities are in essence the product of radiometric quantities, describing the radiation field, and the interaction coefficients, describing their interaction with matter. However, dosimetric quantities are not defined as those products but as directly measurable quantities.

The energy imparted [4], ε, by ionising radiation to matter in a volume, is:

$$\varepsilon = R_{in} - R_{out} + \Sigma Q$$

Where R_{in} is the radiant energy incident on the volume, i.e., the sum of the energies (excluding rest energies) of all those charged and uncharged ionising particles which enter the volume, R_{out} is the radiant energy emerging from the volume, i.e., the sum of the energies (excluding rest energies) of all those charged and uncharged particles which leave the volume, and ΣQ is the sum of all changes (decreases: positive sign; increases: negative sign) of the rest energy of nuclei and elementary particles in any interactions which occur in the volume. The unit is J. The energy imparted, ε, is a stochastic quantity and the expectation value is a non-stochastic quantity called mean energy imparted [4], $\bar{\varepsilon}$.

The absorbed dose [2,3], D, is the quotient of $d\bar{\varepsilon}$ by dm, where $d\bar{\varepsilon}$ is the mean energy imparted by ionising radiation to matter of mass dm, thus:

$$D = \frac{d\bar{\varepsilon}}{dm}$$

The unit is J kg^{-1} with the special name gray (Gy).

The linear energy transfer [3,4] or linear electronic stopping power, L, of a material, for a charged particle, is the quotient of dE by dl, where dE is the mean energy lost by the particle, due to collisions with electrons, in traversing a distance dl, thus:

$$L = \frac{dE}{dl}$$

The unit is J m^{-1} and may be expressed in eV m^{-1} or some convenient submultiple, such as keV μm^{-1}.

The distribution of absorbed dose in linear energy transfer [4], D_L, is the quotient of dD by dL, where dD is the absorbed dose contributed by primary charged particles with linear energy transfers between L and L+dL, thus:

$$D_L = \frac{dD}{dL}$$

The kerma [2,3], K, is the quotient of dE_{tr} by dm, where dE_{tr} is the sum of the initial kinetic energies of all charged ionising particles liberated by uncharged ionising particles in a material of mass dm, thus:

$$K = \frac{dE_{tr}}{dm}$$

The unit is J kg^{-1} and the special name for the unit of kerma is gray (Gy). For uncharged ionising radiation of energy E (excluding rest energy), the relationship between energy fluence, Ψ, and kerma, K, may be written as:

$$K = \Psi \left(\frac{\mu_{tr}}{\rho}\right) = \Phi \left(E \left(\frac{\mu_{tr}}{\rho}\right)\right)$$

Where μ_{tr}/ρ is the mass energy transfer coefficient.

The exposure [2,3], X, is the quotient of dQ by dm where the value of dQ is the absolute value of the total charge of the ions of one sign produced in air when all the electrons (negatrons and positrons) liberated by photons in air of mass dm are completely stopped in air, thus:

$$X = \frac{dQ}{dm}$$

The unit is C kg^{-1} and the old special unit of exposure, röntgen (R), may be used temporarily. To convert the old unit to the SI unit, one can use 1 R = 2.58 x 10^{-4} C kg^{-1} exactly. The definition of exposure in terms of radiometric and interaction coefficient quantities is:

$$X = \Psi \frac{\mu_{en}}{\rho} \frac{e}{W}$$

Where Ψ is the energy fluence, μ_{en}/ρ is the mass energy absorption coefficient in air, e is the elementary charge of 1.60219 10^{-19} C and W is the mean energy expended in air per ion pair formed. For photons in dry air the value for W is about [5] 33.85 ± 0.15 eV.

Radioactivity

The decay constant [2,3], λ, of a radioactive nuclide in a particular energy state is the quotient of dP by dt, where dP is the probability of a given nucleus undergoing a spontaneous nuclear transition from that energy state in the time interval dt, thus:

$$\lambda = \frac{dP}{dt}$$

The unit is s^{-1}. The quantity (ln 2)/ λ is commonly called the half-life, $T_{1/2}$, of the radionuclide, i.e., the time it takes for an amount of radioactive nuclides to fall to half its initial number.

The activity [2,3], A, of an amount of radioactive nuclide in a particular energy state at a given time is the quotient of dN by dt, where dN is the expectation value of the number of spontaneous nuclear transitions from that energy state during time interval dt:

$$A = \frac{dN}{dt}$$

The unit is s^{-1} and the special name for the unit of activity is becquerel (Bq) with 1 Bq = 1 s^{-1}. The old special unit of activity is the curie (Ci) with 1 Ci = 3.7×10^{10} Bq exactly.

RADIATION DETECTORS

A dosimeter consists of a radiation-sensitive part (the detector) and an electronic device used to transform the signal from the detector into an electrical signal to the reading device. Different types of detector and their suitability for use in photon and electron beams applied for the sterilisation of tissues are discussed.

Calorimeters

Calorimetry is a fundamental dosimetry method to measure the absorbed dose, i.e., the absorbed energy in matter due to radiation divided by the mass of the matter. This is done by measuring the increase in temperature due to the absorbed energy of the radiation and comparing it with a calibrated heat source. National calibration institutes and research laboratories apply it, for the determination of absorbed dose. The sensitive part of the calorimeter, the absorber, has been reduced to small masses and can be made of tissue-equivalent plastic, water or graphite. Calorimeters are not commercially available and are not used for routine applications.

Chemical dosimeters

Chemical dosimeters, especially ferrous sulphate Fricke dosimeters, have long been used as reference dosimeters. The solution used as a detector may be considered as water equivalent, both in atomic composition as in density. The reading in optical density of the dosimeter can be directly converted to absorbed dose, but it is advisable to check the response of the whole system, including the densitometer, using a calibrated radiation source such as a cobalt-60 γ beam. The absorbed dose range which may be determined with adequate accuracy is high, i.e. 50 to 200 Gy for the conventional Fricke dosimeter [5]. The minimum amount of solution is about 2 cm^3 and special care for the cleaning of the vessel and the preparation of the solution is needed. For tissue sterilisation the absorbed doses are in the range of tens of kGy and the Fricke dosimeter may be too sensitive. Other chemical dosimeters are able to measure such high doses, e.g. the ceric-cerous dosimeter.

Ceric-cerous sulphate dosimetry is applicable as a high dose measurement system. Radiation leads to the reduction of ceric ions (Ce^{4+}) to cerous (Ce^{3+}) ions. The dose range is between 0.5 kGy and 50 kGy and the dose rate should be below 10^6 Gy/s. The dose is determined by performing conventional spectrophotometric analysis in the ultraviolet region, or by measuring the difference in the electrochemical potential between the irradiated and unirradiated solution with an electrochemical potentiometer. For irradiation temperatures between 0-62 °C the temperature coefficients are known.

Gas-filled ionisation chambers

Ionisation chamber dosimetry is the most common method used for absorbed dose measurements in radiation beams of X-ray generators and cobalt-60 sources. Compared with other detectors, ionisation chambers are readily available, and are simple to use as field instruments. They are precise, reproducible, the oldest radiation detector used, can be small and can easily be related to the national standards. However, ionisation chambers require certain precautions and corrections before the readings can be interpreted in terms of a dosimetry quantity. Corrections should be applied for atmospheric conditions, electronic characteristics, chamber design, such as measuring the radiation sensitive volume, and material being air, which differs from water or tissue both in atomic composition and in density. The large difference in density for air (1.205 x 10^{-3} g/cm^3) and for water (1 g/cm^3) results in a perturbation of the particle fluence when even a small volume of medium is replaced by the air-filled cavity.

Since most ionisation chambers are unsealed and open to the atmosphere, the mass of air inside the fixed volume of the cavity depends upon the temperature t (in °C) and the pressure p (in kPa) of the ambient air. If the calibration conditions are as usual, 22 °C (in some countries 20 °C is standard) and 101.325 kPa (76 cm Hg pressure or 1 atm), the mass of air under the t and p conditions is:

$$m_{t,p} = m_{22,101.325} \cdot \frac{273.15+22}{273.15+t} \cdot \frac{p}{101.325}$$

As the ionisation is inversely proportional to the mass of air, the charge measured must be multiplied by:

$$k_{t,p} = \frac{273.15+t}{273.15+22} \cdot \frac{101.325}{p}$$

The chamber is also sensitive to humidity as the air composition changes with the addition of water. The correction is complex but within stated humidity intervals small. Therefore, generally no correction is applied for humidity but only the working conditions are specified. The saturation of the charge collected in the chamber [5], i.e., the recombination of ions of opposite sign before they reach the electrodes, reduces the collection efficiency. For photons and electrons, ion recombination is dependent on the geometrical specifications of the ionisation chamber, the voltage applied, the dose rate if the radiation is continuous or the dose per pulse if the radiation is pulsed. For photon irradiation and the Nuclear Enterprise NE-2571 [6] 0.6 cm^3 ionisation chamber, the maximum continuous dose rate is about 100 Gy/min for 99% collection efficiency.

The leakage current is composed of a dose independent and a dose dependent part. The dose independent part can be measured in absence of radiation (including background) and is due to leakage in the cable, connectors (dirt and moisture) and the first (low noise, low bias) amplifier. This leakage current can be subtracted from the current during irradiation. The dose dependent part is difficult to measure but can be kept at low levels by reducing radiation to the stem of the ionisation chamber and to the cable, connectors and electronic equipment. Also the use of triax cable with a guarded wire can reduce the leakage current.

Radiographic dosimeters

Radiographic dosimetry, i.e. collecting dose information on film material, is an excellent practical method for relative dose measurements in photon and electron beams with the highest spatial resolution in the plane of the film. The response is highly dependent on the photon energy, especially in the low and medium-energy photon range such as cobalt-60 γ rays. A poor reproducibility in the signal to dose ratio exists that makes the use of films calibrated in a batch necessary.

GafChromic film [7,8] changes to a blue colour when exposed to ionising radiation. GafChromic films are not radiochromic dye films but contain microcrystals of a monomer. Ionising radiation causes partial polymerisation of the monomer into a blue polymer. The GafChromic film MD-55 has a dose range from 1 Gy to 1 kGy and the GafChromic DM-1260 from 10 Gy to 10 kGy [8]. They are read spectrophotometrically at certain wavelengths corresponding to multiple broad radiation-induced absorption bands and on their shoulder, which allows dose readings over a relatively broad dose range. The GafChromic film MD-55 has its maximum sensitivity wavelength at 676 nm [7]. GafChromic film is dependent on the temperature during irradiation and read-out and little dependent on relative humidity. GafChromic film can be used for measurement of both gamma ray and electron dose distributions.

Solid-state dosimeters

The most commonly used solid-state detectors are thermoluminescent dosimeters (TLDs) and semiconductor junction detectors. Both have an atomic composition that differs from water and tissue and have a density that is higher than for water. This causes energy dependence compared to absorbed dose in water and perturbs the particle fluence in a water phantom. Thermoluminescent dosimeters are convenient because of their small volume, no associated voltage and their availability. However, they need an individual calibration, are read out long after irradiation and they need careful conditioning. Semiconductors are read out immediately. However, not only are they energy dependent but also dose rate and temperature dependent and they may be damaged after a cumulated dose of about 100 Gy.

DOSIMETRY AT LOW TEMPERATURES

For the sterilisation of tissues the irradiation temperatures are either room temperature (290 K), or the temperature (195 K) of dry ice (solid CO_2), or the temperature of liquid nitrogen (77 K). Performing dosimetry at the low temperatures may be a problem. One solution is to thermally isolate the detector from the rest of the equipment and use a heater to keep it at a constant higher (near room) temperature. Another method is to measure first in an experimental set-up at room temperature, with tissue substitutes and coolant substitutes. An even better approach is to keep the coolant out of the beam. A monitor chamber is present at room temperature, or the dwelling time is used as a monitor. The ratios of the measurements inside the irradiation box and the monitor dosimeter are used to derive the doses for the irradiation at low temperature. Yet another method is to perform the measurements at the low temperatures with special equipment.

Alanine electron paramagnetic resonance (EPR) dosimeters can be used for this purpose. It is known that the signal to dose ratio decreases with decreasing temperatures. Coninckx et al. [9] have performed measurements for five different alanine EPR dosimeter systems at room and liquid nitrogen temperature and determined the relative dose response (see Figure 2). They showed that the response at low temperatures was lower than at room temperature and that above 10 kGy the signal was decreasing with increasing dose. In addition, it was reported that at least some of these dosimeters showed an increasing signal if the time between irradiation and reading was increased. This is different from the behaviour at room temperature when fading was reported.

Hategan et al. [10] showed that the frozen Fricke dosimeter could be used at 77 K. However, the signal of these dosimeters was one order of magnitude lower compared to room temperature irradiation with the same absorbed dose. For a temperature of 268 K, the signal of the Fricke dosimeter was about half the value of the same absorbed dose at room temperature (293 K). The authors explain these differences by a lower rate of migration of the induced water radicals and their local recombination. In addition, there is no oxygen present in the frozen sample, which reduces the radiochemical yield as well. With these explanations, the authors speculate that at liquid nitrogen temperature the Fricke dosimeter can be used for doses up to 8 kGy. Before the Fricke dosimeter can be used as a dosimetry system at these temperatures, however, reproducibility and accuracy have to be tested.

Ramos-Bernal et al. [11] demonstrated that LiF co-doped with Mg, Cu and P, and $CaSO_4$ doped with Dy can be used as dosimeters at liquid nitrogen temperatures. The response was linear at low doses (< 0.4 kGy) and supra linear at higher doses (< 1 kGy).

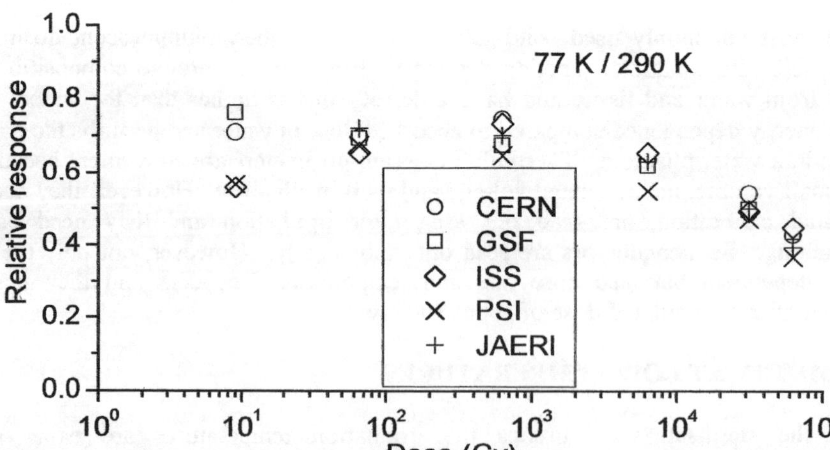

Figure 2. The relative responses of alanine electron spin resonance dosimeters of five different laboratories, namely CERN, GSF, ISS, PSI and JAERI, at liquid nitrogen temperature (77 K) compared to room temperature (290 K) as a function of the dose, according to Coninckx et al. [9].

Their result shows a strong dependence of the thermoluminiscence response on irradiation temperature. They are still investigating the response at higher doses. These crystals require a series of pre-irradiations and thermal annealings before a reproducible signal is established. Biramontri et al. [12] studied the dose responses of undyed and dyed PMMA dosimeters at 2 kGy and 25 kGy between 77 K and 318 K. At 25 kGy the undyed PMMA dosimeter, Radix RNI5, has a linear dependence with a coefficient of +0.15%/K and +0.25%/K for irradiation temperature below and above 258 K, respectively. The dyed PMMA dosimeters, Red 4034 and Amber 3042 have a smaller temperature dependence in this range. Red 4034 has relatively high responses between 77 and 195 K. At a dose of 2 kGy, the dose response of Amber 3042 increases nearly linearly with a coefficient of about 0.5%/K, for temperatures above 195 K. The dose response of Gammachrome YR is almost 30% higher at 195 K compared to 293 K and decreases with increasing temperature above 195 K, with a temperature coefficient of –0.3%/K. In this study batch to batch variations and other influences, apart from temperature, were not investigated.

QUALITY CONTROL

Quality control of dosimetry and exposure arrangements in sterilisation of tissues by ionising radiation has similarities to quality control of these aspects in radiobiology. In the present section experience from dosimetry and exposure arrangements in radiobiology are summarised.

For exposure of animals in radiobiological experiments, long-term stable arrangements are needed. At installation, after a major revision and at regular intervals the whole irradiation arrangement has to be checked [13]. Such a check includes the determination of the half value layer (HVL), when possible the peak tube voltage, tube current and exposure time and the determination of the organ or whole animal dose. For routine checks, this is too time-consuming. Therefore, simple checks have to be applied to establish the performance of the equipment. The radiation output is the single most important parameter needed to assess the overall performance of the X-ray tube and high voltage generator. In addition this parameter is relatively easy to assess. The X-ray output is dependent on, for example, tube voltage, tube current, exposure time, beam filtration and focus-to-detector distance. Methods to determine the X-ray output are given in the next section. The absorbed dose to the target area should be homogeneous. If during a single exposure more than one tissue sample is irradiated, all tissue samples should get comparable doses. This demands that the radiation field used is homogeneous. Methods to reach and check field flatness will be introduced.

For every irradiation, it is necessary to state the location, date, irradiation geometry, and setting of exposure parameters, like tube voltage, tube current and exposure time. If the exposure is terminated at a predetermined level of radiation, the monitoring of the exposure time is essential to establish the performance.

X-ray tube and ^{60}Co γ-ray source output

The X-ray tube output is defined as the air kerma per unit of tube-current exposure-time product at a quoted focus-detector distance, usually 1 m. An equivalent definition is the air kerma rate per unit of tube current at a fixed distance from the focus. The tube-current exposure-time product is also referred to as the focal spot charge. The unit of X-ray tube output is Gy/mAs. The output is dependent on the tube voltage, filtration, collimator, scatter material if any and position of the detector in the beam, including phantom material if any. Changes in any parameters will be reflected in the output.

There are two common experimental arrangements for the determination of the X-ray tube output. In both arrangements an ionisation chamber is positioned at a fixed distance from the focus of the X-ray tube in a well-collimated beam. The ionisation chamber and electrometer system should be calibrated at an appropriate radiation quality and the chamber should have a flat response over the range of energies used. In the first arrangement the measurement is performed in air, whereas in the second arrangement the measurement is performed inside a phantom. A sheet of lead can be used in the central beam behind the ionisation chamber or phantom to obtain standard backscatter conditions. The alignment of the ionisation chamber and the phantom in the beam and the focus-to-detector distance and eventually the depth in the phantom should be fixed, following strict procedures. In the second arrangement, the phantom affects the energy distribution of the X-ray beam. If the calibration factor of the ionisation chamber is energy dependent for these radiation qualities, one has to correct for this. The second method may be preferred, if conditions are more comparable to the actual experimental conditions. For the following conditions the output can be determined:

For constancy checks the X-ray tube output is measured repeatedly with constant exposure factors, e.g., tube voltage, tube current and exposure time. A too high output may indicate inadequate filtration of the X-ray beam. A too low value may indicate problems with the tube voltage waveform. The tube voltage dependency of the radiation output is measured for fixed tube current and exposure time. The X-ray tube output is approximately proportional to the square of the tube voltage [14]. The radiation output constancy should be examined with varying tube currents while keeping the tube voltage and exposure time constant. Any variation may be due to problems in the tube current calibration. The X-ray tube output constancy should be examined with varying exposure times while keeping the tube voltage and tube current constant. Some high voltage generators need a while to establish the tube voltage and tube current set. For these generators the X-ray output will decrease with decreasing exposure times. By plotting the X-ray output exposure time product against the exposure time the effective "dead time" can be estimated (see Figure 3).

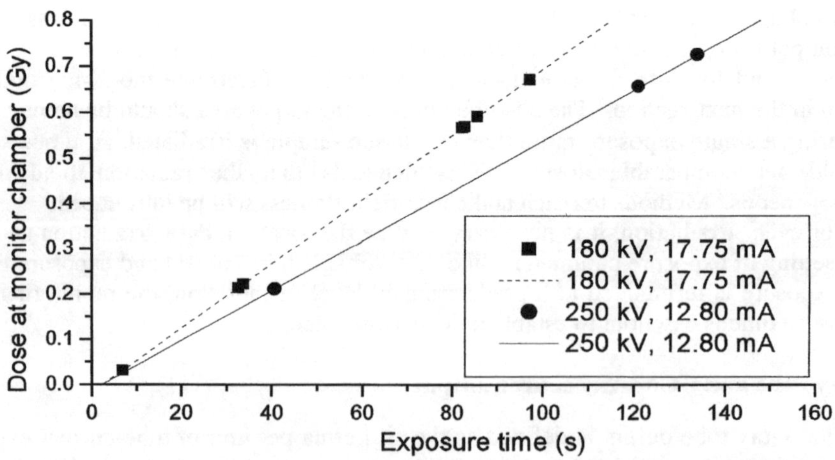

Figure 3. The X-ray tube output is measured as a function of the exposure time for a Philips XMG 300 BG operated at 180 kV or 250 kV tube voltages at the maximum tube currents. The dead time of the equipment is where the lines cross the x-axis.

These dead time effects can also occur with radioactive sources since the movement of sources, samples or shutters will take some time. Similarly, it is important to measure occasionally the yield from radionuclide sources. In radionuclide sources there may be different isotopes present with different decay times, e.g. ^{57}Co in a ^{60}Co source or ^{134}Cs in a ^{137}Cs source.

Field flatness

For the great majority of sterilisation experiments homogeneous irradiations need to be applied. However, the output at the centre of a beam may be different from that at its edges. For X-ray sources, this effect is influenced by changes in angular direction between incident electrons and emitted bremsstrahlung photons, the heel effect in anode-cathode direction, finite dimensions of the focus and wear of the anode. In addition, environmental effects influence the homogeneity of the field such as distance to the focus, beam filtration, back- and side-scatter material. As the scattered radiation is dependent on the field size and field shape, care has to be taken that these parameters are controlled. Checks on field flatness can initially be done with X-ray film, but if a greater accuracy than ± 10 per cent is required [15], more accurate detectors should be used. Erroneous results can be obtained with film since it can be much more sensitive to scattered (lower energy) than to primary radiation, masking the flatness effect.

If the homogeneity of the field is inadequate, a field flatness filter can be designed to overcome these troubles. As the contribution of primary and scattered radiation to a point in the field and the influence of the contribution due to scattered radiation, in the presence of a field flatness filter are probably not well known, a trial and error method has to be applied. This necessitates checking the homogeneity of the field, when the flatness filter is applied. At least the dose on the axes parallel and perpendicular to the anode cathode direction and intercepting the central beam position should be measured at regular distances. If the homogeneity distributions in both directions are similar and there is symmetry around the central beam position, then a circular symmetric beam flatness filter is a good choice. The homogeneity of the field is shown without (-Fil.) and with (+Fil.) field flatness filter in Figure 4.

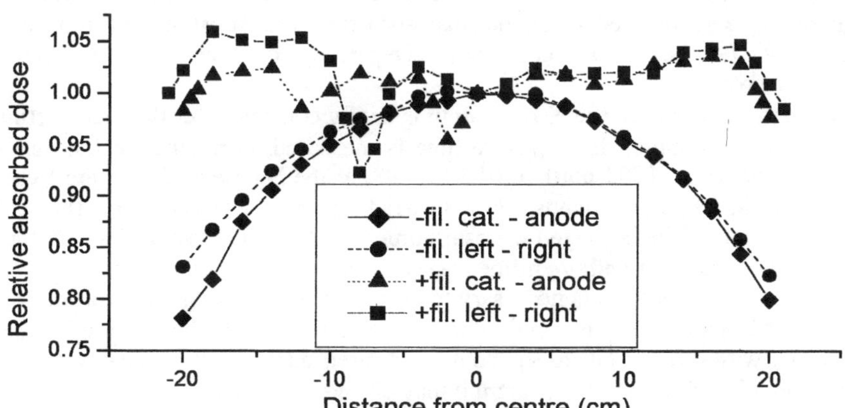

Figure 4 The relative absorbed dose distribution was measured through the centre of the beam, along lines parallel (cath. - anode) and perpendicular (left - right) to the cathode - anode direction, before (- Fil.) and after (+ Fil.) the homogeneity filter was applied. The valleys in the absorbed dose (+ Fil.) are due to solder used to keep the filter together.

With the field flatness filter the X-rays in the beam centre are generally more attenuated than that at the beam edges. Therefore the beam quality will change along the beam axis (see Table 1). To estimate the influence of this primary radiation effect on the depth dose distribution the half value layer (HVL) must be measured on the central beam axis and on the edge of the beam. It is, however, more informative to measure the doses at different depths and locations in order to get an impression of the dose homogeneity in the target volume. For radionuclide sources, e.g. ^{60}Co, field flatness is also affected by source to surface distance, and variation in scatter conditions. Similar to X-ray sources, filters can be applied to improve homogeneity. On the other hand, it is possible to optimise the source distribution to improve the field homogeneity.

Table 1. First and second half value layer at the central beam axis and at the border of the field, measured at a tube voltage of 300 kV, for the XMG 300 BG X-ray tube.

Central beam axis		Border beam axis	
First HVL (mm Cu)	Second HVL (mm Cu)	First HVL (mm Cu)	Second HVL (mm Cu)
2.75	3.68	2.67	3.52

Monitoring

The output may depend on the age of the X-ray tube and also of the warming up procedure followed. Variations may also be expected when new tubes are installed, or after repairs of the equipment. Therefore, X-ray monitors should be positioned permanently in the X-ray beam to reveal any variation. Especially useful are transmission chambers, because they are less sensitive to changes in the location in the field, they monitor also the field size and they do not cast a shadow in the field. It is, however, important to measure the HVL, X-ray output and field flatness with the transmission chamber installed. If another dosimeter is used as monitor, it is important to position the chamber reproducible in the beam whilst keeping the "shadow" of the dosimeter outside the target volume. If the monitor is used to control the irradiation, through setting a predefined level, the exposure time or exposure time tube current product should be checked for constancy. Departures from constancy indicate that something is wrong.

For modern X-ray tubes, the X-ray output is nearly constant over the total lifetime of the tube. As an example, the exposure time is measured, corrected for the delivered dose after January 1st, 1993 until April 24, 1998, at the Biomedical Primate Research Centre, Rijswijk, The Netherlands. For all irradiations in this geometry, the relative standard deviation of the corrected exposure time is 1.2%. On October 28, 1994 a new tube was installed, almost without influence on the corrected exposure time.

For radioisotope sources it is necessary to check the proper positioning of the source, collimator and shutter (if present) during irradiation. If the source is highly contaminated with different isotopes from the desired nuclides, it may be necessary to check the radiation quality as the spectrum may change with time.

IN-PHANTOM MEASUREMENTS

According to the International Atomic Energy Agency (IAEA) recommendations [16] an ionisation chamber calibrated for the X-ray quality considered can be used to derive the absorbed dose in water in a water phantom by the following relation:

$$D_w = M_u \; N_K \; (\bar{\mu}_{en} \; / \; \rho)_{w,a} \; k_{a,w}$$

Where D_w is the absorbed dose in water, M_u is the dosimeter reading (corrected for temperature, pressure, leakage current, ion recombination, etc.), N_K is the air kerma calibration factor for the X-ray quality considered, $(\bar{\mu}_{en} / \rho)_{w,a}$ is the ratio of the mean mass energy absorption coefficients of water to that of air averaged over the spectral energy fluence distribution of the photons at the point of measurement in the water phantom, and the overall correction factor $k_{a,w}$ is a product of the following components: the correction factor for the energy and angular dependence of the response of the ionisation chamber, the displacement correction factor which accounts for the effect of the displacement of water by the air volume equal to the external size of the ionisation chamber and the correction factor which accounts for the effect of the protective sleeve needed if a non-watertight ionisation chamber is inserted into the water phantom.

For HVLs of 1 and 3 mm Cu, the recommended [16] values of $k_{a,w}$ for both the NE 2571 and PTW M23332 chambers are 1.02 and 1.01, respectively. The quoted $k_{a,w}$ values refer to 5 cm depth in the water phantom and field sizes of 10 cm x 10 cm and the uncertainties are 2 to 3 per cent at one standard deviation. For the NE 2571 ionisation chamber $k_{a,w}$ values of 1.025 ± 0.006 and 1.009 ± 0.006 are presented [17] at HVLs of 1 and 3 mm Cu, respectively. The dependence of $k_{a,w}$ on depth in phantom (2 and 5 cm) and on field size (20 cm^2 to 200 cm^2) is about 1%.

Measurements are often performed in polymethylmethacrylate (PMMA) and the absorbed dose has to be expressed in muscle tissue. Therefore, the following relation is formulated [18]:

$$K_m = M_u \; N_K \; (\bar{\mu}_{en} \; / \; \rho)_{m,a} \; k_{a,PMMA}$$

Where K_m is the kerma in the PMMA phantom expressed in muscle tissue, $(\bar{\mu}_{en} / \rho)_{m,a}$ is the ratio of the mass energy absorption coefficient of muscle tissue to that of air averaged over the spectral energy fluence distribution of the photons at the point of measurement in the PMMA phantom, and $k_{a,PMMA}$ is the overall correction factor for a given ionisation chamber to account for differences between measurements free-in-air and at depth in a PMMA phantom. The numerical values of $k_{a,PMMA}$ and $k_{a,w}$ are expected to be identical as the displacement correction factor for stemless chambers is varying marginally between densities of 1.00 and 1.17 g/cm^3 for X-rays with HVLs of 1 and 3 mm Cu [19].

Conversion from air kerma to absorbed dose in muscle tissue can be made employing mass energy absorption coefficients according to Hubbell and Seltzer [20]. In Figure 5 the mass energy absorption coefficient ratios for water, skeletal muscle, soft tissue and bone to air are shown.

For ^{60}Co it is more practical to use a reference dose in water instead of air [21]. The absorbed dose to water at the reference depth (5 g/cm^2 for ^{60}Co) in water for a reference beam of quality Q_0 and in absence of the chamber is given by [21]:

$$D_{w,Q_0} = M_{Q_0} \; N_{D,w,Q_0}$$

Where M_{Q0} is the reading of the dosimeter under the reference conditions used in the standards laboratory and $N_{D,w,Q0}$ is the calibration factor in terms of absorbed dose to water of the dosimeter obtained from a standards laboratory.

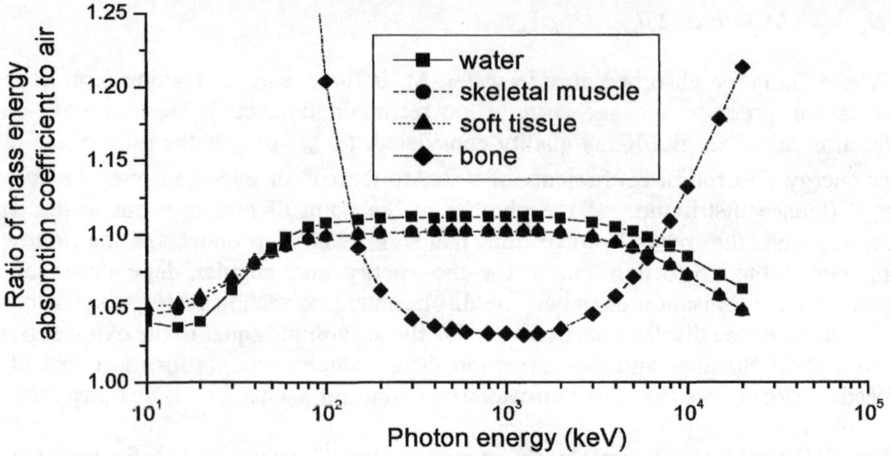

Figure 5. The ratio of the mass energy absorption coefficient of water (solid squares, dash point point line), skeletal muscle (solid circles, solid line), soft tissue (solid triangles, dashed line) and bone (solid diamonds, dash point line) to air against the photon energy, according to Hubbell and Seltzer [20].

When a dosimeter is used in a beam of quality Q, different from Q_0 used in its calibration, the absorbed dose to water is given by [21]:

$$D_{w,Q} = M_Q \; N_{D,w,Q_0} \; k_{Q,Q_0}$$

Where the factor $k_{Q,Q0}$ corrects for the effects of the difference Q_0 and Q, and the dosimeter reading M_Q has been corrected to the reference values of influencing quantities, other than beam quality, for which the calibration factor is valid. The IAEA Technical Reports Series 398 supplies values for the correction factor $k_{Q,Q0}$. For high energy photons produced by clinical accelerators the beam quality Q is expressed by the tissue phantom ratio $TPR_{20,10}$. This is the ratio of the absorbed doses at depths of 20 and 10 cm in a water phantom, measured with a constant source to centre of detector distance of 100 cm and a field size of 10 cm × 10 cm at the plane of the chamber. For high-energy electron beams R_{50}, the half-value depth in water (in g/cm^2), is used as beam quality index.

FACTORS AFFECTING DOSE HOMOGENEITY

For sterilisation of tissues by radiation, it is important to deliver a well-known dose homogeneously, avoiding local under- and over-exposure. Under-exposure means that the radiation will not achieve the desired log microorganism kill. In the case of over-exposure the damage to the extracellular tissue matrix may deteriorate the efficacy of the use of the tissue. Therefore a homogeneous dose in the tissue is necessary. For an isotropic point source of photons, the dose to a point in the tissue is determined by the source to point of interest distance according to the inverse square law, by the attenuation material between the source and the point of interest and by the build-up and scatter contributions. Build-up occurs near interfaces of different materials or density due to the lack of secondary particle equilibrium (see Figure 6). If the photon energy increases the build-up region becomes larger and the effects of attenuation decreases, at least for photon energies below about 50 MeV.

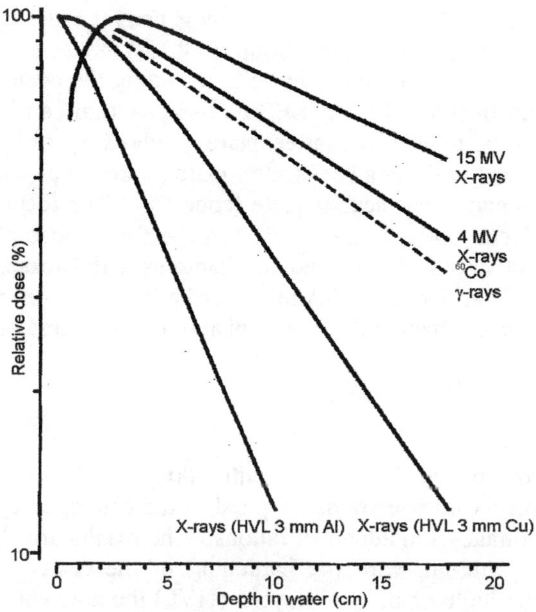

Figure 6. Relative dose is shown as a function of depth in water for various photon beams.

Electrons have a finite range and scatter when they collide with material. Their depth dose distributions show (see Figure 7) an increase in dose (comparable to the build-up for photons), then a decrease with depth (electrons scatter out of the beam) and finally a sharp drop at a depth comparable with the range of the electrons. Due to the generation of bremsstrahlung photons by the electrons in the material there is a small tail in the depth dose distribution for depths beyond the range of the electrons. When the energy of the electrons increases the build-up region and the apparent range of the electrons increases.

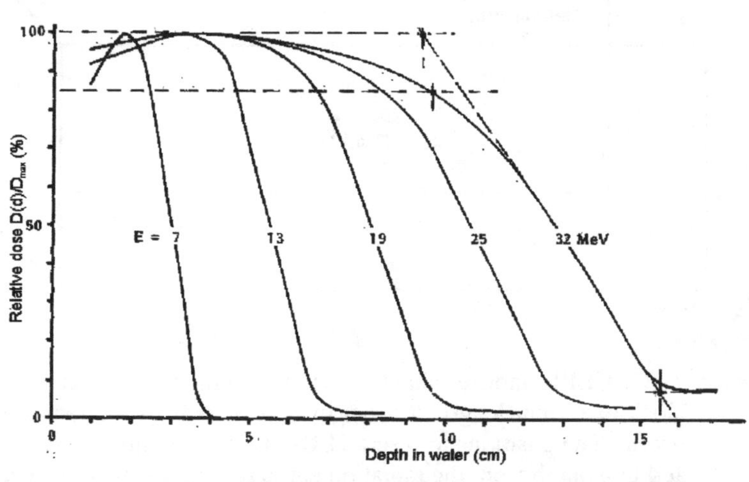

Figure 7. The relative dose is shown as a function of the depth in a water phantom for broad beams electrons of various energies.

To illustrate the effects of scatter on the homogeneity of the dose distribution, consider the European Late Effects Project Group (EULEP) mouse phantom Davies [22]. The EULEP mouse phantom is shown in Figure 8, including the positions to insert LiF TLDs for purpose of dosimetry. The EULEP mouse phantom can be irradiated as a single mouse phantom without backscatter plate (code 1-), with 2 other mouse phantoms on both sides and without a backscatter plate (code 5-) or with 2 other mouse phantoms on both sides and a backscatter plate (code 5+). The focus to centre of the phantom distance is 68 cm, both for the single and the five mouse phantoms without backscatter plate, and 200 cm for the five mouse phantoms with backscatter plate.

The backscatter plate is an 8 cm PMMA block (radiation direction) x 30 cm x 30 cm and is positioned immediately behind the mouse phantoms. The gradient is defined as:

$$Gradient = \left\{ \frac{D_4 - D_{21}}{D_{12.5}} \right\} 100\%$$

Where D_4 is the entrance dose, $D_{12.5}$ is the central dose and D_{21} is the exit dose. The irradiations are performed with spectra as supplied in the catalogue of Seelentag et al. [23] with various tube voltages and added filtrations. The results are shown in Table 2. For the single mouse phantom, with the hardening of the X-ray beam (more tube voltage and filtration and higher half value layers (HVL) the gradient decreases and the irradiation becomes more homogenous. For the 5-geometry the gradient decreases due to increased side scatter. The effect is however limited. If the backscatter plate is added (5+geometry) the gradient decreases from about 20% to 3%. This means that a backscatter plate makes the dose distribution more homogenous and can partly compensate for the attenuation. Tomljenovic et al. [24] have shown (see Figure 9) that the backscatter factor is dependent on the beam quality and that a maximum exists at about 1 to 2 mm Cu HVL. In addition it is shown that when the field size increases the backscatter factor also increases.

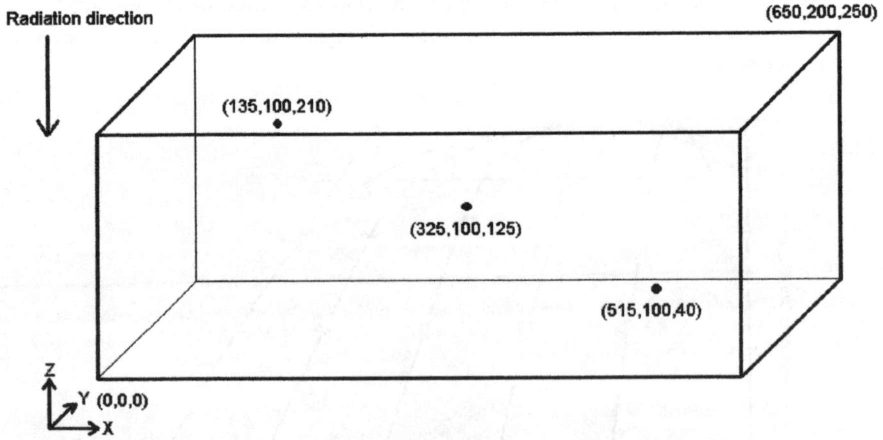

Figure 8. The EULEP mouse phantom is a polymethylmethacrylate (PMMA) block of 65 cm length, 2.5 cm thickness (radiation direction) and 2 cm width. The positions to insert TLDs are in the centre of the phantom and at 4 mm depth from the radiation entrance and exit planes. The TLDs are positioned at 13.5 mm away from the length direction from the edge to decrease the attenuation from the other TLDs. The co-ordinate numbers indicate distances in tenths of mm.

Table 2. The gradient is shown for various irradiation geometries (code 1-, 5- and 5+, where the number indicates the number of mouse phantoms and + or – indicate with or without backscatter plate) and X-ray spectra, including catalogue number, tube voltage (kV) and added filtration according to Seelentag et al. [23].

Geometry:						1-	5-	5+
Seelentag catalogue	Tube voltage (kV)	Added filtration			HVL	Gradient		
		mm Al	mm Cu	mm PMMA	mm Cu	(%)	(%)	(%)
C91	150	3.67			0.33	32.3	27.2	3.4
C102	180	5	0.42	4.8	1.03	27.6	22.8	2.9
C117	230	3.67	0.42		1.29	25.9	21.5	3.0
C121	250	4	1.6		2.44	23.8	20.5	3.6
C122	250	4	2.7		2.88	22.5	18.6	2.6
C133	300	2	3.05		3.49	21.3	17.4	4.0

To improve the dose distribution in the phantom it is possible to irradiate from opposite directions instead of from one direction. If only irradiation from one direction is possible, rotation of the sample can make the dose distribution more homogeneous. This was demonstrated by Zoetelief et al. [25] irradiating a cylinder of water with a diameter of 12.5 cm with 300 kV X-rays (HVL = 3.2 mm Cu) and a focus to centre of phantom distance of 90 cm, both in uni- and bi-lateral direction (see Figure 10). With the central dose normalised to 1, for an unilateral irradiation the dose varies between 0.50 and 1.52. If bilateral irradiation is performed with two opposing X-ray tubes the dose distribution becomes more homogeneous with relative doses between 1 and 1.02.

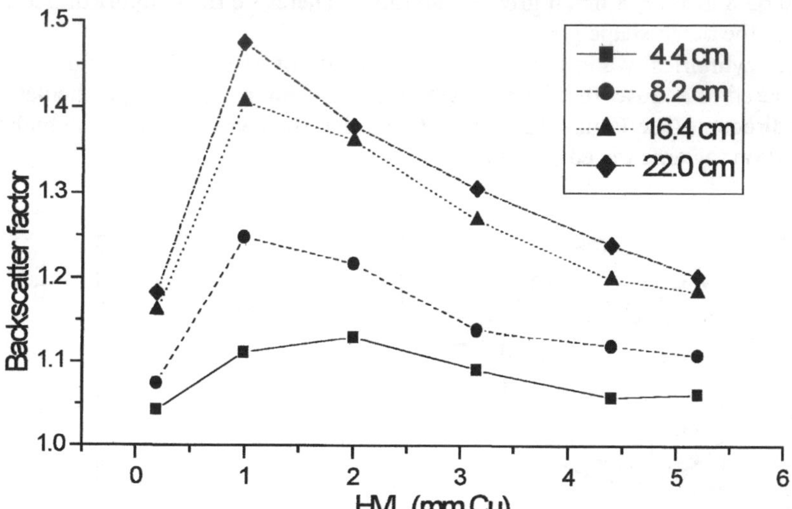

Figure 9. The backscatter factor for wide beam orthovoltage photon spectra on a water phantom with dimension of 15 cm x 30 cm x 30 cm and a source to surface distance of 100 cm, as a function of the beam quality in terms of HVL (mm Cu). The field size is supplied in the legend as diameter in cm at the entrance surface of the water phantom, according to Tomljenovic et al [24].

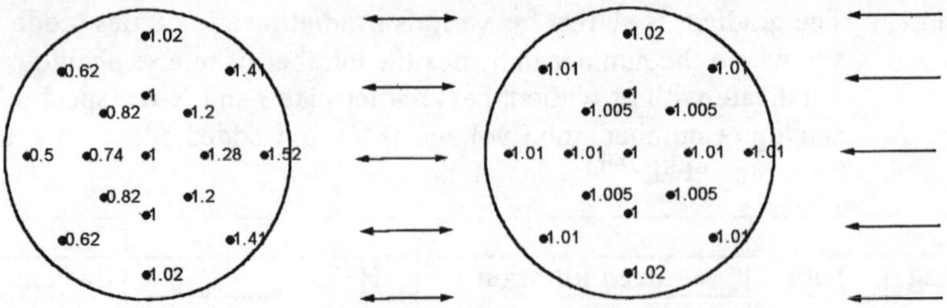

Figure 10. The relative dose distribution is shown in the central plane of a cylindrical water phantom with a diameter of 12.5 cm irradiated by 300 kV X-rays with a HVL of 3.2 mm Cu and a focus to centre of phantom distance of 90 cm. On the left-hand side the situation is shown for the uni-lateral irradiation and on the right-hand side for the bi-lateral irradiation, according to Zoetelief et al. [25].

However, if the diameter of the cylinder is increased from 12.5 cm to 20.1 cm the dose homogeneity decreases. For the large cylinder the focus-to-centre of phantom distance is increased from 90 to 150 cm in order to make the dose more homogenous. The dose distribution changes from 1 to 1.02 for the small cylinder, from 1 to 1.14 for the large cylinder (compare Figure 10 with Figure 11). At this point it was decided to rotate the large cylinder to further improve the homogeneity from 1 to 1.11.

Also the shape of the phantom has influence on the dose distribution, as shown in Figure 11. A cylindrical water phantom with a diameter of 20.1 cm is compared to a rectangular water block with a cross section of 20 cm x 24 cm. For the cylinder the relative dose distribution is between 1 and 1.14 whereas for the block phantom it is between 0.65 and 1.17, a much greater variation. Therefore the cylindrical geometry is preferred to the block shape geometry.

For both cylindrical water phantoms, with a height of 31 cm, Zoetelief et al. [25] have also measured the dose distribution along the cylinder axis, perpendicular to the radiation direction (see Figure 12). It is shown that the central dose is the highest and the doses drop towards the edge of the phantoms.

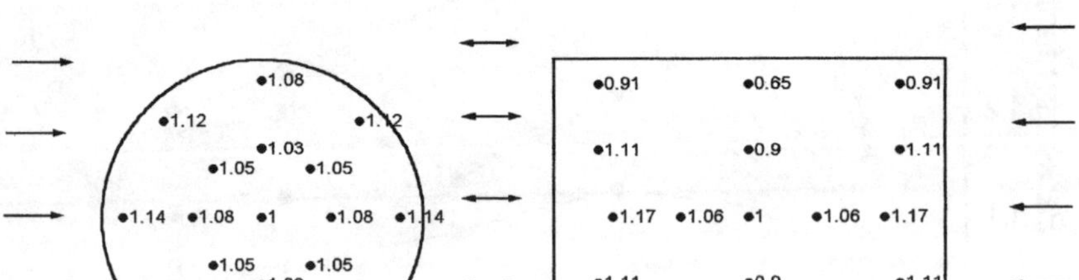

Figure 11. The relative dose distribution is shown in the central plane of a cylindrical water phantom with a diameter of 20.1 cm (left-hand side) and in the central plane of a block of water phantom of 20 cm x 24 cm x 73.8 cm (right-hand side). The bilateral irradiation is performed with 300 kV X-rays with a HVL of 3.2 mm Cu and a focus to centre of phantom distance of 150 cm, according to Zoetelief et al. [25].

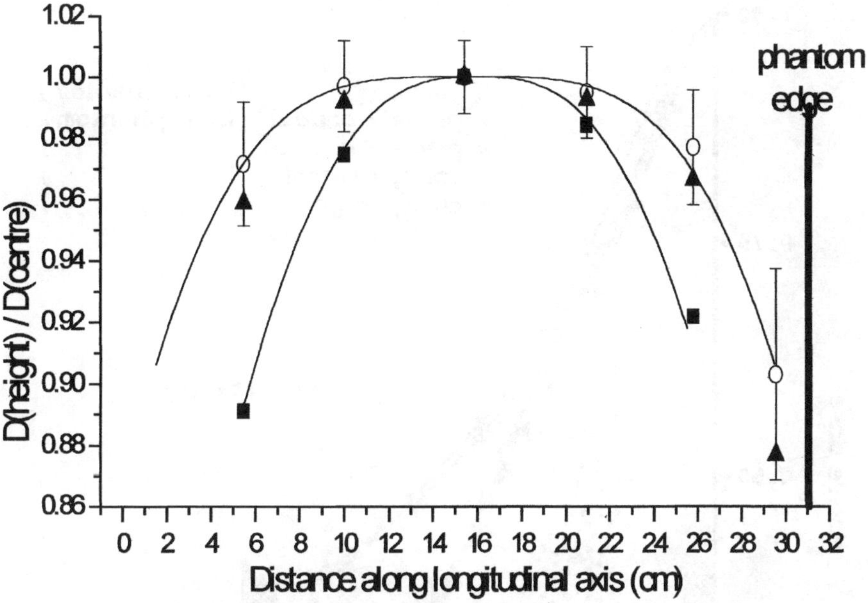

Figure 12. The relative dose distribution is shown on the cylinder axis for a water phantom with a diameter of 12.5 (solid triangles) or 20.1 cm (solid squares) and a height of 31 cm. For the small diameter also the average plane dose is measured and the open circles indicate the result. The bilateral irradiation is performed with 300 kV X-rays with a HVL of 3.2 mm Cu and a focus to centre of phantom distance of 90 or 150 cm, for the small or large diameter, respectively, according to Zoetelief et al. [25].

For the cylinder with the larger diameter the drop is larger than for the smaller diameter. In addition an average plane dose is measured (see Figure 12) and the drop for the average plane dose is less than for the dose on the cylinder axis. This could be expected because the side scatter contribution is less near the edge of the phantom, in all cases considered. To make the dose distribution more homogeneous scatter material has to be added to both sides of the cylinder. The thickness should be about 5 and 10 cm for the 12.5 and 20.1 cm diameters, respectively.

Inhomogeneous phantoms have an effect on the dose distribution. This was illustrated by Zoetelief et al. [25]. A lung in a PMMA phantom is simulated by pressed cork with a density of 0.3 g cm^{-3}. The depth dose distributions for various situations are shown in Figure 13. The open circles show the depth dose curve for the central beam axis in a homogeneous PMMA phantom. It is a smoothly decreasing curve with increasing depth. Measuring the depth dose for the lung phantom at a location where only PMMA is present, that is the caudal (solid squares) and cranial (crosses) sides in Figure 13, both curves start below the homogeneous curve due to lack of backscatter from the simulated lung. With increasing depth this difference becomes smaller. For large depths both curves are above the homogeneous curve due to the strong scatter contribution of the PMMA behind the cork. The cork attenuates the beam much less than solid PMMA and this is the cause for the strong scatter contribution at greater depths.

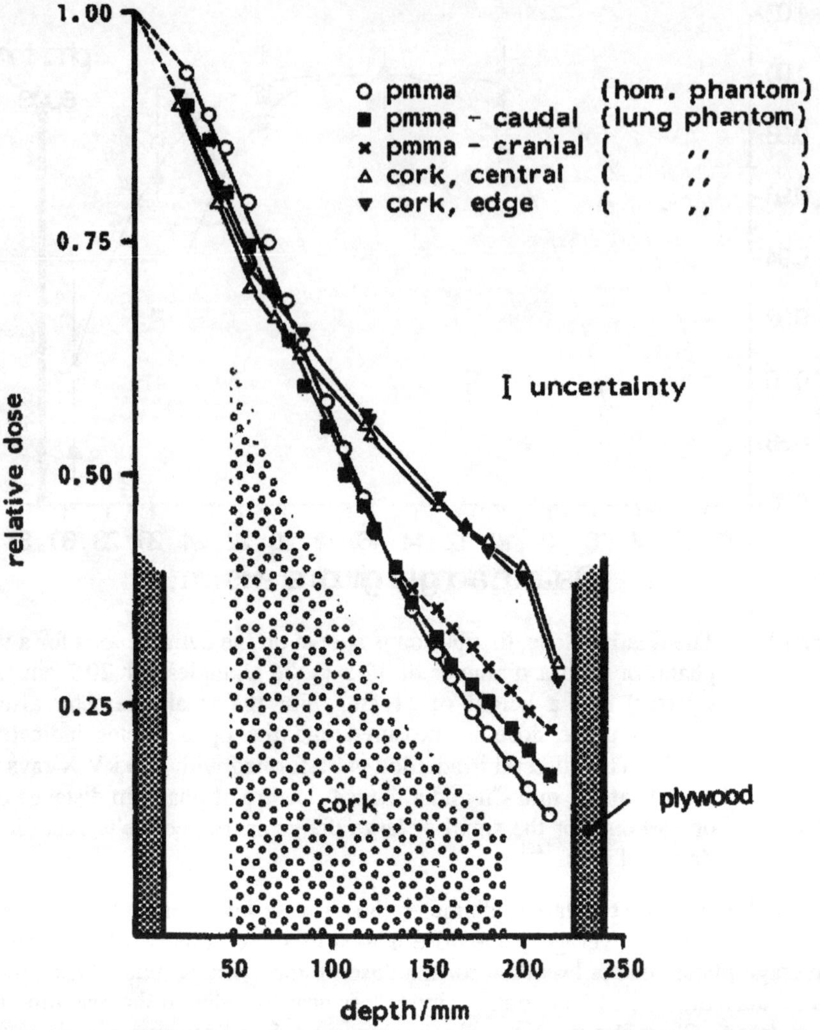

Figure 13. The relative depth dose distribution in a homogeneous PMMA phantom on the central beam axis and for the lung phantom at various locations. The homogeneous PMMA phantom is a block of 22.5 cm x 24.2 cm x 73.8 cm. The lung phantom has similar external dimensions and a cork insert of 14 cm x 10.7 cm x 14 cm to simulate the lung. Unilateral irradiation is performed with 300 kV X-rays with a HVL of 3.2 mm Cu and a focus to centre of phantom distance of 150 cm, according to Zoetelief et al. [25].

The depth dose curves through the cork at the central location (open up triangle) or along the edge (closed down triangle) in Figure 13, both start below the homogeneous and even the PMMA + cork curves, due to the lack of backscatter from the cork. The location at which the phantom is traversed is of little importance. Past the border of PMMA and cork the curves become more flat, due to less attenuation in cork compared to PMMA. At the next border, from cork to PMMA, the curves steepen again due to increased attenuation. The dose is now twice the value for the homogeneous phantom.

BIOLOGICAL FACTORS AFFECTING THE EFFECT OF IRRADIATION

When absorbed doses in tissues are the same, there still can be a considerable difference in the effect due to biological factors. One biological factor is the presence or absence of oxygen during the irradiation and in Figure 14 cell survival curves are shown for both conditions, according to Barendsen et al. [26]. For the radiation with a low linear energy transfer, the radiation in presence of oxygen is much more effective in killing cells than in absence of oxygen. This effect is attributed to the formation of radicals associated with oxygen, that are effective in damaging the deoxyribonucleic acid (DNA). In addition, in Figure 14 it is shown that with increasing linear energy transfer, the effect of oxygen becomes less. At a linear energy transfer of 166 keV/μm the oxygen effect has disappeared.

Temperature is also a biological factor and for sterilisation the low temperature deserves additional attention. In the literature low temperature can both enhance and reduce the effects of irradiation. An explanation for a reduction in biological effect is the reduced range of the radicals formed by the radiation. Thus the radicals are less likely to damage the DNA, resulting in an enhanced survival. On the other hand, at low temperature the repair mechanism of the cell is working at a lower level and the damage is less likely to be repaired. This will result in an enhanced cell killing at low temperatures. In most publications the effects of the radiation are reported to be reduced at low temperatures.

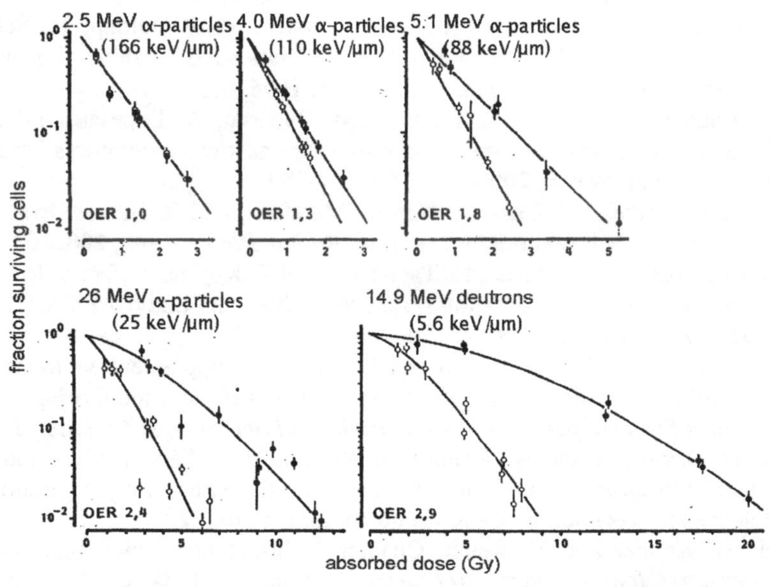

Figure 14. The cell survival fractions are shown as a function of absorbed dose, according to Barendsen et al. [26]. The irradiation is performed with alpha particles or deuterons of different energy resulting in different linear energy transfer given between brackets. Irradiation is performed in presence (open circles) or absence of oxygen (solid circles). The oxygen enhancement ratio (OER), defined as the ratio between the doses necessary for an equal effect of the irradiation in the absence and presence of oxygen, is supplied in the different sections of the Figure.

REFERENCES

1 J. B. Kowalski, Y. Aoshuang and A. Tallentire, Radiation sterilization - evaluation of a new approach for substantiation of 25 kGy, *Radiat. Phys. Chem.*, **2000**, 58, 77-86.

2 International Commission on Radiation Units and Measurement (ICRU), *Radiation Quantities and Units*, ICRU, Bethesda (MA), **1980**, ICRU Report 33.

3 International Commission on Radiation Units and Measurement (ICRU), *Fundamental Quantities and Units for Ionizing Radiation*, ICRU, Bethesda (MA), **1998**, ICRU Report 60.

4 International Commission on Radiation Units and Measurement (ICRU), *Quantities and Units in Radiation Protection Dosimetry*, ICRU, Bethesda (MA), **1993**, ICRU Report 51.

5 A. Dutreix and A. Bridier, Dosimetry of Photons and Electrons, In: *The Dosimetry of Ionizing Radiation Volume I*, K. R. Kase, B. E. Bjärngard and F. H. Attix (eds.), Academic Press, Orlando, USA, **1985**, pp. 163-228.

6 Nuclear Enterprise Technology Limited, *Instruction manual for 0.6 cc ionisation chamber (guard stem) type 2571*, NE Technology Limited, Reading (UK), **1989**.

7 N. V. Klassen, L. van der Zwan, J. Cygler, GafChromic MD-55: Investigated as a precision dosimeter, *Med. Phys.*, **1997**, 24, 1924-1934.

8 W. L. McLaughlin and M. F. Desrosiers, Dosimetry systems for radiation processing, *Radiat. Phys. Chem.*, **1995**, 46, 1163-1174.

9 F. Coninckx, A. Janett, T. Kojima, S. Onori, M. Pantalonp, H. Schönbacher, M. Tavlet and A. Wieser, Responses irradiations of alanine dosimeters to at cryogenic temperatures, *Appl. Radiat. Isot.*, **1996**, 47, 1223-1229.

10 A. Hategan, D. Martin, C. Butan, L.M. Popescu, A. Popescu, and C. Oproiu, Results on electron irradiated Fricke solutions at low temperatures, *Nucl. Instrum. Methods Phys. Res. B*, **2000**, 161-163, 387-389.

11 S. Ramos-Bernal, E. Cruz, A. Negrón-Mendoza, and E. Bustos, Irradiation dose determination below room temperature, *Radiat. Phys. Chem.*, **2002**, 63, 813-815.

12 S. Biramontri, I N. Haneda, H. Tachibana and T. Kojima, Effect of low irradiation temperature on the gamma-ray response of dyed and undyed PMMA dosimeters, *Radiat. Phys. Chem.*, **1996**, 48, 105-109.

13 J. J. Broerse, J. Th. M. Jansen and J. Zoetelief, Supplement V: Measurement of the half value layer, X-ray output and field flatness and monitoring. In: *EULEP-Eurados Protocol for X-ray Dosimetry in Radiobiology*. J. Zoetelief, J. J. Broerse, R. W. Davies, M. Octave-Prignot, M. Rezvani, J. C. Sáez Vergara and M.P. Toni (eds.), European Commission Community Research, Project Report Nuclear Science and Technology EUR 19606 en, Luxemburg, **2000**.

14 H. M. Kramer and H. Reich, Chapter 6, Strahlungsquellen und Spektren, In: *Dosimetrie Ionisierender Strahlung*, H. Reich (ed.), B. G. Teubner, Stuttgart, Germany, **1990**.

15 A. Dixon-Brown, Supplement VIII: Factors affecting X-ray dosimetry for radiobiology and some pitfalls. In: *Protocol for X-ray Dosimetry EULEP*, J. Zoetelief, J. J. Broerse and R. W. Davies (eds.), Commission of the European Communities, Radiation Protection Report EUR 9507 en, Luxemburg, **1985**.

16 International Atomic Energy Agency (IAEA), *Review of data and methods recommended in the international code of practice: IAEA technical report series No. 277 on absorbed dose determination in photon and electron beams*, IAEA, Vienna, Austria, **1996**.

17 J. Seuntjens and F. Verhagen, Dependence of the overall correction factor for a cylindrical ionization chamber on field size and depth in medium-energy X-ray beams, *Med. Phys.*, **1996**, 23, 1789-1796.
18 J. Zoetelief and J. T. M. Jansen, Calculated energy response correction factors for LiF thermoluminescent dosemeters employed in the seventh EULEP dosimetry intercomparison, *Phys. Med. Biol.*, **1997**, 42, 1491-1504.
19 C. Ma and A. E. Nahum, Calculations of ion chamber displacement effect corrections for medium-energy X-ray dosimetry, *Phys. Med. Biol.*, **1995**, 40, 45-62.
20 J. H. Hubbell and S. M. Seltzer, *Tables of mass attenuation coefficients and mass energy absorption coefficients 1 keV to 20 MeV for elements Z = 1 to 92 and 48 additional substances of dosimetric interest*, NISTIR 5632, National Institute for Standards and Technology (NIST), Gaithersburg, USA, **1995**.
21 International Atomic Energy Agency (IAEA), *Absorbed dose determination in external beam radiotherapy. An international code of practice for dosimetry based on absorbed dose to water*, Technical Reports Series 398, IAEA, Vienna, Austria, **2000**.
22 R. W. Davies and J. Zoetelief, Supplement IV: Animal phantoms for radiobiological dosimetry. In: *EULEP Protocol for X-ray Dosimetry*. J. Zoetelief, J. J. Broerse and R. W. Davies (eds.), Commission of the European Communities, Radiation Protection Report EUR 9507 en, Luxemburg, **1985**.
23 W. W. Seelentag, W. Panzer, G. Drexler, L. Platz and F. Santner, *A Catalogue of Spectra for the Calibration of Dosemeters*. Gesellschaft für Strahlen- und Umweltforschung mbH (GSF), GSF Bericht 560, Munich, Germany, **1979**.
24 I. I. Tomljenovic, M. M. Ninkovic, D. Bek-Uzarov, S. J. Stankovic and M. Kovacevic, Water phantom backscatter factors for X-rays in the 60 kV to 300 kV region, *Phys. Med. Biol.*, **1999**, 44, 2193-2200.
25 J. Zoetelief, G. Wagemaker and J. J. Broerse, Dosimetry for total body irradiation of rhesus monkeys with 300 kV X-rays, *Int. J. Radiat. Biol.*, **1998**, 74, 2, 265-272.
26 G. W. Barendsen, H. M. Walter, J. F. Flower and D. K. Bewley, Effects of different ionizing radiations on human cells in tissue culture. III. Experiments with cyclotron accelerated alpha particles and deuterons. *Radiat. Res.*, **1963**, 18, 106-119.

IMPROVED METHOD FOR GAMMA IRRADIATION OF DONOR TISSUE

**Ruth Garcia [1], Anthony Harris [1], Martell Winters [2*], Betty Howard [1*]
Paul Mellor [1*], Deepak Patil [1*] and Jason Meiner [1*]**

[1] *STERIS Isomedix Services, 7828 Nagle Avenue, Morton Grove, IL 60053, USA.
{E-mail: Ruth_Garcia@STERIS.com, Anthony_Harris@STERIS.com,
Betty_Howard@STERIS.com, Paul_Mellor@STERIS.com,
Deepak_Patil@STERIS.com, Jason_Meiner@STERIS.com}*

[2] *Nelson Laboratories, 6280 South Redwood Road, Salt Lake City, Utah 84123-6600, USA.
{E-mail: MWinters@nelsonlabs.com}*

ABSTRACT

Recently there has been a major concern about the sterility of donor tissue. Gamma irradiation has proven to be an effective way of processing tissue for sterility [4]. It has been proven that tissue, irradiated in the frozen state, sustains much less degradation than samples processed at higher temperatures. An easy and common way of maintaining tissue in the frozen state during shipment and irradiation involves packaging the tissue with dry ice. All dosimetry systems commonly used in the irradiation industry are temperature sensitive. Therefore, samples packaged with dry ice must take these dosimeter temperature effects into consideration. STERIS Isomedix Services has evolved a method to provide a standardised process, which ensures a high degree of accuracy for absorbed dose determination of low temperature radiation processing, and specifically to support the irradiation of donor tissue.

KEYWORDS

Gamma irradiation; tissue banking; donor; allograft; dosimeter; BI; low temperature

INTRODUCTION

In 1881, the first human bone transplant was performed under aseptic conditions. In 1999, tissue banks in the United States distributed approximately 650,000 musculoskeletal allografts, compared with 350,000 in 1990 [5]. More recent reports indicate usage of more than 800,000 musculoskeletal allografts in the US during 2000 [3] and this number continues to increase. The fact that an allograft is tissue transferred between two genetically different individuals, of the same species, raises concern about the introduction of bacteria to the recipient and the immune sensitisation of recipient to donor antigens. The increase in the number of allografts performed brought new concerns about the sterility of grafts. As of March 2003, CDC had received reports of 62 allograft-associated infections [5]. Maintenance of sterility is a major concern whether a graft is fresh or preserved [1].

BACKGROUND

To ensure the highest quality specimen, some form of preservation is desired. The most common form for preservation of tissue is freezing [1]. (Another form of preservation involves freeze-drying samples, which is not discussed in this paper.)

Historically, grafted tissue was processed using aseptic techniques, (to prevent the introduction of additional contamination), or by sterilisation methods. Soaking in antibacterial and antifungal solutions may be used in addition to aseptic recovery to further reduce any microflora normally found associated with tissue specimens. Both EO (ethylene oxide) gas and gamma radiation have been used as tissue sterilisation methods, but each has its drawbacks. EO leaves chemical residuals on specimens that can cause inflammation upon implantation; also the gas may not penetrate the tissue sufficiently to address non-surface contamination. Gamma irradiation kills many pathogens, but high radiation doses may impact the functionality of some tissue types, especially soft tissue. Dr Steven L. Solomon of the CDC notes that, "the potential risks associated with the transplantation of aseptically processed tissues suggest that existing sterilisation technologies used for sterilising allografts, such as gamma irradiation, or new technologies with increased effectiveness against bacterial spores should be considered whenever possible" [2]. A modern processing methodology for terminal sterilisation of grafts after preservation is gamma irradiation *of frozen specimens*. Gamma irradiation is very effective at killing bacteria at absorbed doses less than 20 kGy [1]. At temperatures of -20°C to -147°C damage to biological and physical functions of the grafts are minimal [1]. Freezing in conjunction with gamma irradiation is an ideal way for processing, preservation, and sterilisation of grafts. The two methods combined are less time consuming and more cost efficient. This paper will suggest a method to accurately deliver an absorbed radiation dose to decontaminate donor tissue using dosimetry techniques designed to abrogate the skewing effects of low temperature environments on existing dosimetry systems.

THE CHALLENGE

Given that freezing is a desirable way to preserve tissue and given that irradiation is a proven method of microbial reduction in tissue samples, the irradiation of frozen tissue would seem to be an ideal solution. One of the primary challenges of irradiating a frozen sample is determining the absorbed dose delivered to the frozen sample. Most commonly used dose-measuring or dosimetry methods are influenced by temperature. Therefore, under some conditions, frozen product can influence a dosimeter response resulting in a less accurate or skewed absorbed dose analysis. This makes the placement of a dosimeter in the volume of contaminated frozen tissue impractical. Although limited information on temperature correction factors is available in the literature, there are batch-to-batch variations in dosimeter response and the irradiation process itself is such that the temperature is not constant throughout the entire process, making the application of a single correction difficult. In his book *Dosimetry For Radiation Processing*, Dr. William McLaughlin notes that, "Irradiation temperature is, in fact, the most important environmental factor contributing to errors in absorbed dose estimation, and in radiation processing it is sometimes poorly determined and difficult to correct for". Accurate absorbed dose measurement is key to the overall quality for frozen tissue irradiation as advocated by the IAEA, a strong proponent of health safety through tissue irradiation.

Specimen density is also a challenge for the determination of irradiation absorbed dose applied. Gamma rays emitted from cobalt $^{-60}$ (the isotope typically used in industrial and some medical applications) are very penetrating with 1.17MeV and 1.33 MeV photons. Gamma rays nonetheless will lose their energy as they pass through dense material. Frozen tissue specimens are generally packed in dry ice to maintain a low temperature and since dry ice typically used in processing is approximately 0.47 g/cm^3, and tissue is approximately 1.0 g/cm^3, concerns are raised that the target irradiation dose is not delivered to the internal (centre) area of the sample carton.

PROPOSED SOLUTION

The proposed solution is based on a dosimetric technique involving an alternate dose monitoring position with an established mathematical relationship to other dose zones represented throughout a standard sized sample. The assumptions were:

1. Dosimeter calibrations were performed at ambient temperature (≅ 25°C).
2. Dosimeters placed on the external surface of a carton of frozen material with a layer of insulating material between the dosimeter and the temperature compromised carton surface would remain between 0-25°C, a temperature range at which little or no temperature correction was required for the dosimeter used in testing.
3. The most challenging area for dose penetration is in the centre of the densest product mass. In homogeneous products of similar dimensions, repeated experimentation and volume characterisation has shown that the 'geometric centre' of a homogeneous square or compact rectangle cardboard/carton sample provides the greatest density challenge and is generally the location of the low dose zone.
4. A standard sized insulated carton would be used for all experiments.
5. A generic brand of dog food pellets was identified for use to simulate dry ice, having a density comparable to that of dry ice.

The hypothesis was that the numeric ratio between an external monitoring position and the most difficult to sterilise zone (the centre internal area of the specimen) in ambient temperature 'surrogate' material would be analogous to the ratio for 'real' material, in this case simulated tissue material surrounded by dry ice. An additional test was performed using biological indicators to confirm that bacterial kill was achieved based on the calculated absorbed dose ratio and to evaluate differences between the D_{10} values of the spore strips placed in surrogate material verses dry ice.

The following steps will be described in this paper:

1. Selection of carton for experiment,
2. Testing external surface temperature,
3. Determining if simulated product (dog food) mimics the physical properties of dry ice,
4. Establishing delivered dose ratio for the tested carton,
5. Biological testing

EXPERIMENTATION

Step 1

The first task was to select a standard sized, insulated carton, appropriate for packaging, transport and processing of low temperature samples. STERIS Isomedix Services selected Polyfoam Packer Corporation 22" x 14.5" x 17.5" (hereafter referred to as insulated carton) carton for testing. This carton was selected because of its purported exceptional insulation properties.

Step 2

Next the insulating properties of the carton were evaluated to support assumption #2 (above): "Dosimeters placed on the external surface of a carton of frozen material, with a layer of insulating material between the dosimeter and the temperature compromised carton surface, would remain between 0-25°C, a temperature range at which little or no temperature correction was required for the dosimeter used in testing."

To verify if the external surface temperature would remain above 0°C, the Digi-Sense® scanning thermometer for data collection was used. The insulated carton was filled with dry ice and then sealed. Thermocouple at location "A" was used to monitor the room temperature, thermocouple at location "B" was placed inside the dosimeter sachet (Harwell Red 4034 Perspex® dosimeters were used) between the dosimeter and the external surface of insulated carton, and thermocouple at location "C" was placed between two Harwell 4034 dosimeters on the external surface of carton (see Figure 1).

Temperature data was collected every 3 minutes for 64 hours (a total 1279 measurements) to determine if the external surface temperature would remain above 0°C. Data shows that the external surface temperature consistently remained between 10°C and 15°C. The fluctuations in the temperature data appeared to follow ambient temperature movement, i.e., the room air conditioning turning on and off. Thermocouple B was selected over thermocouple C because thermocouple B most closely represents how the dosimeter would actually be exposed to the external surface of the carton (see Figure 2).

Step 3

Once the carton is selected and characterised and found capable of maintaining external surface temperature above 0°C, the next step was to determine if the physical properties of the simulated dry ice (dog food) actually mimic the physical properties of real dry ice. Normally this could be done by placing a dosimeter inside the volume of interest and placing a dosimeter outside the carton in a proposed monitoring position. Since the routine dosimeter used in this testing was the Harwell 4034 and the response of Harwell 4034 dosimeters is both dose rate and temperature dependent, ceric-cerous dosimeters were used for this segment of the experiment. The response of ceric-cerous dosimeters is not dose rate dependent and the effects of temperature can be easily compensated for when the irradiation temperature is known. Ceric-cerous can be exposed to temperature between 0°C and 60°C for determining delivered dose. Ceric-cerous used at STERIS Isomedix Services (provided by MDS Nordion) are calibrated at 25°C.

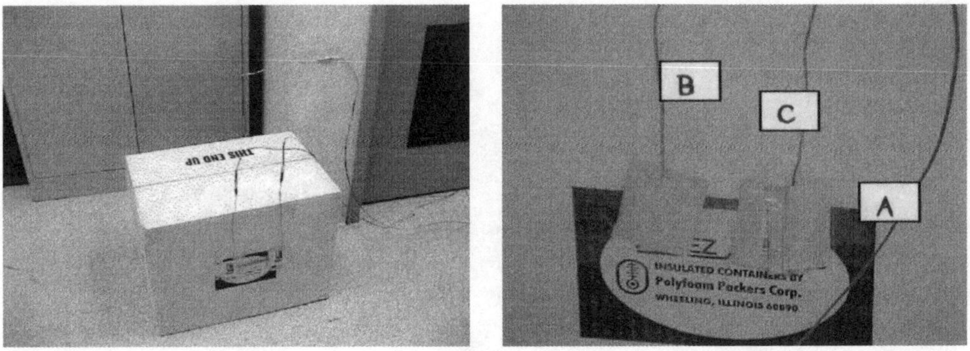

(A) (B)

Figure 1. (A) Carton placement in laboratory environment;
(B) Thermocouple locations A, B, and C.

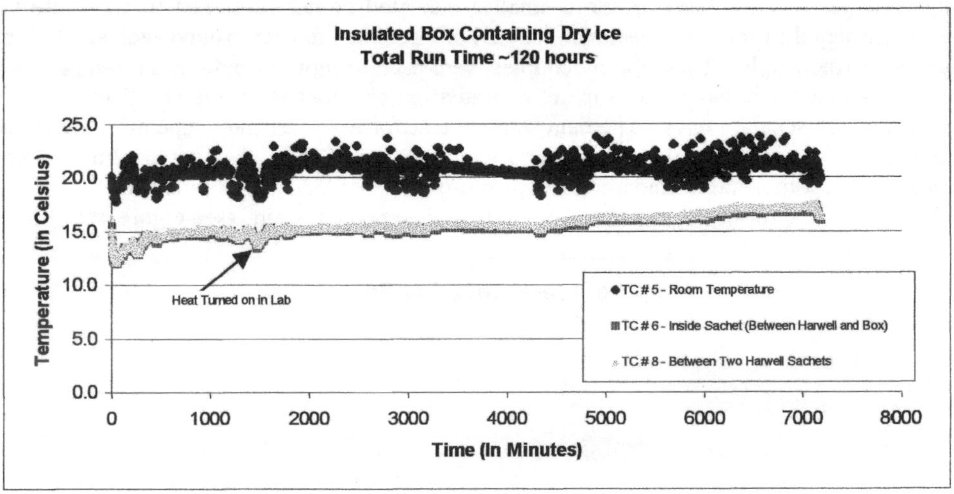

Figure 2. Monitoring external surface temperature using the Digi-Sense® scanning thermometer.

To determine if simulated dry ice (dog food) was truly an accurate representation of real dry ice, two tests were performed to see if the density was within an acceptable range. The first test was performed with ceric-cerous and dry ice and the second test was performed with ceric-cerous and simulated dry ice (dog food). The temperature of dry ice is typically between -80°C to -147°C, well outside the operating range for ceric-cerous. To overcome this obstacle, a self-contained battery powered heater (hereafter referred to as battery heater) designed by Patco Inc. was used to maintain the temperature of ceric-cerous dosimeters used within a constant range (see Figure 3).

Figure 3. Patco Inc. self-contained battery.

The self-contained battery powered heater was designed to run on lead acid batteries with an internal non-electrical thermostat. Three tests were performed for reproducibility to determine if temperature remained at a consistent level. The self-contained battery powered heater was placed inside a smaller insulated cooler (designed to fit inside the larger insulated carton with clearance for real or simulated dry ice around each side) along with a thermocouple. Three thermocouples were used to monitor outer room temperature, and a thermocouple was used to monitor irradiation chamber temperature. Each test was performed on separate days. The data showed that for each test the temperature remained between 21°C and 29°C, with an average temperature of 25°C. This temperature is well within the recommended irradiation temperature range of 0-62°C, as noted in ISO/ASTM 51205, "Practice for use of a ceric-cerous sulfate dosimetry system" (see Figure 4).

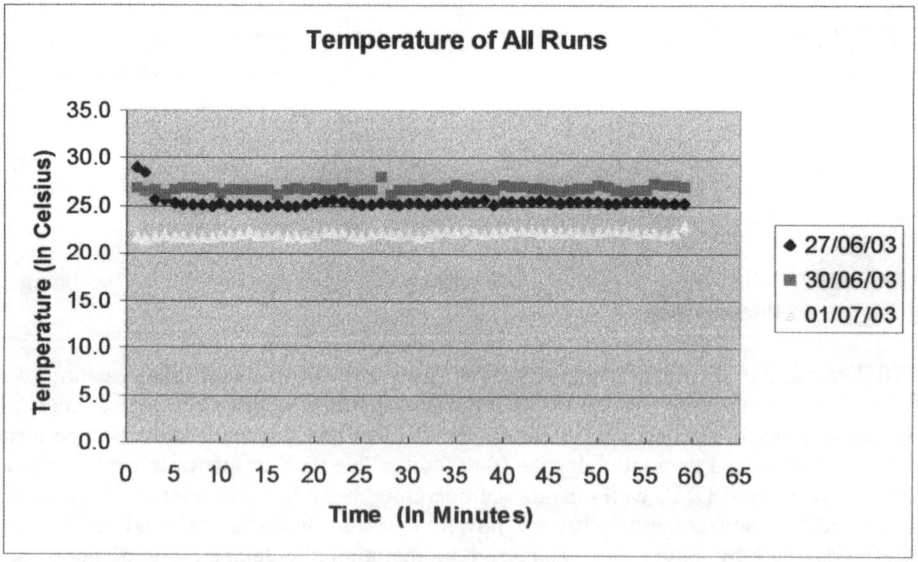

Figure 4. External temperature from three different runs.

Step 4

To establish dose ratio from the centre of the insulated carton to an external monitoring position on the surface of the carton, four test runs were executed. Run A was performed with eleven bundles of ceric-cerous with thermo-labels. This test was performed to create a baseline for establishing a delivered dose ratio from an external monitoring position to a range of positions internally. Runs B, C, and D were performed with three bundles of ceric-cerous with thermo-labels. Run B was performed with the battery heater turned on and sealed within a smaller insulated cooler with a bundle of ceric-cerous dosimeters and a thermo-label (see Figure 5). The small cooler was then placed within the insulated carton and packed in dry ice. The other two dosimeter bundles were placed on the external surface of the carton and designated FC and EC. Run C was performed almost the same as run B but with the heater off and with simulated dry ice (dog food) instead of real dry ice. Run D was performed the same as run C but without the heater (see Table 1).

The purpose for performing runs B and C was to determine if the physical properties of the simulated dry ice (dog food) mimic the physical properties of real dry ice. If the delivered dose ratios for the simulated dry ice (dog food) and the real dry ice are within an acceptable range then dog food may be used to simulate dry ice and the test may continue. Runs C and D were completed in order to determine if the Patco self-contained battery powered heater would create shielding of the ceric-cerous dosimeter.

Run B resulted in a ratio of 1.16 calculated from the external reference position to the internal centre. Run C resulted in a ratio of 1.18 calculated from the external reference position to the internal centre. Run D resulted in a ratio of 1.10 from the external reference position to the internal centre. Comparing results from run B to run C indicates a 2% difference between simulated dry ice (dog food) and real dry ice. This 2% difference is acceptable for using the dog food as a substitute for dry ice without the need for a correction factor.

Figure 5. Layout of how STERIS Inc. overcame the obstacle of using ceric-cerous dosimeters at dry ice temperature.

Table 1. Summary of dose ratio experiments.

Run ID	Configuration (w/in insulated carton)	Purpose	Ratio
A	Dog food with min. dosimeter profile.	Establish baseline internal absorbed dose profile.	N/A
B	Dry ice w/battery heater.	Compare to Run C to show dog food can simulate dry ice.	1.16
C	Dog food w/battery heater	Compare to Run B to show dog food can simulate dry ice.	1.18
D	Dog food w/out heater	Generate ratio between internal and external dosimeter readings without any shielding from battery heater. (Most like actual frozen product.)	1.10

{w/ = with; w/in = within; w/out = without}

Step 5

To determine if this ratio could be confirmed with biological samples, *Bacillus pumilus* strips were used to simulate product with a microbial load being processed using gamma irradiation. *B. pumilus* with a population of 1.1 x 10^6 was selected for testing with an estimated D_{10} value of 1.4 kGy. The objective for testing was to 1) simulate biological lethality in the insulated box configuration, and 2) evaluate any differences between the D_{10} values of *B. pumilus* at room temperature versus dry ice temperature. Any differences in D_{10} values with this organism may reflect similar differences in D_{10} values of other organisms. The required dose for a 6-log reduction should be 6 x 1.4 kGy = 8.4 kGy, so doses above and below 8.4 kGy were used for this evaluation.

Five incremental doses (7, 8, 9, 10, and 11 kGy) were selected for gamma irradiating for statistical analysis. Each dose point was performed twice. The first set of tests used fourteen *B. pumilus* strips for each dose packed in dry ice and the second set of tests had 14 *B. pumilus* strips for each dose packed in simulated dry ice material (dog food). Each dose point was adjusted to reflect the established ratio using this calculation:

Minimum requested dose x Established Ratio = (adjusted) Minimum target dose for exterior

Using this calculation, the five doses of 7, 8, 9, 10, and 11 kGy were adjusted to 7.7, 8.8, 9.9, 11, and 12.1 kGy. The *B. pumilus* strips were sent to Nelson Laboratories Inc. for sterility testing.

Three supplemental irradiations were used for testing of the *B. pumilus* under dry ice conditions. Doses of 12, 13, and 14 kGy were chosen and adjusted to 13.2, 14.3, and 15.4 kGy for testing of the *B. pumilus*.

The supplemental dose points were used to extend the irradiation range thus ensuring sufficient information for calculating a D_{10} value.

BIOLOGICAL PROCEDURES

Procedure for Fraction Negative and Limited Spearman-Karber testing

Ten BIs of each dose were individually transferred to tubes of soybean casein digest broth. The strips were incubated at 30-35°C for 7 days then scored as positive or negative for growth.

Procedure for Survivor Curve testing

Three BIs of each dose were tested for population verification by pooling three BIs together, vortexing the strips in sterile water with sterile glass beads until macerated, performing serial dilutions, and plating onto soybean casein digest agar. The plates were incubated at 30-35°C for 2 days then enumerated.

CALCULATIONS

The D_{10} value of each organism was determined using the Limited Spearman-Karber, Fraction Negative, and Survivor Curve methods. The calculations follow:

Limited Spearman-Karber

Details on performing this calculation are in ANSI/AAMI/ISO 11138, Annex D. For the Limited Spearman-Karber calculation, it was necessary to use doses with exact intervals in the calculations (e.g. 9.0, 10.0 kGy) rather than the actual doses delivered (e.g. 9.1, 10.3 kGy). This is a restriction required by the calculation and cannot be avoided.

Fraction Negative

$$= \text{Dose} / (\log N_0 - \log \text{MPN})$$

Where:

N_0	=	Number of organisms on BI pre-irradiation
MPN	=	ln (# BIs tested / # BIs negative)

Survivor Curve

$$= \text{Dose} / (\log N_0 - \log N_f)$$

Where:

N_0	=	Number of organisms on BI pre-irradiation
N_f	=	Number of organisms on BI post-irradiation

BIOLOGICAL RESULTS

Results of sterility and population verification testing:

Table 2. Sterility and population verification results.

Dry Ice	9.1	10.3	11.1	12.1	13.5	14.4	15.6
# Negative	0	0	0	1	3	10	10
CFU/Strip	UFA	~17 [1]	UFA	UFA	UFA	UFA	UFA

{UFA = the results were outside of the statistically accurate range for a plate count}
{1 In this case even the 10.3 kGy results were below the desired range. This dose was used for the D_{10} determination because it was the lowest dose, which resulted in any growth due to the dilutions used for testing the 8.8 kGy strips}

Table3. Number of strips negative for growth out of 10 tested and colony forming units (CFU) per strip after receiving the specified dose in kGy.

Dog Food	8.3	9.5	10.7	11.8	12.8
# Negative	0	1	4	10	10
CFU/Strip	~1.1 [2]	UFA	UFA	UFA	UFA

{UFA = the results were outside of the statistically accurate range for a plate count}
{2 In this case even the 8.3 kGy results were below the desired range. This dose was used for the D_{10} determination because it was the lowest dose, which resulted in any growth from the strips}

Summary Table - Calculated D-Values for Dry Ice and Dog Food

Table 4. Calculated D-Values.

	Fraction Negative	Survivor Curve	Limited SK	Limited SK 95% Confidence
Dry Ice	2.20 kGy	2.14 kGy	2.13 kGy	2.07 – 2.19 kGy
Dog Food	1.72 kGy	1.38 kGy	1.68 kGy	1.62 – 1.75 kGy

SUMMARY AND DISCUSSION

In general, this series of experiments demonstrates that irradiation of materials in the configuration described above provides a reliable method of determining an external monitoring position and corresponding dose ratio to use in microbial reduction of low temperature samples. The ambient temperature spore strips (BI's – biological indicators irradiated in dog food surrogate) using the establish dose ratio, accurately reflected the expected D_{10} value; spore strips irradiated in the frozen state exhibited a measurably higher D_{10} value. This is consistent with the cryo-preservative effects of low temperatures on biological systems as demonstrated previously by Block and others [7]. If tissue is irradiated in a frozen state to preserve cellular qualities, it follows that bacteria will also benefit from this cryo-preservative effect. This difference in D_{10} value is an interesting result of the experimentation, but is independent of the ratio establishment and does not exclude using surrogate (simulated) material for dose mapping. In fact, the use of surrogate material for dose mapping is a well-accepted irradiation industry practice that has been once again confirmed in this study with the use of the heating element. The biological indictor results do emphasise the importance of considering the effects of temperature when determining a sterilisation dose for temperature-compromised product. In these situations, it may not be appropriate to base a sterilisation dose on published D_{10} values that may have been established under very different temperature conditions.

Using surrogate material data and biological spore strip data, STERIS Isomedix Services has demonstrated a procedure for establishing a dose ratio with a standard box that provides a useful method of quickly and accurately determining the irradiation dose to frozen tissue specimens. The standardised carton can be provided to customers for specimen packaging, ensuring a quick and easy way for processing product under low temperature conditions.

REFERENCES

1. Stevenson, SDVM, Autograft/allograft biology of bone grafts, *Orthopedic Clinics of North America*, **1999**, 30 (4), 543-5.
2. Solomon, SL, CDC response to infections related to human tissue transplantation, *US Senate Committee on Governmental Affairs*, **2003**, May, 14.
3. Joyce, MJ, Musculoskeletal allograft tissue safety, *American Academy of Orthopaedic Surgeons*, **2003**, Feb, 5-9.
4. Boyce, T, Edwards, J and Scarborough, N, Allograft bone. The influence of processing on safety and performance, *Orthopedic Clinics of North America*, **1999**, 30 (4), 571-81.
5. McLaughlin, WL, Boyd, AW, Chadwick, KH, McDonald, JC, and Miller, A, *Dosimetry for Radiation Processing*, Taylor and Francis, Philadelphia, PA, USA, **1989**, p. 189.
6. *United States Pharmacopeia* 26 & *National Formulary* 21, United States Pharmacopeial Convention, Rockville, MD, USA, <71>, <55>, <1035>, and <1211>.
7. Block, SS, *Disinfection, Sterilization, and Preservation, 5th Ed.*, Chapters 6 (calculations) and 37 (D_{10} value discussion), Lea & Febiger, Philadelphia, PA, USA, **2001**.

RAPID HEAT TRANSFER DYNAMICS AND COLD GAMMA STERILISATION METHODS FOR SOFT TISSUE ALLOGRAFTS

M. Hayzlett *, S. Griffey and G. Greenleaf

LifeCell Corporation, One Millenium Way, Branchburg, New Jersey 08876, USA
{E-mail: mhayzlett@lifecell.com}

ABSTRACT

The sterilisation of soft tissue allografts presents a particular challenge. Methods must be sufficiently robust to effect bacterial and viral kill while maintaining native structure as well as the biochemical and biomechanical properties of the grafts. The use of cold temperature gamma irradiation has been described as a method for sterilising tissue while minimising deleterious effects. In our experience, cold temperatures alone did little to prevent matrix damage of a freeze-dried dermal product when exposed to sterilising gamma doses. Our studies confirmed that a negligible change of product temperature was detected during gamma processing (photo-heating). Assuming tissue damage was not instantaneous, some accumulation of energy from all sources could contribute to a local damage 'event' over time. We focused on optimising the design and composition of the gamma container to increase target density and improve heat transfer. We theorised that increasing the *rate* of energy transfer away from the tissue would prevent harmful chemical reactions by preventing the required energy thresholds from being attained. Numerous tests have produced good results even with 15 kGy of accumulated dose absorbed by the tissue. Localisation of the reaction kinetics by cold temperatures may limit the number of opportunities for adverse reactions at any given time during gamma exposure, by nature of the probability of a given number of photons intercepting molecules of interest. This rate may fall within the capability of special materials designed at LifeCell to disperse energy to a –80° C cold reservoir. Gamma processing of human sheet dermis, tendon and other tissue derived products has been accomplished with these special methods. Preliminary tests indicate good results employing accumulated doses up to 40 kGy upon osteoinductive materials. Current research is investigating the application of 12 kGy upon cryoprotected skin prior to processing, followed by a second 12 kGy exposure of the same tissue in protective packaging in the same environment.

KEYWORDS

Sterilisation; gamma; energy transfer; soft tissue; dermis

INTRODUCTION

LifeCell Corporation has been processing soft human tissues for over ten years, predominantly preparing acellular dermis as a natural scaffold for clinical use in treating burns, soft tissue deficits and supplemental repair of connective tissue. Efforts to improve the safety of these products without affecting their functional characteristics resulted in a method of gamma irradiation treatment presented in concept within the following paper. There are any number of gamma-induced biochemical events that are expected to take place in the matrix and surrounding frozen solutions, and it is also expected that precise understanding of all the mechanisms taking place at any given time during and after irradiation would be very challenging to characterise. Since the object of the investigations undertaken by LifeCell scientists was to leave the acellular dermis unchanged in all the characteristics that we have found to be relevant to success as an allograft, our efforts focused upon the design of an irradiation method that supported that end result.

Therefore, LifeCell Corporation wished to share the findings of our experimentation regarding the concept of rapid heat transfer mitigating the typical gamma damage that often abrogates the use of gamma irradiation to reduce microorganisms in human derived tissues for transplant. The exact treatment procedure requires a custom design for the process specific to the tissue type, thus, the materials and designs of the equipment utilised for LifeCell products is proprietary and not disclosed herein. However, it is hoped that with the general knowledge of how rapid heat transfer affects the state of materials, other scientists could undertake their own trials of gamma irradiation upon tissue they wish to sanitise and benefit from our experience.

DISCUSSION

Research into the feasibility of using gamma irradiation to affect the bioburden reduction in soft tissues started with the premise that the irradiated condition of the tissue would be in its final packaged form. LifeCell Corporation has a proprietary process for removing the epidermis and cells from human skin, while preserving the dermal matrix and biochemical components necessary for regeneration as an implant. During the last several years, this product has been primarily marketed as Alloderm®. Motivation for reducing the bioburden stemmed from the attempt to qualify the product as a Medical Device for use as a repair tissue for dura-mater under Section 510K pre-market approval guidelines from the US Food & Drug Administration, which required Terminal Sterilisation of the graft. The final form of the graft as it is distributed is freeze-dried, packaged in a heat-sealed Tyvek® bag as well as a plastic-coated foil bag that is also heat-sealed. The Alloderm® has a storage condition of 2 years at 2-10°C, and has a matrix condition that is within quality standards verified by histological scoring for several weighted characteristics of collagen bundle condition.

Various methods of loading these individual product units into coolers were attempted, using ambient and –80°C conditions. Standards for experiments were to observe how well the treated samples compared to controls using the aforementioned histology standard. A significant premise for these trials was that the use of any process upon the tissue must leave the dermal matrix unchanged in form and function. The morphology of the dermal matrix does not have a direct relationship to the efficacy of this particular type of graft in animal models, as several examples of what LifeCell would consider 'damaged' did in fact revascularise and stimulate cell infiltration without significant inflammation. However, the company maintains a conservative stance towards minimal manipulation of the matrix to ensure the consistent efficacy that the product has demonstrated in over 500,000 human grafts.

Many of the samples with an absorbed gamma dose of 11 kGy did not compare favourably with the controls, particularly in the quality of the reticular and papillary regions of the matrix, and separation of collagen bundles (see Figure 1).

The transition to a new method of preparing the tissue for gamma irradiation was based upon both the unique processing steps of Alloderm® and theoretical concepts of what may have caused the alteration of the matrix. Most of the processing steps for the tissue involve solutions, with the use of a proprietary mixture of a starch-based cryogenic preservation additive (CPA) prior to any freezing event. The CPA is a complex carbohydrate that acts to dehydrate the tissue such that subsequent freezing of the tissue in a –80°C environment results in amorphous ice formation at a glass-transition temperature (GTT) of approximately –17 C. The mass of the tissue in its frozen form prior to lyophilisation is approximately four times the mass of the dried product, with most of this mass as free water. Once in this frozen form, it was postulated that the structure of the matrix would be rigidly braced from alteration as a result of gamma absorption. Preparation of the tissues was routine with tissue exposed to gamma irradiation (13 kGy) at –80°C in a dry ice environment, without any support or contact with any material other than the dry ice. There was a significant improvement in the histological condition of the tissue, with all test samples comparing favourably to controls after all tissue was lyophilised and rehydrated for sectioning several days after the gamma processing. Samples of the same donor tissue retained for 30-day accelerated storage at 37°C revealed a reversal of the favourable condition, with all irradiated samples drastically altered and failing the same histological standard.

Concurrent with the trials of that time were test vials of the CPA and alternate formulations without tissue. Various temperatures were included for an analysis of any evidence of radical formation as a result of gamma irradiation. It was postulated that interaction of potential radical species with lipids would lead to lipid peroxidation. A by-product of such a reaction would be malondialdehyde (MDA), which could be detected using a thiobarbituric acid (TBA) test.

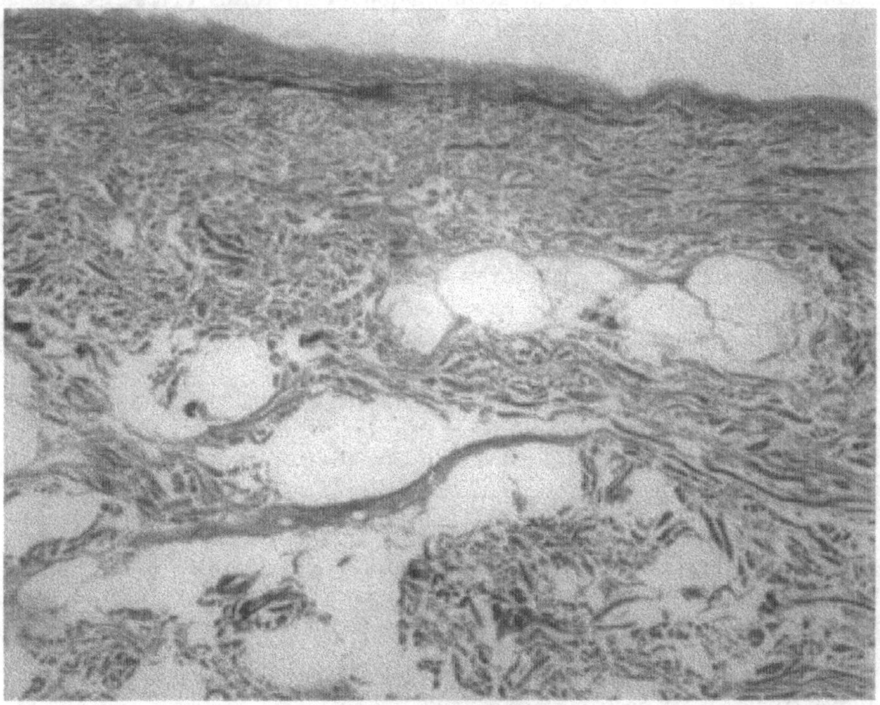

Figure 1. 20 x view of human dermis [1] section using Hemotoxylin and Eosin (H & E) staining. Compressed papillary region at the top has a distinct boundary to the remaining reticular region, and collagen bundles have an amorphous appearance.

Peak absorbance in spectral analysis was based upon a published assay procedure of Stocks and Dormandy (*Br. J. Haematol.* **20**:95-111, 1971). Evidence of peroxidation may have a relationship to the reactivity of the CPA formulation and the temperature of the solution during gamma irradiation. Using the results of these tests, it was decided that changes were occurring in the chemical nature of the CPA. The effect of the changes apparently did little to alter the GTT of the CPA solutions in bench testing, but the biochemical effects have not been investigated further.

The design of the support system for the tissue was based upon the hypothesis that biochemical changes in the environment of the tissue during irradiation had the potential for matrix damage long after irradiation was completed. Up to this point the system had prevented immediate changes, perhaps by limiting the diffusion of what may have been radical species during the gamma exposure. A theory was postulated that instead of delaying the biochemical changes, the energy threshold required to develop the (probable) radicals could be diminished though heat energy removal at a rate comparable to the energy provided by gamma absorption.

It must be made clear that this is not to say that the pure energy from gamma is translated into any form that can be dealt with using heat transfer. It is well established that photo heating is not a characteristic of gamma irradiation. The hypothetical concept is that if a cascade of radical-based reactions were initiated within the given localised environment and materials of the LifeCell product, their capacity to propagate may be based upon exothermic by-products that can be diminished with high rates of heat transfer away from the tissue. If the activation energy of any secondary 'biochemical event' were to be an accumulation of all forms of energy available, draining at least the thermal portion of that energy at a rapid rate could limit gamma-induced by-products to low energy 'events'. This concept is complemented by the theory that if the CPA provided sufficient targets for low energy 'events', such an environment could reduce the probability that the tissue (specifically collagen bundle cross-linking) will be involved. The implied assumption that collagen cross-linking is a relatively high energy 'event' has not been substantiated.

The arrangement of the gamma container evolved to a proprietary design that incorporated special materials to contact the tissue in such a manner to be gamma transparent, yet provide a path for heat flux to develop to a cold sink. Tissues processed with 15.5 kGy with this technique have not only demonstrated preservation of the matrix in accelerated storage, but also real-time storage of 1½ years at 4°C. An example of tissue processed in this manner is rendered in Figure 2.

The histological success using the high heat transfer gamma configuration resulted in parity with controls over numerous repeats of experiments on a variety of donor lots (tissue rejected for production use). Further in-vitro tests included:

- Glycosaminoglycan analysis with HPLC
- Type IV and Type VII Collagen staining
- Mechanical testing (suture retention, creep and relaxation, tensile)
- Cytotoxicity (agar diffusion test, mouse cell)
- Hemolysis
- Residual Moisture Analysis (upon lyophilised product)

Animal studies have been performed with subcutaneous implants and subsequent histological sectioning of explants for ten 7-day *in* vivo tests with Sprague-Dawley albino rats (inflammatory response) and twenty 21-day tests with immunocompromised mice (cell infiltration and revascularisation). All evaluations by two blinded scorers resulted in no statistical difference between test (up to 9 kGy) and control articles [3, 4].

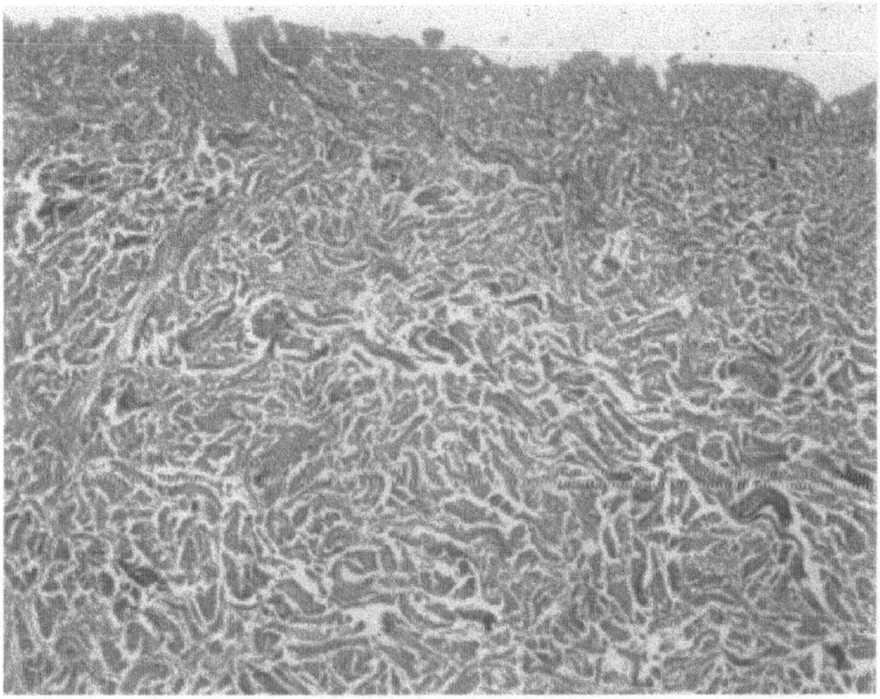

Figure 2. 10 x histology section, H & E staining, human dermis [2]. Tissue sectioned after 30-day-accelerated storage. 15.5 kGy absorbed gamma dose using special heat transfer techniques. Note smooth transition from reticular to papillary regions and retention of collagen bundle structure.

Based upon the concepts that have brought success to the evaluation of Alloderm® processed with gamma irradiation prior to lyophilisation, other collagen-based products have been similarly configured to enable bioburden reduction with a wide range of absorbed doses of gamma irradiation. These include human tendon, micronised placental matrix and micronised dermis (Cymetra®).

CONCLUSIONS

The successful gamma processing of allograft soft tissues is dependent upon the criteria selected to specify form and function throughout processing. This imperative is clearly set forth in a number of regulatory documents in the U.S., with respect to 'minimal manipulation'. The autologous use of these grafts is clinically proven to be a useful tool for reconstructive medicine as long as there is a good demonstration of efficacy with those standards of form and function in place. The development of the gamma processes described in this research has evolved within the structure of the established processing method, which retains the form and function of a dermal matrix. Although this process is proprietary to LifeCell Corporation, the concepts of including rapid heat transfer designs in the irradiation configuration may have applications to other allograft processing methods. Benefits of low temperature gamma processing are widely recognised, and the opportunity to enhance this technique with high heat transfer rate materials may provide an alternative or supplement to chemical stabilisation efforts to minimise gamma induced tissue damage.

REFERENCES

1. Tissue histology slide from Run # 4410-31, LifeCell Corporation.
2. Tissue histology slide from Run # 4767-24, LifeCell Corporation.
3. Toxicon GLP Rat Study #01-0250-G2 with evaluation by S. Livesey, S. Griffey and M. Hayzlett, March 1, **2001**.
4. Toxicon GLP Nude Mouse Study #02-4199-G1 with evaluation by S. Livesey, S. Griffey and M. Hayzlett, November 26, **2002**.

COMPARISON OF DIFFERENT THAWING METHODS ON CRYOPRESERVED RABBIT AORTA

Sung Bo Sim*, Young Min Oh and Sun Hee Lee

Department of Thoracic & Cardiovascular Surgery, Saint Mary's Hospital, Catholic University of Korea, 62 Yoido-dong, Youngdeungpo-gu, Seoul, 150-713, South Korea {E-mail: sungbo@catholic.ac.kr}

ABSTRACT

The studies on cryopreserved arterial allografts have been focused on cooling methods, pre-treatment, cryoprotectant agents, and preservation temperature. But recently, several studies have reported that thawing methods also play an important role in the occurrence of macroscopic and microscopic cracks. This study was designed to investigate the expression of apoptosis after thawing, using a rabbit model to clarify the effect of thawing methods on cryopreserved arteries.

Segments of the rabbit aorta were obtained and divided into 3 groups (n=60) according to whether the specimens were fresh (control, n=20), cryopreserved and rapidly thawed (RT) at 37°C (n=20), or cryopreserved and subjected to controlled, automated slow thawing (ST) (n=20). Cell damage was established using the TUNEL method and the morphological and microstructural changes were also evaluated.

In the group that was rapidly thawed, the expression of TUNEL (+) cells increased significantly more than in the slowly thawed group. In addition, the endothelial denudation, microvesicles and edema were significant in the rapidly thawed group compared with the microstructural changes in the slowly thawed group.

Our study suggests that the rapidly thawing method may be the major cause of cellular damage and delayed rupture in cryopreserved arterial allografts. The expression of TUNEL (+) cells and structural changes were significantly lower in the slowly thawed group, which might have contributed to the improvement of graft failure after transplantation.

KEYWORDS

Cryopreservation; allograft; thawing; cellular injury

INTRODUCTION

An appropriate substitute for an artery to be used in vessel reconstruction is the subject of continued research. Prosthetic vessels and autograft arteries have been used widely, and yet still hold many limitations in the actual process of reconstruction. Increasing interest is paid to the recent methods of using cryopreserved allograft arteries, and the method is especially useful for patients with infective diseases or prosthetic instruments. With the progresses in the use of cryoprotectant agents and preservation methods, the results of cryopreserved aorta allograft transplantations have been good. However, the macroscopic and microscopic fissures and decrease in viability found in the cryopreserved grafts affect the prognosis of the surgery, and many studies regarding such problems arising after the cryopreservation of the tissues have been carried out. There have been reports that most of the cellular damage occurs in the early stages of acquisition, sterilisation, and preservation processing [1]. Other researchers have reported that increasing the preservation temperature prevents the occurrence of fissures [2]. Recently, under the belief that the thawing process is an important factor in the development of fissures, studies have suggested that the thermal stress within the instruments containing the arterial graft after cryopreservation cause the fissures [3], and that methods of slow thawing up to -100°C can prevent fissures [4]. Pascual et al. (2001) [5] concluded that slow thawing methods can better preserve the viability and structural characteristics of cryopreserved vessel tissues. By using aorta grafts cryopreserved using the same method, applying two different thawing methods (rapid and slow), and comparing the histological picture and incidence of DNA fragmentation, we wished to investigate effects of thawing methods on graft tissues.

MATERIALS & METHODS

1. Animal Experiments

Male New Zealand White Rabbits of approximately 2,500 g were injected with thiopental (2 mg/kg) and intubated to induce general anesthesia. Sodium thiopental, pancuronium bromide, and phentanilum chloride were injected to maintain the anesthetic condition. Median incision was made to expose the heart and the aorta graft was taken. The obtained aorta graft was put into 4°C minimal essential medium (MEM).

2. Sample Grouping

The obtained aorta grafts were randomly divided into three different groups and thawed accordingly. The control group (C, n=20) was first transported to the 4°C minimal essential medium (MEM) and preserved for one hour in the MEM + 10% dimethylsulphoxide (DMSO) solution. The group moved first to the 4°C (MEM) and then preserved at -196°C/min in the MEM + 10% DMSO solution before being rapidly thawed was designated as the rapidly thawed group (RT group, n=20). The comparison group moved to the 4°C (MEM), preserved at -196°C/min in the MEM + 10% DMSO solution, and then slowly thawed was designated as the slowly thawed group (ST group, n=20).

3. Thawing Methods

The aorta graft immersed in the MEM was put into a cryotube containing a mixture of MEM and DMSO at the ratio of nine to one. In order to reduce the toxicity, DMSO concentration was progressively increased to 2.5, 5, 7.5, and 10% at five-minute intervals. After adding the cryoprotectant, using a programmable freezer (85-1.7p, Scientemp, Adrian, MI, U.S.A.), each tissue was frozen at the rate of -1°C/min until the temperature of -120°C was reached. After a day, the graft was preserved for 7 days in -196°C liquid nitrogen. Past the preservation period, the RT group was thawed in a water tank (37°C, 5 minutes). The programmable freezer was used for the ST group to increase the temperature at a rate of 1°C/min, up to 38°C during the thawing step.

4. Histological Evaluation

The tissues used in the experiment were H-E stained and observed by light microscopy for the evaluation of their structural morphology.

5. Measurement of DNA Fragments

Using the TUNEL staining method, marked nucleotide was tagged to the 3'-OH end of cut chromatin. Cells with fragmented DNA were magnified 200 times under a light microscope (Axioskop 40®, Carl Zeiss, Jena, Germany) and directly checked to compare the degree of cellular damage in each group. TUNEL kit (ApopTag®, S7100, Peroxidase kit, Oncor, Gaithersberg, MD, USA) was used for staining. To describe in more detail, the paraffin tissue blocks were washed with xylene and ethanol to remove the paraffin, activated with proteolytic enzymes, inhibited with 3.0% H_2O_2, and next reacted with Tdt enzyme (55 μL / 5 cm^2). After adding anti-digoxin peroxidase conjugate (65 μL / 5 cm^2) and left to react at room temperature, peroxidase substrate (75 μL / 5 cm^2) was added to stain for 5 minutes. Finally, 0.5% methyl green was used as the contrast stain for 3 minutes. After fixation, the cells staining brown on the slide sample were judged as being TUNEL (+), and the percentage of TUNEL (+) cells among the total number of cells in the tissue sample was used as the parameter of cellular damage. The percentage of TUNEL (+) cells were expressed by the number of TUNEL (+) cells in each arterial graft, counted by photographs taken at 12, 3, 6, and 9 o'clock directions under high power field, over the total number of cells.

6. Statistics

The percentage of TUNEL (+) cells in each group was expressed by the mean value ± standard deviation. The differences between the two thawed groups were compared using the Mann-Whitney U test. Statistical significance was given for P value < 0.05, applying the confidence interval of 95 percent.

RESULTS

A. The occurrence of apoptosis as seen by the percentage of TUNEL (+) cells

Under the same high power field, more TUNEL (+) cells were observed in the RT group than the ST group (Figure 1). The expression of TUNEL (+) cells in each group was 2.6 ±1.0%, 25.2 ± 6.8%, and 38.8 ± 7.9% for the control group, ST group, and RT group, respectively. Compared to the ST group, the expression of TUNEL (+) cells in the RT group was significantly increased (Figure 2).

B. Structural changes of the cells observed under the light microscope

Changes such as cellular edema, formation of microvesicles, and endothelial denudation were seldom observed in the ST group, while such changes could be seen frequently in the RT group (Figure 3).

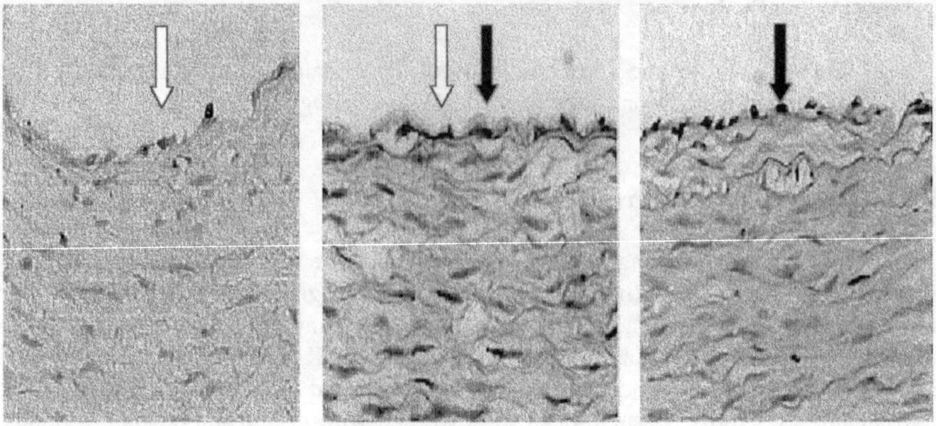

(a) (b) (c)

Figure 1. Light microscopic findings of aortic endothelium from rabbit illustrating TUNEL staining. Open arrows indicate counter-stained TUNEL (-) endothelial cells. Arrows indicate TUNEL (+) endothelial cells. Figures are in the order of (a) control, (b) slowly thawed and (c) rapidly thawed specimens. {Original magnification: x 200}

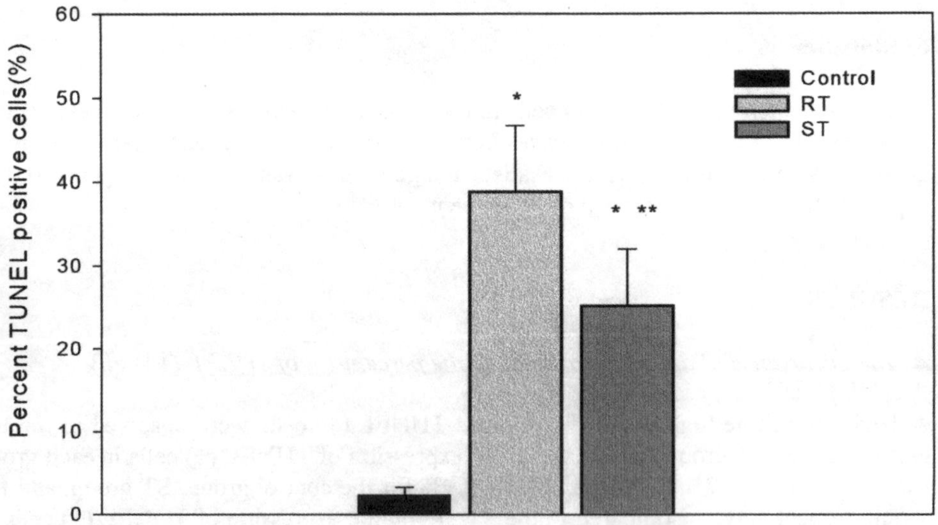

Figure 2. Percentage of TUNEL (+) cells in the three groups.

The expression of TUNEL (+) cells was increased significantly after thawing in two different manners compared to that of control group ($^{*}P < 0.05$).

In the specimen that was rapidly thawed, the expression of TUNEL (+) cells increased significantly more than in the slowly thawed specimen ($^{**}P < 0.05$).

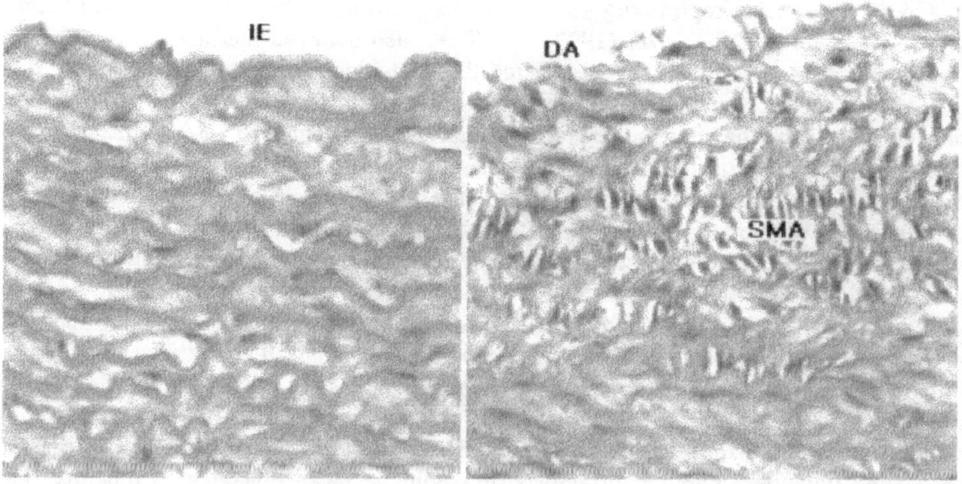

(a) (b)

Figure 3. Light microscopic findings of aortic endothelium from rabbit illustrating H-E staining. (a) Slowly thawed specimen showed minimal endothelial swelling and few anisocytosis of smooth muscle cells. Endothelial lining remained intact (IE). (b) Rapidly thawed specimen showed the loss of endothelial layer creating denuded areas (DA) and disrupted internal elastic lamina. In addition, there were anisocytosis of the smooth muscle cells (SMA) as well as edema in the medial layer. {Original magnification: x 200}

DISCUSSION

As methods of cryopreservation spread widely to preserve tissues, cryopreserved arterial allografts have been used as substitutes in the area of vascular reconstruction. Though these allografts have many advantages to prosthetic vessels and autografts, macroscopic fissures discovered after thawing often made the tissues unsuitable for transplantation, and the microscopic fissures and decrease in viability of the graft endothelium hindered advancements in post-operative success rates. There have been many studies seeking the cause of such problems and methods that could improve the prognosis of cryopreserved allograft transplantation [6-8]. Most of these studies have focused on the process of tissue pre-treatment, cooling methods, and the preservation temperature [9,10].

Hunt *et al.* (1994) [2] reported that using dimethyl sulfoxide as the cryoprotectant agent, freezing slowly, storing at temperature less than 160°C, and then thawing rapidly caused macroscopic fissures in 75% of the cases. But if the maintenance freezing temperature was kept at -80°C, the physical stress brought on by the difference in the temperatures is decreased, and thus did not bring about fissures. Likewise, Wassenaar *et al.* (1995) [3] stated that physical stress from the difference in the temperatures is an important factor in causing fissures, and that such stress arises from within the container holding the graft tissues. Lim *et al.* (1997) [1] thought that the diminished viability of the vessel endothelium arises not from the freezing and thawing processes, but from earlier stages of preparing for the preservation, like graft acquisition and sterilisation.

On the other hand, reports suggesting that the thawing process is an important factor were first made by Pegg *et al.* (1997) [4]. They also believed thermal stress to be an important cause of fissures, but they found that changing the freezing temperature did not affect the incidence of fissures following cryopreservation. The temperature change during the thawing process was thought to be the major cause of the fissures. However, in evaluating the occurrence of tissue fissures or changes, this study judged the presence of fissures macroscopically, and the function of vessel endothelium through in vitro model norepinephrine response [11]. Whether or not cryopreserved vessel tissues retain the characteristics unique to the tissue prior to freezing have been researched extensively through numerous methods. Trials include macroscopic observation of the tissue for endothelial fissures, direct evaluation of physical properties, and predicting fissures and ruptures by connecting the thawed tissue to a circulatory circuit. However, these methods cannot check for the loss of physical properties or the presence of fissures before they actually occur. So for evaluating tissue samples like the cryopreserved arteries, in which the physical properties and viability are directly associated with the operation prognosis, it would be more objective to quantify the degree of cellular damage through methods such as the TUNEL staining method. TUNEL staining can pick out DNA fragments, a feature of apoptosis, through immunochemical methods. Due to the fact that cellular necrosis and DNA replication also stain positive, there is controversy regarding the specificity of the test, but it is the best method for quantifying the fraction of cellular death among all the cells present [12].

Based on the result of this study where more TUNEL (+) cells were observed in the RT group compared to the ST group, method of slowly thawing cryopreserved tissue must be taken into consideration in practice to preserve tissue viability and prevent complications. But contrary to the expectation that there won't be a great difference between the ST group and the control group, there exists a meaningful difference between them. This fact suggests that other elements play a role in injuring the cryopreserved tissue besides the thawing method. As suggested in previous studies, such elements responsible for tissue injury may be found in the preparation process, cooling step, the freezing temperature, or duration of preservation. Still, as seen in our study, arterial grafts frozen and preserved under the same conditions show different histological characteristics and degrees of cellular damage as seen through the TUNEL staining, depending on the thawing methods. Thus, it seems reasonable to say that the thawing processes greatly influence the prognosis of the tissue after transplantation. With vessel tissues, especially cryopreserved arterial grafts, the endothelium plays the controller between the blood and the vessel smooth muscle, and also a very important role in deciding the viability of the tissue and prognosis of the transplantation. So, ongoing studies should be made to reveal the causes of vessel de-endothelialisation observed after the thawing of cryopreserved tissue grafts. In addition, as revealed in many studies, the endothelial and media edema from ice crystals, decreased smooth muscle cells, and formation of perinuclear microvesicles in smooth muscle cells occur in common to all rapidly thawed vessel tissues, and are causes of acute or delayed graft failure and should be the subject of future studies.

As shown above, it would be beneficial to use slowly thawing methods when using cryopreserved tissues, especially cryopreserved arterial grafts. Furthermore, slowly thawing methods should be considered when cryopreserving tissues other than vessels to maximise post-transplantation viability and to diminish graft failures. In the future, studies should be made to find out the ideal thawing method by designing and comparing various models utilising slowly thawing methods for cryopreserved tissues.

REFERENCES

1. C.Y. Lim & E. K. Hong, 'Flow cytometric analysis of endometrial cell viability in arterial allograft', *Korean J. Thorac. Cardiovasc. Surg.*, **1997,** 30, 553-558.
2. C. J. Hunt, Y. C. Song, E. A. J. Bateson & D. E. Pegg, 'Fractures in cryopreserved arteries', *Cryobiology*, **1994,** 31, 506-515.
3. C. Wassenaar, E. G. Wijsmuller, L. A. V. Herwerden, Z. Aghai, L. J. Corina, V. Tricht & E. Bos, 'Cracks in cryopreserved aortic allograft and rapid thawing', *Ann. Thorac. Surg.*, **1995,** 60, S.165-167.
4. D. E. Pegg, M. C. Wusteman & S. Boylan, 'Fractures in cryopreserved elastic arteries', *Cryobiology*, **1997,** 34, 183-192.
5. G. Pascual, N. Garcia-Honduvilla, M. Rodriguez, F. Turegano & J. Bujan, 'Effect of the thawing process on cryopreserved arteries', *Annals of Vascular Surgery,* **2001,** 15, 619-627.
6. E. Rosset, A. Friggi, R. Rieu, P. Rolland, G. Novakovitch, R. Choux, J. F. Pellissier, R. Pellissier & A. Branchereau, 'Mechanical properties of the arteries. Effects of cryopreservation', *Chirurgie,* **1996,** 121, 285-297.
7. E. Rosset, A. Friggi, G. Novakovitch, P. Rolland, R. Rieu, J. F. Pellissier, P. E. Magnan & A. Branchereau, 'Effects of cryopreservation on the viscoelastic properties of human arteries', *Ann. Vasc. Surg.,* **1996,** 10, 262-272.
8. F. Pukacki, T. Jankowski, M. Gabriel, G. Oszkinis, Z. Krasinski & S. Zapalski, 'The mechanical properties of fresh and cryopreserved arterial homografts', *Eur. J. Vasc. Endovasc. Surg.*, **2000,** 20, 21-24.
9. M. Rigol, M. Heras, A. Martinez, M. J. Zurbano, E. Agusti, E. Roig, J. L. Pomar & G. Sanz, 'Changes in the cooling rate and medium improve the vascular function in cryopreserved porcine femoral arteries', *J. Vasc. Surg.*, **2000,** 31, 1018-1025.
10. B. Lehalle, C. Geschier, G. Fieve & J. F. Stoltz, 'Early rupture and degeneration of cryopreserved arterial allografts', *J. Vasc. Surg.*, **1997,** 25, 751-752.
11. J. Bujan, G. Pascual, N. Garcia-Honduvilla, M. J. Gimeno, F. Jurado, A. C. Martin & J. M. Bellon, 'Rapid thawing increases the fragility of the cryopreserved arterial wall', *Eur. J. Vasc. Endovasc. Surg.*, **2000,** 20, 13-20.
12. A. Negoescu, P. Lorimier, F. Labat-Moleur, C. Drouet, C. Robert, C. Guillermet &, E. Brambilla, 'In situ apoptotic cell labeling by the TUNEL method: improvement and evaluation on cell preparations', *J. Histochem. Cytochem.*, **1996,** 44, 959-968.

PART 3

EFFECTS OF RADIATION ON BONE, TISSUES, AND THEIR COMPONENTS

EFFECTS OF GAMMA IRRADIATION ON BONE – CLINICAL EXPERIENCE

William W. Tomford

Massachusetts General Hospital Bone Bank
Department of Orthopaedic Surgery
55 Fruit Street
Boston, MA 02114, USA

ABSTRACT

The use of radiation for sterilisation of musculoskeletal tissue allografts began more than fifty years ago. Based on irradiation of vascular tissues, the development of radiation as a sterilising method for bone has paralleled the popularity of bone transplantation. Historical research suggests that radiation at doses of 25-30 kGy has a minimal adverse effect on bone strength and healing, although effects on osteoinductivity are not fully researched. Future use of radiation may employ high doses to kill viruses and resistant bacteria with pre-treatment of grafts.

KEYWORDS

Radiation; sterilisation; bone

INTRODUCTION

This is an age of increasing use of human tissues for transplantation. In the case of musculoskeletal transplants, such as bone and ligament, there are currently an estimated 750,000 deposits or pieces of tissues used annually. This number is almost twice the number of transplants that were used only five years ago.

With an increasing use of tissue transplants, there is also an increasing concern of transmission of viral disease agents in grafts procured from humans. The number of individuals who are HIV-1 positive or HCV positive has never been higher. Due to the development of drugs that thwart the devastating effects of these types of disease agents in humans, survival in spite of infection by these agents is now commonplace and even expected. Likewise, bacteria and exotic disease agents such as prions now exist for which there is no effective antibiotic treatment. Increasing resistance to medicinals is occurring, and the spread of these agents due to ease of global travel and shipping is uncontrollable.

For all of the above reasons, an effective method of sterilisation of tissues is important, necessary, and warranted. Gamma irradiation is an effective sterilising method, yet many who currently use this method know little about the history and clinical experience of its use. Gamma irradiation has advantages and disadvantages, and these can best be realised through a review of its historical use and present applications.

HISTORICAL USE OF GAMMA RADIATION AS A STERILISING AGENT

In 1919, Grober and Pauli first demonstrated the ability of cathode rays to kill bacteria [1]. However, their use of an electron beam accelerator to provide cathode rays caused problems in application of their methods to objects requiring sterilisation such as medical devices.

Their machines were very large and extremely unwieldy. In addition, because of the comparative low energy of electron beam radiation, there was inadequate penetration of tissues greater in thickness than a few millimetres. Thus their methods required prolonged exposure time in order to achieve sterilisation. In this circumstance, organic tissues such as a bone or ligament could not be sterilised using this method.

In 1939, Trump and Van de Graaff produced a compact pressure insulated electrostatic X-ray generator [2]. Their research at Massachusetts Institute of Technology (MIT) led to the development of methods of sterilisation of food. In fact, a Department of Food Technology was founded at MIT based upon their work in sterilisation of food. Once the Van de Graaff accelerator proved to be useful in a technical sense in sterilising every day objects, rather than simply a research device, the sterilisation of medical products began to be performed extensively.

In 1948, Dunn, also working at MIT, conclusively demonstrated bactericidal action of cathode rays in food sterilisation experiments [3]. In 1951, Proctor and Goldblith also published work on sterilisation of food and suggested a dose of 1-1.5 million rep [4]. Finally, Moriarty, also working at MIT, published tables of sensitivities of organisms to radiation in an attempt to benefit the food sterilisation industry [5]. His work on the effect of cathode ray irradiation on various microorganisms led to the knowledge that *Bacillus subtilis* and *Clostridium spirogenes* were the most resistant organisms to radiation.

The application of radiation to sterilisation of tissues really began with the publication of work in 1948 by Trump and Van de Graaff [6]. Their research on the irradiation of biological materials by high-energy Roentgen rays and cathode rays again proved their ingeniousness at practical applications of radiation. In 1951, working at Children's Hospital in Boston, Meeker and Gross wanted to transplant aortic grafts in children, but had no method of sterilising the grafts [7]. They hit upon the use of cathode rays and persuaded Professor Trump to provide them the ability to use his apparatus at MIT to sterilise allograft aortic grafts. Meeker and Gross noted "the cathode ray machine at MIT was made available to us through the kindness of Professor Trump. This apparatus . . . has beneath it an endless conveyor belt on which materials can be placed (even in a frozen state) and passed through the electron beam". Meeker and Gross performed experiments involving contaminated dog aorta grafts in an attempt to answer the question of whether or not cathode ray irradiation could sterilise these grafts without denaturing or degrading the organic vascular tissue. In their initial trial, they irradiated four aortas at 1.5 million rep at room temperature. This dose was chosen on the basis of experiments in sterilising food previously done at MIT. After transplantation, all four of these grafts failed with intense inflammation and occlusive thrombosis. The authors concluded that it is necessary to irradiate tissues in a frozen state, and they used carbon dioxide at -55°C in future experiments.

At a subsequent trial experiment, Meeker and Gross evaluated sixty unsterile and contaminated aortic grafts which were irradiated at -55°C [8]. They noted only two infections after irradiation of tissues at this dose in a frozen state. However, they were concerned about even two infections and therefore did further experimentation evaluating higher doses of radiation. In their next experiment, they irradiated twenty grafts at 2.0 million rep, eleven grafts at 3.0 million rep, and another eleven grafts at 4.0 million rep. They found doses greater than 3.0 million rep produced irreparable damage to the collagen such that the grafts could not be used. They concluded on the basis of their work that the dosage of radiation should be between 2.0 and 2.5 million rep. At this dose, sterilisation was successful and tissue damage was minimal. They also reiterated their belief that tissue should be irradiated in a frozen state to prevent what they considered an extensive inflammatory response to the irradiated collagen.

From the work of Meeker and Gross on soft tissues, radiation was applied to sterilisation of bone. Historical methods such as boiling, autoclaving, and chemical disinfection, all of which had been used up to that time, were associated with a high incidence of wound infections and slow and inadequate healing. Therefore the idea of using radiation for sterilisation, given its success in the hands of Meeker and Gross, was appealing to many researchers in the early 1950's. Working at the US Navy Tissue Bank in Bethesda, Maryland, Kreuz and Hyatt in 1952 first mentioned the idea of using cobalt rays for sterilising bone grafts in an article on the use of preserved tissues in orthopaedic surgery [9]. At that time, they were actually doing research on the use of cobalt rays but had not published their data. The Navy had a source for providing cobalt rays, and bone grafts procured at the US Navy Tissue Bank were being irradiated in an experimental fashion.

Meanwhile at the Children's Hospital in Boston, having heard about the work of Meeker and Gross and talked with them because they were colleagues at the same hospital, Jonathan Cohen began work on cathode ray sterilisation of bone grafts [10]. In 1955, Cohen reported experimental results of grafts sterilised by cathode rays and transplanted in felines. The bone grafts were frozen and irradiated at 2.0 million rep at MIT in the Van de Graaff accelerator, the same accelerator used by Meeker and Gross. In fact, Cohen chose the dose of 2.0 million rep based upon Meeker and Gross' work. In his experiments, Cohen deliberately contaminated twenty cat humeri and irradiated them at 2.0 million rep. He found no growth on any of the twenty specimens. His transplants included twenty-four transplanted cat humeri irradiated at the same dosage. He found no inflammation and the results of transplantation of irradiated grafts were similar to those of transplantation of non-irradiated. The only limitation noted, however, was that the maximum bone thickness that could be penetrated by the Van de Graaff accelerator, which is an electron beam source, was 1.4 cm, with an average penetration of 0.8 cm. To solve the problem of low penetration, Cohen administered radiation from two sides which required two passes through the accelerator.

The first group to use cobalt radiation as a sterilising method was headed by Macris at the University of Michigan [11]. In 1954 he began the use of a 10,000 Curie source at the University of Michigan based upon experiments that he did from 1952 to 1954 with 2.0 million rep. In these experiments, he showed the bactericidal effects of gamma radiation. With an interest in cardiac surgery, Macris also first experimented with sterilising aortic homografts and looked at the effect of gamma ray radiation on the structural integrity of the tissue. Like Meeker and Gross, he found that 2.0 million rep was an ideal dose.

Similar to what had occurred with the collaboration of Cohen and Meeker and Gross at the Children's Hospital in Boston, Devries, in 1955, working at the University of Michigan with Macris, experimented on the use of cobalt radiation sterilisation of bone [12]. He purposefully contaminated dog humeri with clostridium and treated them with 2.0 million rep to find that they were all sterilised. Devries also experimented with the use of irradiation as a method of preservation. He developed a hypothesis that irradiated bone stored at room temperature might not have to be frozen for storage. However, his experiments showed that irradiated bone stored at room temperature developed crystal deposits and what he called "loss of marrow" similar to bone which had been refrigerated for long periods of time. Following his experiments, he concluded that irradiation could not be used solely as a method of preservation, and freezing or lyophilization was necessary for prolonged storage. Devries also evaluated the effects of irradiation on allograft bones. After transplantation of irradiated bones in eighteen animals, all wounds healed primarily suggesting there were no unusual systemic reactions to the sterilised bones. He also reported on the gradual healing of the bone allografts with what he noted as a disappearance of the transplanted bone and replacement with new bone.

The first reported clinical use of irradiated allograft bone was by Bassett in 1955 in New York City [13]. Bassett reported on the clinical use of cathode ray sterilised grafts using cadaver bone as transplants. He cut pieces of rib, iliac crest, and tibial shafts taken from cadavers into pieces no thicker than 1.5 cm, froze them to -35°C, shipped them to MIT to be put into the Van de Graaff accelerator, and irradiated them at 2.0 million rep in dry ice. He prepared 189 irradiated grafts using this method, which he transplanted into 100 patients. Average follow up was 6.8 months. Two of the hundred patients developed wound infections that he felt were unrelated to the grafts. In his final report, he followed up on thirty-one of one hundred patients in whom he found no untoward clinical radiographic effects. He found that the cancellous grafts healed by four to five months. He ended up exploring two patients and found the grafts had united to the host bone with revascularisation. He noted no foreign body reaction or inflammation around the grafts.

In 1958 Devries followed up on Bassett's work using radioactive cobalt instead of cathode rays as a sterilising source [14]. Devries reported on the radiation sterilisation of allografts utilising radioactive cobalt by procuring strips of iliac crest, sealing them in Pyrex, freeze-drying them, and irradiating them at 4.0 million rep. His source included one hundred aluminium jacketed cobalt rods with tissues exposed from six to twenty-four hours. He transplanted the grafts into eighty-two patients in the 104 operations. He noted five wound infections, none of which he believed were related to the graft. One of these five wound infections had a positive bone culture when he retrieved the bone. He also had seven positive cultures from incisions, but he felt this was related to skin contamination. Most of his grafts were used in spinal fusions and he anecdotally reported them as satisfactory.

In the 1980's, several papers were produced from Paris, France, on the use for radiation for the sterilisation of massive bone allografts. In 1986, Hernigou reported on transplantation of thirty massive allografts that included large segments of diaphyseal bone, metaphyseal bone, and cartilage, used in treatment of tumours and non-unions [15]. The grafts were irradiated in a frozen state with beta radiation at 2.5 to 3.5 megarads (Mrads). He reported no infections. Bone union occurred in twenty-six of thirty diaphyseal grafts and eight of nine metaphyseal grafts. Three whole diaphyses, or so-called intercalary grafts, failed. He used bone scans at three months to detect healing.

In 1993 Hernigou reviewed 127 massive allografts frozen and irradiated at 25 kiloGray (kGy) using radioactive cobalt as a source [16]. His follow up was three years. His results showed no infections in forty-four patients who were not on chemotherapy for adjunct treatment of their tumours, and eleven infections in eighty-three patients who received the grafts but were also on chemotherapy as adjunct therapy. In seven patients he noted non-unions and in eight patients he noted fractures. However, he concluded that irradiation of massive allografts does not jeopardise the clinical result.

Working with Hernigou, Loty, also in France, reported in 1994 on infections in the use of massive bone allografts sterilised by radiation [17]. Loty reviewed one hundred and sixty-four consecutive reconstructions and found fourteen skin sloughs or infections. Complications were most common after preoperative radiation was used for treatment of the primary tumour. The grafts, Loty concluded, never appeared to be responsible for the infections or skin sloughs.

The use of irradiated cancellous bone has been analysed by Zasacki who published in 1991 on the clinical use of freeze-dried irradiated cancellous bone grafts [18]. The dosage of radiation was 25 to 30 kGy. He reported on four hundred and thirty-five patients aged three to seventy-four years with a follow up of three years. Surgeries included two hundred and thirty-four spinal surgeries, thirty-six fusions, and eighty-three cases in which cancellous bone was used near a joint.

The results Zasacki reported were 91% satisfactory with incorporation and rebuilding of bone, which was entirely satisfactory clinically in his conclusion. He felt that radiation did not affect the result, but he also noted that intimate contact of the graft and the host were more important than the use of irradiated bone graft.

In 1999, Kwiatkowski reported on the use of frozen allografts of cancellous bone after loosening of hip replacement prostheses [19]. He examined frozen cancellous chips (3 mm^3), which were irradiated at 25 to 30 kGy. He reviewed thirty-five revision hip arthroplasties and found thirty-three were successful. He had two reoperations, but no infections.

In 2000, Leitman reviewed complications of irradiated allografts in orthopaedic tumour surgery [20]. These grafts had been used by Dr. Henry Mankin in treatment of long bone primary bone tumours. Twenty-four patients served as the experimental group. Each of these patients received a massive long bone osteoarticular allograft from 1987 to 1991. The radiation dose of the long bone allograft, which included cartilage, metaphysis, and diaphysis, was 10 to 30 kGy. The follow up average was five years with a range of two to nine years. A control group of two hundred and eighty-two patients who had similar diagnostic problems and received the same type of grafts in which the graft was not irradiated, were used as a control group. The incidence of fracture, non-union and infection was evaluated.

The authors found that nine of twenty-four patients who received irradiated allografts fractured (38%). Fifty-one of 282 patients who received non-irradiated allografts fractured (18%). The difference between irradiated and non-irradiated groups was significant ($p = 0.03$). Two of twenty-four irradiated allografts resulted in non-union (8%) and the use of fifty-four of 282 non-irradiated allografts resulted in non-union (19%). The difference between the irradiated and non-irradiated groups was not significant ($p = 0.27$). None of twenty-four patients who received irradiated allografts became infected. In comparison, thirty-one of 282 patients who received non-irradiated allografts became infected (11%). However, the incidence of infection in comparing the two groups was not significant ($p = 0.15$). The authors concluded that sterilisation of bone for clinical use can be achieved by irradiation. They also found, however, that excessive doses of radiation may have an adverse effect on biomechanical strength of large bone grafts.

In evaluating the radiation effects on bone, several principles are important to keep in mind. Most important is the fact that bone is a combination of mineral and collagen. Collagen sustains the tensile strength of bone. It also harbours the osteoinductive factors or so-called bone growth factors. Therefore changes in bone collagen have biomechanical and osteoinductive effects. Some of the investigation of the effect of irradiation has evaluated the effects on collagen and on osteoinductive factors. In regard to the mechanical strength of collagen, Hamer in 1996 reported on tests of transverse sections of irradiated thin femoral rings [21]. He used 233 samples divided into three groups. The first group was irradiated at 28 to 30 kGy at room temperature. The second group was irradiated at the same dose but frozen to -80°C, and the third group evaluated grafts, which were only frozen but not irradiated. His results showed that freezing had no effect on the elastic or strength properties of these grafts. Irradiation and irradiation plus freezing had no effect on the elastic properties of the grafts, but significantly decreased their capacity to absorb load and sustain maximum load in mechanical testing. He concluded that radiation affects bone collagen, but if loads are within the elastic region, there is a minimal effect. Anderson reviewed the biomechanical effects of irradiation on cancellous bone [22]. In 1992 he reported experiments on cancellous cubes cut from the proximal tibia and irradiated at 10, 31, 51, and 60 kGy. Tested to compression in failure, he found significant differences compared to controls only for specimens irradiated at 60 kGy.

Voggenreiter in 1996 evaluated rat diaphyseal tibial segments irradiated at 1, 5, 25, and 50kGy [23]. He found that 1 and 5 kGy had no effect. However, using 25 kGy, he found that healing was delayed about 50%. Of grafts which were irradiated at 50 kGy, 60% fractured, suggesting that there was an upper limit to the amount of radiation that could be sustained by these grafts and still maintain their normal mechanical properties.

In 2000, Cornu looked at the effects of processing with or without radiation at 25 kGy on femoral heads [24]. Tested in compression, he found that freeze-drying and lipid extraction of grafts produced a 20% reduction in strength whether irradiated or not. Irradiating these grafts resulted in a 42.5% reduction in compressive strength. Tosello in 1994 reported on crush resistance of cancellous bone fragments irradiated at 25 and 50 kGy. He found that there was no effect or minimal effect at 25 kGy. In regard to cortical bone, Curry in 1997 irradiated paired femurs from eight specimens at 17, 29.5, and 94.7 kGy [25]. He found no change in elastic modulus, but a decrease in bending strength, work to fracture and impact energy absorption, similar to the results noted by Hamer in 1996.

In evaluating radiation effects on osteoinduction and osteogenesis, much of our knowledge comes from Poland and the labs of Komender and Dziedzic-Goclawska. In 1991 Dziedzic-Goclawska reported on the effect of radiation sterilisation on the osteoinductive properties and the rate of remodelling in bone implants preserved by lyophilisation and deep freezing [26]. She found effects were related to the amount of irradiation. Komender, however, noted a therapeutic effect on transplantation of these types of grafts [27].

In 1994 Ijiri reported on the effects of radiation on BMP [28]. He investigated partially purified BMP and type I collagen and found that osteoinductivity was reduced by irradiation at 25 kGy, ETO at 37°C for four hours and ETO at 55°C for one hour, but ETO at 29°C at four hours was less deleterious.

Finally, Moreau in 2000 reported on the cytotoxic effects of irradiation of medullary lipids [29]. The problem he noted was that the effects of radiation of lipids are unknown. His objective was to evaluate the cytotoxicity of irradiation of bone with and without lipid on cultures of osteoblasts. He used bone cores from femoral heads with and without lipid extracted and with or without radiation at 25 kiloGray. He noted that peroxidated lipids were two to three times higher in irradiated cores. Fresh cores with no irradiation had no effect on osteoblasts. Cell death dramatically increased around the irradiated cores that had fat in them, but defatted irradiated cores produced no effect. He concluded that defatting procedures were necessary when irradiating.

CONCLUSIONS

Irradiation has been used to sterilise bone for transplantation for over fifty years. The first methods employed cathode rays, but currently the use of gamma irradiation is the predominant approach. Much of the research on irradiation has involved sterilisation of long bone and osteoarticular grafts, which by virtue of their massive size are difficult to sterilise. Radiation at 25 to 30 kGy or less does not appear to have a significant effect on osteoinductivity or on healing of bone grafts, although osteoinductivity may be reduced even at these levels. As a sterilising agent, gamma radiation has ideal qualities of high penetration and rapid effectiveness.

Current research suggests that future use of radiation may employ extremely high doses with pre-treatment of tissues to mitigate adverse effects. Due to the myriad diseases now detectable in humans, radiation will no doubt continue to be a critical adjunct to safe transplantation of musculoskeletal tissues.

REFERENCES

1. J. Grober and W. E. Pauli, Untersuchungen uber die biologische Wirkung der Kathodenstrahlen, *Deutsche Med. Vchnschr.*, **1919**, 45, 841-844.
2. J. G. Trump and R. J. Van de Graaff, A compact pressure insulated electrostatic X-ray generator, *Physiol. Rev.*, **1939**, 55, 1160-1165.
3. C. G. Dunn, W. L. Campbell, H. Fram and A. Hutchins, Biological and photochemical effects of high energy, electrostatically produced roentgen rays and cathode rays, *J. Appl. Hys.*, **1948**, 19, 605-616.
4. B. E. Proctor and S.A. Goldblith, Electromagnetic radiation fundamentals and their applications in food technology, *Advances Food Res.*, **1951**, 3, 119-125.
5. J. H. Moriarty, Effect of cathode ray irradiation on various microorganisms - Progress Report II, *Department of Food Technology, Massachusetts Institute of Technology*, Cambridge, Massachusetts, USA, May 1, **1950**.
6. J. G. Trump and R. V. Van de Graaff, Irradiation of biological materials by high-energy roentgen and cathode rays, *J. Applied Physics*, **1948**, 19, 5990-604.
7. I. A. Meeker and R. E. Gross, Low temperature sterilization of organic tissue by high voltage cathode ray irradiation, *Science*, **1951**, 11, 283-285.
8. I. A. Meeker and R. E. Gross, Sterilization of frozen arterial grafts by high voltage cathode ray irradiation, *Surgery*, **1951**, 30, 19-28.
9. F. P. Kreuz, G. W. Hyatt, T. C. Turner and C. A. L. Bassett, Use of preserved tissues in orthopedic surgery, *A. M. A. Arch. Surg.*, **1952**, 64, 148-152.
10. J. Cohen, Cathode ray sterilization of bone grafts, *A. M. A. Arch. Surg.*, **1955**, 71, 784-789.
11. J. A. Macris, H. Sloan and J. E. Orebaugh, The use of cobalt 60 as a sterilizing agent for aortic homografts, *Surgical Forum*, **1954**, 5, 268-272.
12. P. H. Devries, W. O. Brinker and L. L. Kempe, Utilization of radioactive cobalt in the sterilization of homogeneous bone transplants, *Surgical Forum*, **1955**, 6, 546-549.
13. C. A. L. Bassett and A. G. Packard, A clinical assay of cathode ray sterilized cadaver bone grafts, *Acta Orthop. Scand.*, **1959**, 28, 198-209.
14. P. H. Devries, C. E. Badgley and J. T. Hartman, Radiation sterilization of homogeneous-bone transplants utilizing radioactive cobalt, *J. Bone Joint Surg.*, **1958**, 40A, 187-203.
15. P. Hernigou, G. Delepine and D. Goutallier, Allogreffes massives cryoconservees et sterilisees par irradiaton, *Rev. Chir. Orthop.*, **1986**, 72, 403-413.
16. P. Hernigou, G. Delepine, D. Goutallier and A. Julieron, Massive allografts sterilised by irradiaton (clinical results), *J. Bone Joint Surg.*, **1993**, 75B, 904-913.
17. B. Loty, B. Tomeno, J. Evrard and M. Postel, Infection in massive bone allografts sterilised by radiation, *Int. Orthop.*, **1994**, 18, 164-171.
18. W. Zasacki, The efficacy of application of lyophilized, radiation sterilized bone graft in orthopedic surgery, *Clin. Orthop.*, **1991**, 272, 82-87.
19. K. Kwiatkowski and G. Ratynski, The usage of frozen allografts of the spongy bone in filling the loss of bone after loosening of the hip prosthesis, *Ann. Transplant.*, **1999**, 4, 59-63.

20. S. A. Litman, W. W. Tomford, M. C. Gebhardt, D. S. Springfield and H. J. Mankin, Complications of irradiated allografts in orthopaedic tumor surgery, *Clin. Orthop.*, **2000**, 375, 214-217.

21. A. J. Hamer, J. R. Strachan, M. M. Black, C. J. Ibbotson, I. Stockley and R. A. Elson, Biomechanical properties of cortical allograft bone using a new method of bone strength measurement: a comparison of fresh, fresh frozen and irradiated bone, *J. Bone Joint Surg.*, **1996**, 78B, 363-368.

22. M. J. Anderson, J. H. Keyak and H. B. Skinner, Compressive mechanical properties of human cancellous bone after gamma irradiation, *J. Bone Joint Surg.*, **1992**, 74A, 747-752.

23. G. Voggenreiter, R. Ascherl, G. Blumel and K. P. Schmit-Neuerburg, Extracorporeal irradiation and incorporation of bone grafts. Autogeneic cortical grafts studied in rats, *Acta Orthop. Scand.*, **1996**, 67, 583-588.

24. O. Cornu, X. Banse, P. L. Docquier, S. Luyckx and C. Delloie, Effect of freeze-dried and gamma irradiation on the mechanical properties of human cancellous bone, *J. Orthop. Res.*, **2000**, 18, 426-431.

25. J. D. Curry, J. Foreman, I. Laketic, J. Mitchell, D. E. Pegg and G. C. Reilly, Effects of ionizing radiation on the mechanical property of human bone, *J. Orthop. Res.*, **1997**, 15, 111-117.

26. A. Dziedzic-Goclawska, K. Ostrowski, W. Stachowicz, J. Michalik and W. Grezesik, Effect of radiation sterilization on the osteoinductive properties and the rate of remodeling of bone implants preserved by lyophilization and deep-freezing, *Clin. Orthop.*, **1991**, 272, 30-37.

27. J. Komender, H. Malczewska and A. Komender, Therapeutic effects of transplantation of lyophilized and radiation-sterilized, allogeneic bone, *Clin. Orthop.*, **1991**, 272, 38-49.

28. S. Ijiri, T. Yamamuro, T. Nakamura, S. Kotani and K. Notoya, Effect of sterilization on bone morphogenetic protein, *J. Orthop. Res.*, **1994**, 12, 628-636.

29. M. F. Moreau, Y. Gallois, M. F. Basle and D. Chappard, Gamma irradiation of human bone allografts alters medullary lipids and releases toxic compounds for osteoblast-like cells, *Biomaterials*, **2000**, 21, 369-376.

EFFECTS OF GAMMA IRRADIATION ON THE MECHANICAL PROPERTIES OF HUMAN CORTICAL ALLOGRAFT BONE

Ming H. Zheng [1], Richard A. Power [2], J. Neil Openshaw [3], Roger I. Price [4], Robert E. Day [5], Joyleen Winter [2], Anne Cowie [2] and David J. Wood[1]

[1] *Dept of Orthopaedic Surgery, University of Western Australia, Nedlands, 6009, Australia*

[2] *Perth Bone and Tissue Bank, Hollywood Private Hospital, Nedlands, 6009, Australia*

[3] *Dept of Orthopaedic Surgery, Fremantle Hospital, Fremantle 6160, Australia*

[4] *Dept of Medical Technology & Physics, Sir Charles Gairdner Hospital, Nedlands, 6009, Australia*

[5] *Dept of Medical Engineering & Physics, Royal Perth Hospital, Perth, 6000, Australia*

ABSTRACT

Musculoskeletal allografts are used widely in tumour and reconstructive surgery. Gamma irradiation has been used to sterilise allograft bone but its effect on the biomechanical properties of osseous tissue is not fully elucidated. In this study, we have examined the effect of gamma irradiation on the mechanical properties of human cortical bone. An examination was carried out of the three-point bending, compressive strength, and fracture toughness of human femoral cortical bone irradiated at doses of 15, 25, and 50 kGy, in comparison with non-irradiated control bone. We found that gamma irradiation degrades the mechanical properties of bone as evidenced by three-point bending and toughness assays. A 6% reduction in ultimate three-point bending stress was observed at 15 kGy, 23% at 25 kGy and 30% at 50 kGy. There was a 12% reduction in toughness at 15 kGy, 13% at 25 kGy and 22% at 50 kGy. The effect on ultimate compressive stress was less marked, with no significant effect seen below a 50 kGy radiation dose. In summary, given the low incidence of disease transmission from allograft bone, it is believed that where large structural allografts are concerned, the possible benefits of gamma irradiation need to be balanced against its detrimental effects on the resistance of the bone to catastrophic mechanical failure.

KEYWORDS

Gamma irradiation; tissue banking; bone allograft; three-point bending; toughness

INTRODUCTION

Musculoskeletal tissue allotransplantation in reconstructive surgery is increasingly applied worldwide. It has been estimated that in 2002 alone more than 800,000 grafts were implanted in the United States, and approximately 4000 grafts are applied in Australia each year. Allograft bone is used most commonly in revision joint arthoplasty and limb salvage surgery for musculoskeletal tumours. Although the clinical results have proved satisfactory, concern remains regarding possible disease transmission from the donor, in particular human immunodeficiency virus (HIV) and hepatitis.

In Australia the processing of allograft tissue is regulated by the Therapeutic Goods Administration under the Blood and Tissue Code. As a consequence, the potential for disease transmission and infection have been reduced dramatically. Apart from the introduction of vigorous blood screening of donors, terminal sterilisation of allograft tissue has also been applied. Several methods of secondary processing have been used, including ethylene oxide gas exposure, or gamma irradiation for the final sterilisation of allograft tissues. In particular, the use of gamma irradiation, established in 1956 [1], has now become the most commonly used form of sterilisation. In a recent survey of North American tissue banks accredited with the American Association of Tissue Banks [2], of the fourteen out of fifty member banks that replied, 79% indicated their use of gamma irradiation for some or all of the banked bone. Applied doses varied from 15 to 25 kGy.

Gamma irradiation is known to induce changes in the mechanical properties of many materials. However there is a paucity of literature concerning gamma irradiation of human bone, and the findings are conflicting. A study by Triantafyllou *et al.* [3] demonstrated that a 25% to 50% reduction in three-point bending strength of bovine cortical bone occurred at 30 kGy. Komender [4] described a 30% reduction in three-point bending failure load at 60 kGy and 20% at 50 kGy, but did not test bone irradiated at the lower doses currently recommended. On the other hand, Bright and Burstein [5] tested human bone irradiated with 25 kGy in tension and compression, and found no effect. Another important mechanical property known to be affected by gamma irradiation is fracture toughness. Low fracture toughness is associated with brittle or catastrophic failure [7,8].

In view of the increasing use of structural allograft bone, plus the widespread and increasing use of gamma irradiation, it is considered important to define the mechanical effects at radiation doses currently in use. Therefore an examination of the three-point bending, compressive strength and fracture toughness of human cortical bone gamma-irradiated at 15, 25 and 50 kGy, was carried out.

MATERIALS & METHODS

Specimen Preparation

Diaphyseal cortical bone was obtained from banked femurs at the Perth Bone and Tissue Bank. Bending and compression testing specimens were obtained from a 47-year-old male donor and crack propagation specimens from a 49-year-old male donor, both of whom had had suspicious serology.

The bone was stored at -80°C, in accordance with the standard practice of the Perth Bone and Tissue Bank. Specimens of cortical bone were cut according to specific experimental requirements (see below) using an Isomet low speed diamond saw (Buehler, Lake Bluff, IL, USA), cooled with Ringer's lactate. All cancellous bone and periosteum were removed.

Storage conditions and thawing procedures for all specimens closely followed those used by the Perth Bone and Tissue Bank in clinical practice. The prepared bone specimens were re-frozen. Prior to the start of each experiment, each specimen was thawed to room temperature in normal saline and then wrapped in saline-saturated gauze.

The bone mineral contents (BMC) of the beams were measured by dual-energy X-ray absorptiometry (DXA), using a Hologic QDR 1000W (Hologic Inc, MA, USA). Optical microscopy and scanning electron microscopy were used to assess the features of fracture surfaces.

Gamma Irradiation

For each of the experiments; (i) three-point bending, (ii) compression testing, and; (iii) fracture toughness (see below), the specimens allocated to a particular experiment were divided into four groups. Group 1 was the control group, without gamma irradiation. The remaining groups were irradiated with a ^{60}Co source (Steritech, Melbourne, Australia) using standard Bone Bank practices. Group 2 received 15 kGy; Group 3, 25 kGy; and Group 4, 50 kGy, respectively. Dosimeters were placed adjacent to the specimens to record the actual dose. The delivered doses were accurate to within ± 5 kGy.

Three Point Bending

Sixty-four beams were utilised, each of trapezoidal cross-section, 20 to 30 mm^2 in area, with the saw cuts oriented so that the long axis of a specimen was chosen to be parallel to that of the femur. Each specimen was placed on a three-point bending jig with a 40 mm span between the supports. The testing was performed with an Instron TM/SM (Instron, High Wycombe, UK) universal testing machine. For three-point bending, each beam was loaded at the midpoint at a rate of 25 mm per minute, until failure occurred. A load versus displacement curve was plotted and the applied load in kgF at the moment of failure recorded. The cross-section of a fracture site was then photographed and the boundary digitised manually with a GTCO Digipad 5 (GTCO Corp, Rockville, MD, USA). The load axis was plotted on the digitised figure and the centroid and section modulus calculated using custom designed software. By introducing the failure load the ultimate stress was calculated in MPa.

Uniaxial Compression

This experiment utilised forty parallel-ended beams, 8 mm in length and 10 to 20 mm^2 in cross-sectional area, cut from the femoral cortex contralateral to the material used for the three-point bending specimens. The cross-sections were photographed and digitised, (and the cross-sectional areas calculated) as described above. The compressive load was applied at a rate of 25 mm per minute until failure occurred. A load *vs* displacement curve was plotted and the load at the moment of failure was deduced. The ultimate compressive stress was calculated by dividing this load by the cross-sectional area.

Fracture Toughness

After initial cutting of the femoral diaphysis into approximately 45 mm lengths with a standard orthopaedic air saw, 48 parallel sided beams of dimensions 3 x 6 x 40 mm were cut using the Isomet saw. Each beam had been cut so that its long axis was oriented parallel to the long axis of the femur and with the larger dimension of its cross-section oriented radially. Using the Isomet saw a 3 mm deep cut was placed in the centre of each beam, producing a round tipped notch. At the tip of the notch a starter crack was made using a dermatome blade. The volumetric bone mineral densities (vBMD) of the specimens were calculated using the DXA BMC measurements, plus measured specimen dimensions, ignoring the central notch. No significant variation in mean vBMD was found between the groups.

The beams were placed on a three-point bending jig with a 24 mm span. Testing was performed in the Instron testing machine using a 20 kgF full-scale-deflected load cell with the load applied opposite the notch to yield a deformation of 5 mm per minute; this being the slowest available crosshead speed. Crack-opening displacement (notch width) was measured with a clip gauge attached across the notch. Calibration of the clip gauge confirmed a linear response to displacement of the feet of the gauge. Each beam was tested to failure and load *vs* crack-opening displacement was recorded. A typical curve is shown in Figure 1. The load (P_Q) corresponding to a 2% apparent increment of crack extension is determined by a specified deviation from the linear part of the load versus displacement curve, where the 5% secant line of the linear elastic part of the curve (the line with slope 0.95 that of the linear elastic part of the curve, and passing through the origin) intersects with the curve.

There are several measures of fracture toughness based on linear elastic fracture mechanics. The critical stress intensity factor (K_{Ic}) is the most commonly used index, being an estimate of fracture toughness under plane strain conditions. Where conditions for the valid measurement of K_{Ic} are not met, the specimen strength ratio (R_{sb}) (Ref. 6; Annex 3) can be a useful comparative measure of material toughness in specimens of the same form and size. Crack propagation testing was performed on compact bend specimens using published protocols; K_{Ic} was calculated as follows [6]:

$$K_Q = (P_Q S / B W^{3/2}).\ f(a/W)$$

where:

$$f(a/W) = \frac{3(a/W)^{1/2}\,[1.99 - (a/W)(1 - a/W)(2.15 - 3.93a/W + 2.7(a/W)^2)]}{2(1 + 2a/W)(1 - a/W)^{3/2}}$$

and:

P_Q	=	load at 2% increment of crack growth
B	=	specimen thickness
S	=	span
W	=	specimen width
a	=	crack length (cm)

If $1 \leq P_{max}/P_Q \leq 1.1$, and a and B are both $\geq (K_Q/\sigma_{YS})^2$

where:

P_{max}	=	maximum load sustained by the specimen
σ_{YS}	=	0.2% offset yield strength in tension

then:

$$K_{Ic} = K_Q$$

The compact bend specimen strength ratio (R_{sb}) is calculated from the failure load, the specimen dimensions, and the yield strength in tension of the material:

$$R_{sb} = (6P_{max}W)/B(W-a)^2\ \sigma_{YS}$$

In all experiments, testing of the beams was conducted blind to their irradiation status. For each variable, the results for each group were compared with the control group using the pooled variance Student's test.

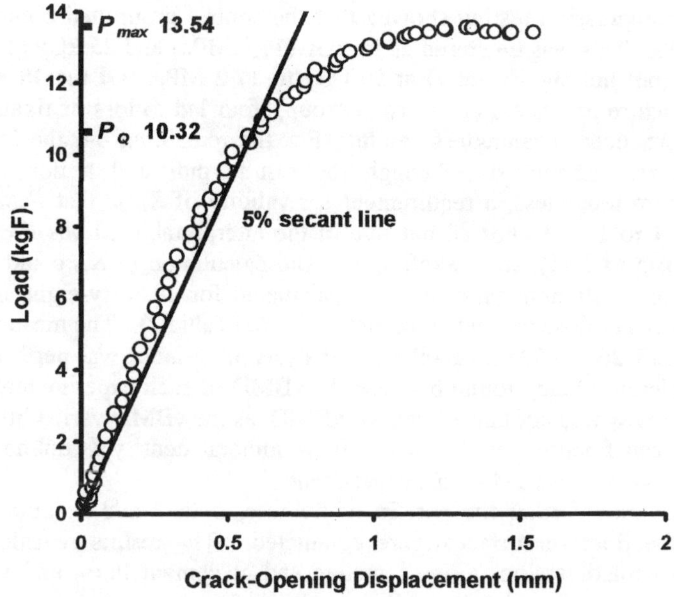

Figure 1. Typical load *vs* crack-opening displacement curve, showing the raw data points; P_{max}, the maximum load sustained; the 5% secant line of the linear elastic part of the curve; and P_Q the sustained load at the point where the secant line intercepts the curve.

RESULTS

Given the fact that spatial variations in the mineral content of a bone specimen, or the sample preparation, might affect the uniformity of measurement of the mechanical properties, DXA images and measurements were used to ensure that each specimen had an approximately homogeneous structure and composition. The DXA results also indicated that the relationship between gravimetric bone mass and BMC was highly linear ($r = 0.99$, 2% standard error of the estimate). Using a three-point bending test, there was a significant reduction in ultimate three-point bending stress of the irradiated groups compared with the control group (Table 1).

Table 1. Effect of gamma irradiation on the ultimate three-point bending stress of human femoral cortical bone.

Radiation Dose (kGy)	3-Point Bending Stress [MPa] (SD)	Reduction (%) *	p-value *
0	269 (12.5)		
15	255 (24)	5.7	< 0.05
25	208 (41)	22.6	< 0.01
50	185 (21)	30.0	< 0.01

{* In comparison with non-irradiated cortical bone}

The uniaxial compression testing showed that the control group had a mean failure stress of 18.0 MPa. This was unaltered at 15 kGy (17.7 MPa) and 25 kGy (17.9 MPa), and fell slightly (but not significantly) at 50 kGy to 17.2 MPa. Of the 48 specimens used to assess fracture toughness (12 in each group), four led to loss of fixation of the clip gauge and produced meaningless results. For the remaining 44, the load at 2% crack extension was calculated. Though the results indicated a dose-dependent variation in fracture toughness, a requirement for validity of K_{Ic} is that P_{max}/P_Q falls within the range 1 to 1.1 [6]. For all but five of the interpretable curves P_{max}/P_Q was greater than 1.1 (up to 1.45), and therefore no valid calculation of K_{Ic} could be made. However the bending strength ratio, R_{sb} was calculated for all forty-eight specimens. There was a significant dose dependent reduction in R_{sb} (Table 2). The mean vBMD of the specimens was 1.26 ± 0.5 [SD] gcm^{-3}. An analysis of variance was performed, with no significant difference being found between the vBMD of each experimental group (p = 0.68). Therefore R_{sb} was not normalised for vBMD, as the vBMD varied little and the relationship between fracture toughness and bone mineral density is unknown at the crack propagation velocities used in this experiment.

In order to examine further the pattern of fracture, optical and scanning electron microscopy of the fracture surfaces were conducted. The results revealed dentate fracture patterns with diversions along lamellae and at cement lines, and with some evidence of a degree of osteonal pull-out (Figures 2 and 3).

Table 2. Effects of irradiation on the fracture toughness of human cortical bone.

Radiation Dose (kGy)	Bending Strength Ratio [R_{sb}] (SD)	Reduction (%)*	p-value*
Control	1.64 (0.13)		
15	1.45 (0.12)	11.5	< 0.001
25	1.43 (0.16)	12.8	< 0.001
50	1.28 (0.13)	21.9	< 0.001

{* In comparison with non-irradiated cortical bone}

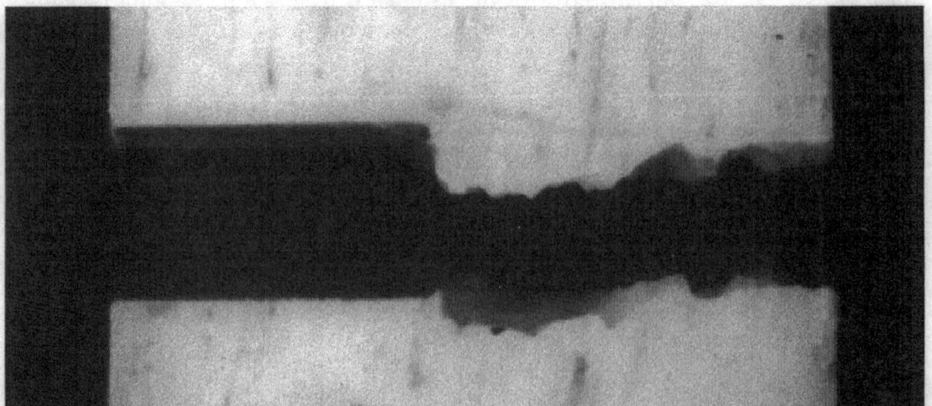

Figure 2. Photomicrograph (x 20) of dentate fracture from crack propagation testing of compact bend specimen. Initial cleavage plane is interlamellar.

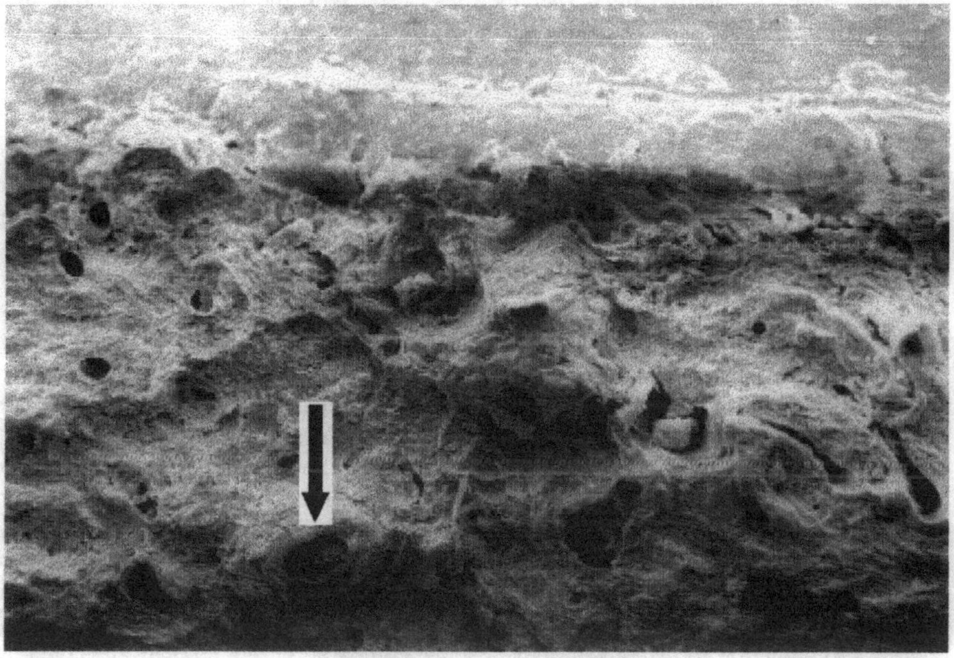

Figure 3. Scanning electron micrograph (x 82) showing a fracture surface resulting from crack propagation testing. Surface roughness indicates that crack diversion has occurred at the interfaces between different microstructural elements of the bone. The arrow indicates where an osteon has pulled out of the fracture surface.

DISCUSSION

Massive irradiated allografts have been shown to perform satisfactorily in the short term, despite a 6% allograft fracture rate [9]. A minimum dose of 15 kGy is considered necessary to achieve a sterility assurance level (SAL). However, several reports have indicated that gamma irradiation cannot be regarded as a significant virus inactivation method for bone allografts [10-12]. Given that there is also concern over its long-term mechanical integrity arising from radiation damage, the question arises at to whether allograft bone should be routinely irradiated.

In this study we have shown that even at the lower dose of 15 kGy there was a significant effect on the strength of human cortical bone. A 6% reduction in ultimate three-point bending stress was observed at 15 kGy, 23% at 25 kGy and 31% at 50 kGy. There was a 12% reduction in toughness at 15 kGy, 13% at 25 kGy and 22% at 50 kGy. The effect on ultimate compressive stress was less marked, with no effect seen below 50 kGy. Our data are consistent with those reported by Komender [4] and Hamer *et al.* [13], but are contradictory to the data of Bright and Burstein [5] who indicated no effect of gamma irradiation in either tension or compression. On the other hand, it is not possible on the basis of this study to draw conclusions as to the effect of irradiation on toughness at low crack velocities. This may be due to the fact that the crack propagation test used in our study differed from those reported by Wright and Hayes [4], Behiri and Bonfield [15,16] and Norman *et al.* [8].

In our study, a compact bend specimen was used with the crack propagating in a radial plane, rather than a compact tensile specimen with longitudinal crack propagation. In addition, the minimum crosshead speed of 5 mm per minute in this study allowed only a single load versus crack-length measurement at fracture initiation, due to the high crack velocity. These methodological differences may account for our failure to determine the stress intensity factor (K_{Ic}).

The P_{max}/P_Q ratio was consistently greater than acceptable for a valid determination of K_{Ic}. This may indicate that the phase of plastic deformation occurring by microfracture and delamination is large. This suggests that a large proportion of the energy was absorbed before ultimate failure. In turn, this indicates a high resistance to crack propagation, or a different mode of initiation of failure.

Although a valid K_{Ic} value could not be calculated, the values derived were less than expected in light of previous reports of longitudinal crack propagation in human bone, and both longitudinal and transverse crack propagation in bovine bone [8]. This is possibly a result of the relatively high crack velocity. The report of Norman *et al.* [8] does not state whether P_{max}/P_Q satisfied the conditions for validity of K_{Ic}, as multiple estimates were possible at different crack lengths because of the slow crack velocity, avoiding the need for calculation of P_Q.

In a material such as cortical bone, resistance to crack propagation is intuitively likely to be higher in the transverse direction than longitudinally. A crack front encounters many more sites at which crack arrest or diversion are possible when it crosses perpendicular to osteons or cement lines, compared with a crack front travelling generally parallel with the osteonal axes.

In spite of the fact that we could not measure the effect of irradiation on toughness at low crack velocities, the bending strength ratio R_{sb} is a relative measure of the resistance to crack propagation in specimens of approximately the same form and size. Although no exact comparison of the results of this study with previous estimates of bone fracture toughness is possible, our data showed clearly that irradiation has a significant dose-related deleterious effect upon the toughness of human cortical bone at high crack velocities.

CONCLUSIONS

Gamma irradiation of bone for allotransplantation causes significant decreases in three-point bending strength and toughness of human cortical bone. Gamma irradiation of bone should be avoided where possible, especially in massive structural allografts in young patients.

By adhering to prescribed strict screening and graft procurement procedures, the risk of disease transmission may be sufficiently low that the detrimental effects of irradiation on the mechanical properties of allograft bone would outweigh the benefits of sterilisation.

Indeed, for protection against viral contamination, these benefits are reportedly limited at conventional radiation doses. If sterilisation is considered obligatory, an alternative method may be worth considering.

ACKNOWLEDGMENTS

This work was supported by the National Health & Medical Research Council, the Western Australian Institute of Medical Research, and the Perth Bone & Tissue Bank.

REFERENCES

1. T. C. Turner, C. A. L. Bassett, J. W. Pate, *et al.*, Sterilization of preserved bone grafts by high-voltage cathode irradiation, *J. Bone Joint Surg.*, **1956**, 38A, 862-884.
2. American Association of Tissue Banks: *Standards for Tissue Banking*, Arlington, American Association of Tissue Banks, **1991**.
3. N. Triantafyllou, E. Sotiropoulos and J. N. Triantafyllou, The mechanical properties of the lyophylized and irradiated bone grafts, *Acta Orthop. Belg.*, **1975**, 41 (Suppl. 1), 35-44.
4. A. Komender, Influence of preservation on some mechanical properties of human haversian bone, *Mater. Med. Pol.*, **1976**, 8, 13-17.
5. R. W. Bright and A. H. Burstein, Material properties of preserved cortical bone, *Trans. Orthop. Res. Soc.*, **1978**, 3, 210.
6. American Society for Testing and Materials. ASTM E399-83 Standard Test Method for Plane-Strain Fracture Toughness of Metallic Materials. *ASTM Annual Book of Standards.* Designation E 399-83, Philadelphia, USA, **1983**, 487-511.
7. W. Bonfield, J. C. Behiri and B. Charalambides, Orientation and age-related dependence of fracture toughness of cortical bone. In: *Biomechanics: Current Interdisciplinary Research* (S. M. Perren and E. Schneider, eds), Dordrecht, Martinus Nijhoff, **1985**, 185-189.
8. T. L. Norman, D. Vashishth and D. B. Burr, Fracture toughness of human bone under tension, *J. Biomech.*, **1995**, 28, 309-320.
9. P. Hernigou, G. Delepine, D. Goutallier and A. Julieron, Massive allografts sterilised by irradiation, *J. Bone Joint Surg.*, **1993**, 75B, 904-913.
10. S. J. Withrow, S. A. Oulon, T. L. Suto, *et al.*, Evaluation of the antiretroviral effect of various methods of sterilizing/preserving corticocancellous bone, *Orthop. Res. Soc.*, **1990**, 226.
11. B. M. Fideler, C. T. Vangsness, T. Moore, Z. Li and S. Rasheed, Effects of gamma irradiation on the human immunodeficiency virus. A study in frozen human bone-patella ligament-bone grafts obtained from infected cadavera, *J. Bone Joint Surg.*, **1994**, 76A, 1032-1035.
12. D. G. Campbell and P. Li, Sterilization of HIV with irradiation: relevance to infected bone allografts, *ANZ J. Surg.*, **1999**, 69, 517-521.
13. A. J. Hamer, J. R. Strachan, M. M. Black, C. J. Ibbotson, I. Stockley and R. A. Elson, Biochemical properties of cortical allograft bone using a new method of bone strength measurement. A comparison of fresh, fresh-frozen and irradiated bone, *J. Bone Joint Surg.*, **1996**, 78B, 363-368.
14. T. M. Wright and W. C. Hayes, The fracture mechanics of fatigue crack propagation in compact bone, *J. Biomed. Mater. Res.*, **1976**, 10, 637-648.
15. J. C. Behiri and W. Bonfield, Crack velocity dependence on longitudinal fracture in bone, *J. Mater. Sci.*, **1980**, 15, 1841-1849.
16. J. C. Behiri and W. Bonfield, Fracture mechanics of bone – the effects of density, specimen thickness and crack velocity on longitudinal fracture, *J. Biomech.*, **1984**, 17, 25-34.

THE EFFECT OF COLD GAMMA RADIATION STERILISATION ON THE PROPERTIES OF DEMINERALISED BONE MATRIX

Arthur A. Gertzman [1*], Moon Hae Sunwoo [1], Denise Raushi [1] and Michael Dunn [2]

[1] *Musculoskeletal Transplant Foundation, Research & Development Division*
125 May Street, Edison, NJ 08837, USA
{ E-mail: Arthur_Gertzman@MTF.ORG*

[2] *Div. of Orthopaedic Surgery, UMDNJ, Robert Wood Johnson Medical School*
51 French Street, New Brunswick, NJ 08903-0019, USA

ABSTRACT

Demineralised bone in a viscous hyaluronan carrier was processed at sterilising and sub-sterilising doses of gamma radiation (10.2, 17.2 and 29.5 kGy). The pH, viscosity and osteoinductivity were measured. Osteoinductivity (OI) measured by implantation in an athymic (nude) mouse demonstrated significant loss at a dose of 17.2 kGy; OI scores declined 53% ($p = 0.024$). Viscosity also declined with increasing radiation dose. pH was unaffected.

INTRODUCTION

Tissue forms based on demineralised bone matrix (DBM) are widely used for initiating new bone growth in osseous defects. This application arises from the osteoinductive behaviour of DBM resulting from the exposure of the natural bone morphogenic proteins (BMPs) present in the bone when the calcium-based mineral component is removed [1]. Most currently available DBM products are processed and packaged using aseptic clean room technology to achieve sterility. It is believed that sterilisation by gamma radiation (cobalt 60) or other high energy or high-energy dose rate processes (electron-beam) would degrade the BMPs present in the DBM and reduce or eliminate its osteoinductive potential. Earlier studies by Stevenson *et al.* [2] demonstrated that irradiation of intact, cleaned, cortical bone at non-sterilising doses (15 kGy) under cold conditions (-70°C) preserved the osteoinductivity of the resultant graft. In that work, cortical struts were implanted in 8 mm middiaphyseal segmental rat defects. To achieve 10 sterility assurance level (SAL), it would be ideal to perform a terminal sterilisation, i.e., sterilisation with a penetrating radiation source of the processed and fully packaged DBM. DBX® (a registered trademark of MTF, Edison, NJ, USA) is a DBM dispersed in hyaluronan and formulated at physiologic conditions. It is processed using aseptic technology and is not sterilised by radiation or chemical means. In an effort to extend the work of Stevenson *et al.* [2], we evaluated the effect of cold gamma sterilisation on DBX in its fully processed and packaged form.

MATERIALS AND METHODS

The purpose of this study was to evaluate the effect of gamma irradiation sterilisation (15 kGy) on the physical and osteoinductive properties of three different lots of MTF demineralised bone matrix (DBM) formulation. Osteoinductivity of the gamma-irradiated lots was compared to that of control MTF DBM (positive control) and heat-inactivated MTF DBM (negative control).

When implanted into normal animals, human DBM is xenogeneic and is expected to provoke an immune response that may compromise the analysis of osteoinduction. To avoid this, we used the athymic mouse model. The athymic mouse lacks a thymus gland and, therefore, cannot mount a humoral immune response to the human DBM implants. Precedence of the use of an athymic mouse *(nu/nu)* model for studying the osteoinductive potential of demineralised bone allograft was noted by Schwartz *et al.* [3]. Samples of DBM were implanted bilaterally into the mouse hamstring muscle. The hamstring muscle group (*biceps femoris*) is a large, easily accessible muscle, which is commonly used as an implant site to evaluate heterotopic bone formation. Histological evaluation of the control and test articles was conducted 28 days after implantation to assess osteoinduction.

DBX putty from three different donors (Table 1) was stored in dry ice and subjected to 15 kGy gamma irradiation. The actual dose received was 17.2 kGy (16.1 - 18.2). Temperature of dry ice conditions was maintained throughout radiation and shipping. DBX Putty is a commercial bone defect filler produced by MTF. It is formulated of 32% w/w DBM (200-800 microns) ("Putty") in sodium hyaluronan of 700,000 Daltons molecular weight in a pH 7.2 phosphate buffered saline (PBS). The hyaluronan is at a 4% solution.

Table 1 Donor Selection.

DONOR	AGE (Years)	GENDER	WEIGHT (lb)	CAUSE OF DEATH
I	50	Male	300	Myocardial Infarction
II	41	Male	280	Chronic Obstructive Pulmonary Disease
III	22	Female	120	Motor Vehicle Accident

IDENTIFICATION OF SAMPLE GROUPS

Group I: MTF DBX from donor # I (putty) terminally sterilised with 15 kGy gamma irradiation. A total of five (5) samples were implanted.

Group II: MTF DBX from donor # II (putty) terminally sterilised with 15 kGy gamma irradiation. A total of five (5) samples were implanted.

Group III: MTF DBX from donor # III (putty) terminally sterilised with 15 kGy gamma irradiation. A total of five (5) samples were implanted.

Group IV: Control MTF DBX from donor # I in Carrier I (putty). A total of five (5) samples were implanted.

Group V: Control MTF DBX from donor # II in Carrier I (putty). A total of five (5) samples were implanted (putty).

Group VI: Control MTF DBX from donor # III in Carrier I (putty). A total of five (5) samples were implanted.

Group VII: Heat-inactivated Negative Control +100°C/24 hours: MTF DBX from donor # II. A total of five (5) samples were implanted.

Materials comprised of 35 samples of DBM (7 groups x 5 samples per group) packaged in individual syringes for implantation, plus several extra samples from each group. Each sample contained approximately 10 mg. of the test article. The samples were randomised and implanted bilaterally in the hamstring muscles of athymic nude mice. Intramuscular implantation of active DBM is expected to induce cartilage and then bone formation within the implants, a process termed osteoinduction. Animals were sacrificed at four weeks post-implantation. Decalcified histology was then performed on the explanted samples; 5 histological slides with 2 sections per slide were prepared for each sample (10 sections total per sample). Slides were stained with hematoxylin and eosin (n = 2 per group) or Masson's trichrome (n = 3 per group) and samples were evaluated for osteoinductivity. A semi-quantitative scoring system was utilised to assess osteoinduction.

The relative amount of osteoinduction was evaluated semi-quantitatively by the study investigator, using the scoring system described below; the observer was blinded to the identification of the implant. Osteoinductive scores were based on the degree to which new bone, bone cells, osteoid, calcified cartilage remnants, and marrow elements are present. To be consistent with proposed standards in the industry the following scoring system based on the method of Draft Standard ASTM F04 Division; B. Boyan, *et al*, was utilised:

0	No evidence of new bone formation
1	1-25% of the section is covered by new bone
2	26-50% of the section is covered by new bone
3	51-75% of the section is covered by new bone
4	> 75% of the section is covered by new bone

The overall score for each implant was obtained by averaging the highest five scores from the histological slides; scores for each experimental group were determined by pooling the overall scores of the individual implants. The results of semi-quantitative scoring are presented as a mean ± standard deviation. Significant differences between groups were determined by the use of a Students-t test. P values < 0.05 were considered significant.

Images of histological slides from each experimental DBM group were also captured and stored using a digital camera and computer system (*Image Pro Plus* ™ imaging software).

For testing of pH and viscosity effects, two lots each of commercial grade DBX putty and paste were subjected to gamma irradiation doses of 10.8, 17.2 and 29.5 kGy each at –70°C. These four lots of DBM in hyaluronan were measured for pH using a glass electrode designed for a semi-solid (Hamilton, "TripTrode"). This permits direct pH measurement without need for sample dilution. Each lot was tested in triplicate.

Additionally, viscosity of the four DBX samples was measured using a penetration method adopted from ASTM D-1403-96. The method presses a metal cone of known shape and weight over a fixed sample volume and records the penetration depth of the cone in a standard time. Consequently, viscosity is inversely proportional to the penetration value. Each lot was tested in triplicate.

RESULTS

Osteoinduction

The results of the histological scoring from the athymic mouse implants are shown in Table 2. Values are presented as means and standard deviation (n = 5). The non-irradiated tissue averaged 1.68; the irradiated samples averaged 0.77. The difference between control and irradiated bone scores was statistically significant ($p = 0.024$) based on a t-test. The negative control, heat inactivated at 100°C for 24 hours, scored consistently 0.0. Overall, osteoinduction decreased 53.7% and was consistent for each of the three donors (Table 3).

Table 2 Effect of cold terminal gamma irradiation sterilisation on the osteoinductivity of DBM in a hyaluronan formulation.

Gamma Irradiation (at -40 to -70°C)	Donor ᵗ	Osteoinductive Bone Score (0-4)	
		Mean (n = 5)	Standard Deviation
None (Control)	I	2.27	0.50
	II	1.07	1.10
	III	1.70	1.19
17.2 kGy (16.1 – 18.2)	I	1.40	1.23
	II	0.67	0.58
	III	0.25	0.50
Negative Control (No Gamma) Heat Inactivated	II	0.0	0.0

Table 3 Effect of gamma irradiation (17.2 kGy at -40 to -70°C) on osteoinductive score.

DONOR NO.	% REDUCTION
I	38.3
II	37.4
III	85.3
Mean	53.7

Formulation pH

pH values of DBX in both paste and putty forms were evaluated and found to be unchanged. Both lots tested of each form ranged in pH from 7.38 to 7.62 before irradiation and 7.39 to 7.67 after irradiation (Table 4).

Table 4 pH Of DBX putty and paste before and after irradiation. (results are shown as average (sd) [No.] of samples)

	Control	Test Group I	Test Group II	Test Group III
Lot Number (Description)	**Before Irradiation**	**After 10.4 kGy**	**After 15 kGy**	**After 25 kGy**
054210781072 (Putty)	7.38 (0.02)[3]	7.40(0.01)[2]*	7.40(0.01)[2]*	7.39 (0.01)[3]
048189961032 (Putty)	7.60(0.01)[3]	7.64(0.01)[3]	7.63(0.01)[3]	7.62(0.01)[3]
054208301062 (Paste)	7.67(0.01)[3]	7.66(0.01)[3]	7.66(0.01)[3]	7.67(0.01)[3]
054209131102 (Paste)	7.62(0.01)[3]	7.62(0.01)[3]	7.59(0.01)[3]	7.58(0.01)[3]

{* Average for sample 054210781072 was calculated using only two sets of pH values}

Viscosity

Penetration (cone) values increased sharply with increasing radiation dose. Initial values for two lots each of DBX putty and paste were 57.2 to 70.4 and 67.4 to 71.0, respectively. These penetration values increased (decreasing viscosity) with higher doses of gamma radiation (Table 5). Three of four lots irradiated at 25 kGy had viscosity decrease to a level below the range of the cone penetration test. The marked effect of gamma irradiation on DBX viscosity is summarised in Figure 1.

Table 5 Penetration of DBX putty and paste before and after irradiation. (results are shown as average (sd) [No.] of samples)

	Control	Radiation Dose		
Lot Number (Description)	**Before Irradiation**	**After 10.4 kGy**	**After 15 kGy**	**After 25 kGy**
054210781072 (Putty)	57.2(0.50)[3]	70.5(0.28)[2]	72.2(0.21)[2]	75.6(0.23)[3]
048189961032 (Putty)	70.4 (2.01)[3]	79.9(0.75)[3]	89.8(0.40)[3]	Samples were too soft to measure penetration
054208301062* (Paste)	63.4(0.81)[3]	81.5(2.25)[3]	80.1(0.85)[3]	
054209131102 (Paste)	71(0.30)[3]	81.8(0.71)[2]	79.9(1.15)[3]	
Average Increase in Penetration	-	12.9	15.1	18.4

{* Average for sample 054208301062 was calculated using only two sets of penetration values}

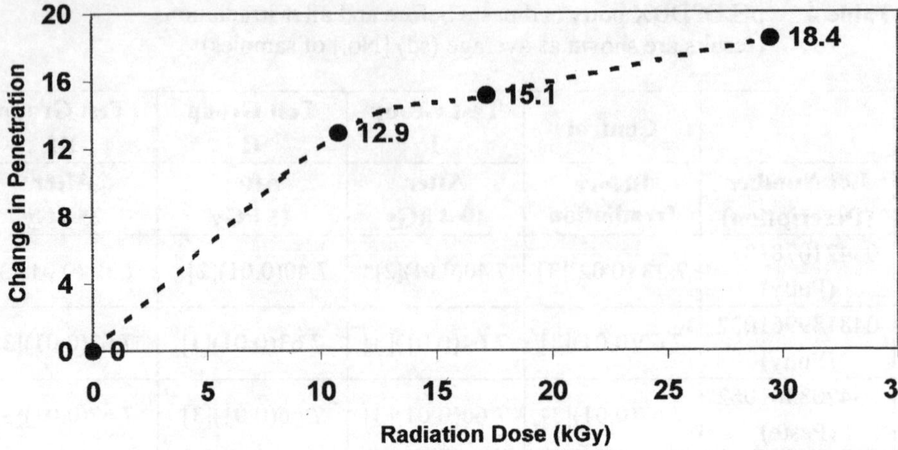

Figure 1. Effect of gamma irradiation on frozen DBX.

DISCUSSION

The pH of the DBX was unaffected by a gamma radiation dose of 17.2 kGy. The DBM in the formulation was processed with a pH 7.2-7.4 phosphate buffered saline and is expected to be unaffected as to pH.

The viscosity decrease is probably due to the loss in molecular weight from the irradiation of the sodium hyaluronan carrier used in the product formulation. The loss in osteoinduction of the DBM is probably due to the degradation of the bone morphogenic proteins (BMP) present in the DBM. While all three lots tested had measurable osteoinductivity after the 17 kGy dose, the biological efficacy of the DBM was significantly compromised.

Whereas product sterility is essential for a surgical implant, this loss of biological performance is not justified as product sterility is achieved by well-established aseptic process techniques.

REFERENCES

1. Urist, M. R., Bone Formation by Autoinduction, *Science*, **1965**, 150, 893-899.
2. Stevenson, S., The effects of processing and low dose irradiation on cortical bone grafts, *Clin. Ortho. Related Res.*, **2000**, 375, 275-285.
3. Schwartz, J., *Periodontal Surgery*, **1998**, 69, 470-478.
4. Boyan B., "Standard Guide for the Assessment of Bone Inductive Materials", 8 May **2000**.

COMPLICATIONS OF STRUCTURAL ALLOGRAFTS FOR MALIGNANT BONE TUMORS

Yong-Koo Kang *, Jin-Young Jeong, Yang-Guk Chung, Won-Jong Bahk and Seung-Koo Rhee

Department of Orthopedic Surgery, St. Vincent's Hospital, The Catholic University of Korea
93 Jee-dong Paldal-gu, Suwon, Kyunggi-do, Republic of Korea, 442-723
{E-mail: ykang@vincent.cuk.ac.kr

ABSTRACT

To compare the complications in non irradiated and irradiated structural allograft for malignant bone tumours, thirty-five patients who had been treated from April 1992 to December 2000 were studied. The duration of follow-up was 1 to 9 years (average 5 years). There were 23 males and 12 females, and age ranged from 1 to 55 years old (average 26 years). Diagnoses were 18 osteosarcomas, 7 chondrosarcomas, 3 malignant giant cell tumours, and 7 in Ewing's sarcoma, plasmacytoma, malignant lymphoma, metastatic tumours. The anatomical locations were 20 femurs, 7 tibiae and 8 humeri. There were 17 deep frozen, non-irradiated allografts supplied by our hospital bone bank. There were 25 irradiated, either deep-frozen or freeze dried allografts. Dosage of the irradiation ranged from 15 to 25 kGy. Reconstruction related complications from all patients developed in 13 patients (37.1%). The complications of the non-irradiated allografts developed in 7 out of 15 (46.6%), included infection in 3, delayed or non-union in 3, and graft fracture in 1. The complications of the irradiated allograft developed in 6 out of 20 (30%), included infection in 1, delayed or non-union in 2 and graft fracture in 3. The complications of irradiated allografts were relatively less, especially infection. However, the graft fracture of irradiated allografts were high. Delayed or non-union and graft fractures were salvageable with rigid internal fixation and autogenous bone graft. However, infections were difficult to salvage.

KEYWORDS

Allograft; bone tumour; sterilisation; irradiation; complication

INTRODUCTION

Recent advances of management of malignant bone tumors have improved survival rate dramatically. Radiographic imaging techniques such as CT, MRI and bone scan have improved, and enable us to better define the anatomic confines of the lesion, and provide critical information in deciding whether amputation is necessary or limb salvage is possible. Chemotherapy has improved most dramatically. Recently many clinical trials with chemotherapeutic regimens have reported adjuvant or neoadjuvant chemotherapy very useful to eradicate micrometastasis and improve survival. Moreover chemotherapy can induce the tumour down staging, and enable more effective limb salvage surgery.

The requirements of wide resection of the malignant bone tumours always lead to extensive resection with normal tissue envelope of the tumour, and cause considerable problems of reconstruction. After wide resection of malignant bone tumours, reconstructive options of bony defect include the autograft, allograft, tumour prosthesis and composite transplantation. Among those reconstructive methods, allograft transplantation is one of popular procedures.

Advantages of allograft transplantation include a biological reconstruction with excellent functional restoration of joints with soft tissue attachment, a lack of donor site morbidity, a capacity to shape the tissue to actual deficit, almost unlimited amount of supply of allograft with maintenance of bone bank. However, it has many disadvantages of the high rates of complications such as disease transmission, infection, non-union, bone resorption and graft fracture. Joint instability and articular cartilage collapse with misalignment of reconstruction, and cartilage fragmentation with secondary degenerative arthritis have been reported in cases of osteo-articular allograft. Among those complications, infection is one of most serious complications.

Various techniques have been attempted to reduce the complications, especially infection through the grafts. In order to minimise the infection through the graft, ionising radiation for sterilisation/decontamination of tissues was introduced in many tissue banks. However, the use of the ionising radiation for the sterilisation of tissues, either in the initial or the final stage of a defined processing method, will affect the mechanical and biological properties of the tissues. These changes may affect final results of allograft transplantation.

The purpose of this study was to review the clinical results in terms of functional results and to compare the reconstruction related complications in non-irradiated and irradiated structural allografts, retrospectively.

MATERIALS & METHODS

Thirty-five patients who had been treated with allograft transplantation for malignant bone tumours from April 1992 to December 2000 were retrospectively reviewed. The duration of follow-up was 1 to 9 years (average 5 years). There were 23 males and 12 females, and age ranged from 1 to 55 years old (average 26 years). Diagnoses were 18 osteosarcomas, 7 chondrosarcomas, 3 malignant giant cell tumors, and 2 in Ewing's sarcoma, and 1 in each MFH, plasmacytoma, malignant lymphoma, synovial sarcoma, metastatic tumour. The anatomical locations were 20 femurs, 7 tibiae and 5 in humeri and 1 in each ileum, metacarpal and metatarsal. There were 15 deep frozen, non-irradiated allografts, which were harvested in sterile technique in operation room, and supplied by hospital bone bank. There were 20 irradiated, either deep-frozen or freeze dried allografts. Dosage of the irradiation ranged from 15 to 25 kGy. Reconstructive procedures were 15 arthrodeses, 10 allograft-prosthesis composites and 7 intercalary grafts and 3 osteoarticular allografts. Internal fixations for reconstructive procedures were 10 medullary nail, 8 nail and plate, 7 plate, and 10 prostheses with or without plate fixation. Autogenous cancellous bone grafting was performed in 10 patients. Freeze-dried allografts were reconstituted in sterile saline solution at the operation room before the operation. All patients received preoperative, intraoperative and postoperative intravenous antibiotics.

The length of allograft ranged from 5 to 26 cm; twenty-five patients received neoadjuvant chemotherapy and 10 did not. Average time for chemotherapy was 10 months. The chemotherapeutic drugs most frequently used were doxorubicin, ifosfamide, high dose-methotrexate and cisplatin for osteosarcoma. Functional results were evaluated by modified ISOLS scores, which contained pain, functional activity, emotional acceptance, use of external support, walking ability and gait for lower extremity and hand positioning, manual dexterity and lifting ability for upper extremity.

RESULTS

The overall survival rate in this group of patients was 75% by Kaplan-Meier's methods. There were eight deaths from wide spread metastases and 5 alive with disease and 22 continuous disease free survival. Two patients had been treated with amputation due to local recurrence. Functional results of limb salvage surgery with allograft transplantation by modified ISOLS score were 63 to 80% (average 72%). There were no differences in survival rates and functional results between irradiated and non-irradiated groups.

There were no cases with disease transmission through the graft. The reconstruction related complications from all patients developed in thirteen patients (31.4%). The complications of the non-irradiated allografts developed in seven out of fifteen (46.6%), included infection in three, delayed or non-union in three, and graft fracture in one. The complications of the irradiated allograft developed in six out of twenty (30%), included infection in one, delayed or non-union in two and graft fracture in three. Three out of four infections were related to immunosuppressed conditions in post-operative chemotherapy. Among those three infections, two (including one patient of irradiated allograft) were initiated with marginal necrosis of surgical wound, and progressed to deep infection during post-operative chemotherapy. Among four patients with infection, one had been treated with amputation at the level of the thigh due to continuous infection, two were salvaged with staged procedures, which were removal of the allograft and followed by vascularised fibular graft, or secondary allograft. In one the infection continued. Five patients with delayed or non-union from both groups were successfully managed with rigid internal fixation and cancerous bone graft. Four out of 5 patients with delayed or non-union had been treated with chemotherapy. Three out of 4 graft fractures were managed with rigid fixation and cancerous bone graft. However, one patient with graft fracture from non-irradiated allograft had been treated with staged procedure with tumour prosthesis. Four patients with delayed or non-union and 2 patients with fracture were successfully salvaged with internal fixation and autogenous bone grafts. The overall failure of allograft reconstruction, which meant removal of allograft, developed in 5 patients (14.3%) who had infection in 4 or graft fracture in 1. In 30 patients who had bone union at graft-host junctions with initial surgery, average union time was 6 months in metaphysis and 10 months in diaphysis. There was no non-union in metaphyseal junction, whereas 5 delayed or non-union out of 30 diaphyseal junctions. There was no difference in allograft-host bone union time between irradiated and non-irradiated allografts.

DISCUSSION

The requirement of the wide resection with normal tissue cuff for malignant tumours leads to extensive tissue defects causing considerable problems of reconstruction. The reconstructive procedures most widely used are bone transplantation and prosthetic replacement. Bone transplantation is a biological reconstruction and can be done by either autogenous or allogenous bone graft.

Utilisation of tumour replacing mega-prosthesis for hip, knee and shoulder supplies an effective means of restoring skeletal continuity. Prosthetic reconstruction restores excellent function and usually provides the patient well satisfied because of the preservation of joint movement. While prosthetic reconstruction may seem best indicated in older patients who have less functional demands on the implants, it has been used to a great extent in young patients with life threatening sarcomas because of the advantage of immediate functional restoration and minimal morbidity. While the early results have been satisfactory, concerns for long-term prosthetic function have limited widespread acceptance. Many mega-prostheses are implanted each year with good results. However, complications are frequent, resulting in re-operation and, at times, even amputation.

Autogenous bone transplantation represents beyond doubt the most physiological procedure for reconstructing skeletal defects. However donor sites of the autograft for extensive structural defect are limited, and when using autogenous grafts with a major cortical component in the reconstruction of long diaphyseal defects, late fatigue fractures often occur in spite of grafted bone healing. When large segments of bone need to be replaced, the vascularised bone transfer is available, most frequently fibula. Although microvascular surgery may improve the results, autogenous bone graft for major structural defect due to extensive resection of the malignant tumours is difficult to offer congruent skeletal substitution. Another important limitation of vascularised bone grafting is the amount of material available.

The transplantation of large bulk allografts is alternative to autograft. It was initiated by Laxer in 1906, who reported his experience with partial or total knee joint replacement using allogeneic tissue for traumatic joint loss. It was not popularised due to the inability of storing the allograft tissue for future use. In 1973, Parrish reported that freezing transplantable tissue to -40 degrees would not only preserve the tissue but also decrease its immunologic competency. Further reports by many authors in the 1970s firmly supported the potential use of structural allograft in orthopedic oncology. Large segment structural allografts can be used intercalary, as allograft prosthesis composites and osteoarticular allografts.

The advantages of allografts include the lack of any donor site morbidity, the capacity to shape the tissue to actual deficit, and the ability to advance peri-articular soft tissues into the allogeneic tissue to allow for not only greater inherent stability but also greater functional outcome. Allograft offers almost unlimited amounts of material for skeletal reconstruction with maintaining of tissue banks.

The disadvantages of allografts include the lack of donors, continued and evolving quality control issues, costs, transmission of disease, and the high complication rates with infection, non-union, fracture and resorption. Non-union and fracture of the allograft are attributed mainly to low osteogenic capability. Osteoarticular allograft includes an inordinately high rate of articular cartilage collapse, subsequent degeneration and instability of the joint leading to additional surgical procedures.

It is evident that complication rate of allograft reconstruction was relatively high in patients with malignant bone tumors. There were reports similar to this study. Delayed or non-union rate in this study was 16%, which is comparable to Mankin's reports. This is, of course, not surprising. It is well known that bone formation is profoundly diminished by chemotherapeutic agents especially methotrexate and doxorubicin. The toxic effects of these drugs are directed against osteoblastic activity. Thus, one would expect delayed bone healing of the graft-host junction and bone resorption. Fortunately, these complications could be successfully salvaged with autogenous cancellous bone graft in this study. However, it might be recommended that an immediate application of autogenous cancellous bone graft in allograft-host junction in patients receiving chemotherapy to enhance the healing and to reduce non-union of the allograft.

Most difficult complication was infection. There were 4 (16%) infections. These infection rates are comparable with other reports, too. These high rates of infection may be related to extensive surgical exposures and soft tissue resection and a long operation time for reconstruction, and also might be related to rather immunologically suppressed conditions of the patients than contamination of the allografts. Since infection of the patients with chemotherapy developed in the early stage of the reconstruction and progressed to failure, the meticulous wound management during the surgery, and close observation and early debridement of the marginal necrosis are important to prevent the deep infection. However, it is a problem that in most instances of malignant bone tumour there is a need to continue chemotherapy to prevent metastasis, and to induce neutropenia.

CONCLUSIONS

The clinical results of irradiated allografts were as good as non-irradiated allografts. The complications of irradiated structural allograft were relatively high, especially infection. However, graft fractures were common in irradiated allograft. Delayed or non-union and graft fracture were salvageable with internal fixation and autogenous bone graft. However, infection was difficult to salvage.

REFERENCES

1. Berrey, B. H., Lord, C. F., Gebhardt, M. C. and Mankin, H. J., Fractures of allografts: frequency, treatment, and end-result, *J. Bone Joint Surg.*, **1990**, 72, 825.
2. Conrad, E. U., Strong, D. M., Obermeyer, K. R. and Moogk, M. S., Tissue donor testing and recipient tracking for hepatitis C. *Communication to American Association of Tissue Banks*, **1994,** Board of Governors, Northwest Tissue Center, Seattle, Washington.
3. Dick, H. M., Malinin, T. I., Mealy R. and Mnaymneh, W. A., Massive allograft implantation following radical resection of high-grade tumors requiring adjuvant chemotherapy treatment, *Clin. Orthop.*, **1985,** 197, 88-95.
4. Enneking, W F., Dunham, W., Gebhardt, M. C., Malawar, M. and Pritchard, D. J., A system for the functional evaluation of reconstructive procedures after surgical treatment of tumors of the musculo-skeletal system, *Clin. Orthop.*, **1993,** 286, 241-246.
5. Enneking, W. F. and Mindell, E. R., Observations on massive retrieved human allografts, *J. Bone Joint Surg.*, **1991**, 73, 1123.
6. Gebhardt, M. C., Flugstad, D. I., Springfield, D. S. and Mankin, H. J., The use of bone allografts for limb salvage in high-grade extremity osteosarcoman, *Clin. Orthop.*, **1991**, 270, 181-196.
7. Gebhardt, M. C., Flugstad, D. I., Springfield, D. S. and Mankin, H. J., The use of bone allografts for limb salvage in high-grade extremity osteosarcoma, *Clin. Orthop.*, **1991**, 270, 181.

8. Horowitz, S. M., Glasser, D. B., Lane, J. M. and Healey, J. H., Prosthetic and extremity survivorship after limb salvage for sarcoman, *Clin. Orthop.*, **1993**, 293, 280-286.
9. Langlais, F., Lambotte, J. C., Collin, P. and Thomazeau, H., Long-term results of allograft composite total hip prosthesis for tumors, *Clin. Orthop.*, **2003**, 414, 197-211.
10. Malinin, T. I., Buck, B. E., Temple, H. T., Martinez, O. V. and Fox, W. P., Incidence of clostridial contamination in donors' musculoskeletal tissue, *J. Bone Joint Surg. Br.*, **2003**, 85, 1051-1054.
11. Mankin, H. J., Gebhardt, M. C., Jennings, L. C., Springfield, D. S. and Tomford, W. W., Long-term results of allograft replacement in the management of bone tumors, *Clin. Orthop.*, **1996**, 324, 86-97.
12. Ohno, T., Shigetomi, M., Ihara, K., Matsunaga, T., Hashimoto, T., Kav Sugiyama, T. and Kawai, S., Skeletal reconstruction by vascularized allogenic bone transplantation: effects of statin in rats, Transplantation, **2003**, 76, 869-71.
13. Ortiz-Cruz, E., Gebhardt, M. C., Jennings, L. C., Springfield, D. S. and Mankin H. J., The result of Transplantation of intercalary allograft after resection of tumors, *J. Bone Joint Surg.*, **1997**, 79A, 97-106.
14. Ottolenghi, C. E., Massive osteo and osteo-articular bone grafts: technique and results of 62 cases, *Clin. Orthop.*, **1972**, 87, 156.
15. Parrish, F. F., Allograft replacement of all or part of the end of a long bone following excision of a tumor, *J. Bone Joint Surg.*, **1973**, 55, 1.
16. Piriou, P., Sagnet, F., Norton, M. R., de Loubresse, C. G. and Judet, T., Acetabular component revision with frozen massive structural pelvic allograft: average 5-year follow-up, *J. Arthroplasty*, **2003**, 18, 562-569.
17. Simon, M. A., Aschliman, M. A., Thomas, M. and Mankin, H. J., Limb-salvage treatment versus amputation for osteosarcoma of the distal end of the femur. *J. Bone Joint Surg.*, **1986**, 68A, 1331-1337.
18. Thompson, R. C., Pickvance, E. A. and Garry, D., Fractures in large-segment allografts, *J. Bone Joint Surg.*, **1993**, 75, 1663.

EFFECTS OF RADIATION ON THE INTEGRITY AND FUNCTIONALITY OF SOFT TISSUE. CURRENT SITUATION: CARTILAGE, HEART VALVES, TENDONS AND OTHER TISSUES. CHANGES WITH INCREASING DOSE/DOSE LIMITS

D. Michael Strong

Northwest Tissue Center / Puget Sound Blood Center
921 Terry Avenue, Seattle WA 98104, USA
{Email: dmstrong@psbc.org}

ABSTRACT

Over the last several decades, numerous studies have been performed to evaluate the effects of irradiation on various soft tissues. Both gamma and e-beam sources have been used with doses ranging from 17 to 80 kGy. A variety of animal models have been employed including dogs, pigs, goats and sheep and several studies have been carried out with human tissue as well. The primary tissue studied has been tendon with bone-tendon-bone as the principle tissue of interest. Various biomechanical and biochemical measurements have been employed comparing irradiated and control tissues. Measurements have included: physicochemical properties, histological changes, collagen structure and cross linking, cyclic stretching, tensile strength, maximum force and strain energy to maximum force, maximum stress, maximum strain and strain energy density to maximum stress, and Young's modulus testing. In general, increasing damage occurs with increasing doses of irradiation. At lower doses (17-20 kGy), there is some disagreement as to significant changes however decreases in strength (20-30%), and modulus (50-60%) have been reported. At doses of 25 kGy and above, biochemical, biophysical and mechanical and material properties were more significantly altered. Limited studies have also been published on irradiation effects on meniscal grafts, cartilage, heart valves, dura mater, skin, fascia, sclera and amnion. Only irradiated tissues such as cartilage and amnion have been considered acceptable since they do not require load-bearing forces and have been used in wound coverage, craniofacial repairs or other non-weight-bearing reconstruction.

KEYWORDS

Irradiation; soft tissues; tendons; dose limits; tissue effects; transplants

INTRODUCTION

The growing popularity in the use of allografts for a variety of surgical applications along with the recent reports of infectious complications has increased the interest in sterilisation procedures to reduce the risk of disease transmission [1-3]. This review will focus on the effects of irradiation on soft tissues including cartilage, heart valves, dura mater, fascia, amniotic membranes, sclera and tendons, with a primary focus on tendons.

CARTILAGE

Although significant adverse effects can be demonstrated on cartilage and extracellular matrix with doses as low as 5-10 Gy [4], a large body of literature exists on the use of irradiated (typically 40 kGy) cartilage in a variety of clinical applications. Irradiated human costal cartilage is used as an implant material for reconstructive procedures because of its strength, homogenous consistency, ease of preparation and light weight (Table 1). It has been used with good results in reconstruction of the tympanic membrane with only a 16% complication rate [5]. It has been used in reconstruction of the face with good to satisfactory results in about 75% [6]. Complication rates for facial contour restoration have been reported in 5-7% of cases [7] and in the follow up of 16-72 months complete resorption in 88% of grafts, many of which had acquired chondrocytes from the host and produced new proteoglycan matrix [8]. In the long term follow up of an average of 9 years, 62 of 145 irradiated allogeneic cartilage grafts had a resorption rate of 75% and 18 of 24 followed from 11 to 16 years were completely resorbed. In spite of this, patients maintained satisfactory facial contours with fibrous tissue replacement of the cartilage [9].

Table 1. Irradiated Costal Cartilage Applications.

- Cleft lip reconstruction
- Facial fractures and contours
- Craniofacial anomalies
- Congenital deformities – ear, chin, orbits
- Orbital reconstruction
- Iliac crest following bone graft harvest
- Tympanic membrane
- Rhinoplasty

Irradiated cartilage has been used extensively in augmentation rhinoplasty with good to excellent results in 83% with follow up of 1-27 months [10]. Revision was required in 17% including 15% for warping. However, a comparison of warping between irradiated and non-irradiate costal cartilage found no significant difference [11]. No resorption has been reported with irradiated costal cartilage used for orbital and periorbital reconstruction [12]. Long term follow up in 21 patients up to 48 months revealed no graft infections, extrusions or clinically detectable graft distortion or resorption [13]. Auricular cartilage has been reported as superior to costal cartilage in laryngotracheal reconstruction [14], however irradiated glutaraldehyde preserved bovine cartilage is reported to have a high rate of graft loss thus questioning the reliability of this tissue [15].

HEART VALVES

The successful use of aortic valve allografts was first reported in 1956 [16]. This was followed by reports from others that led to various means to better process and preserve these grafts [17,18]. An earlier report of the use of irradiation sterilisation of arterial grafts [19] led to the development of similar procedures for aortic valves [20]. Initial results using a dose of 20-22 kGy on frozen grafts were promising with an incidence of 20% diastolic murmurs in 50 patients was an improvement compared to previous reports [21]. As patients continued to be followed, outcomes were considered to be less than satisfactory and cause for concern [22]. Results after at least four years of patient follow-up demonstrated a death rate of 35% with graft failure being responsible for at least half. An additional 20% of patients required removal of the graft. Of the remaining patients, two-thirds of these patients had murmurs of aortic insufficiency leading the authors to conclude that this technique should be abandoned [23].

DURA MATER, SKIN, FASCIA, SCLERA, AMNION, MENISCUS

Standard sterilising doses of 25 kGy have been applied to other tissues that have been used primarily for wound coverings or surgical repair. Dura mater has been used for neurosurgical repair [24] although the reports of Creutzfeld-Jacob disease transmission, an organism resistant to these doses of irradiation, have drastically reduced the use of this tissue [25].

Doses of 21 kGy did not cause any significant change in the permeability of rehydrated freeze-dried chorio-amniotic, dermo-epidermal and fascial grafts prepared for clinical use [26]. The use of irradiated amniotic membranes for wound coverings has been successfully employed throughout Asia [27]. Measurements of stress-strain on skin grafts following these doses demonstrated a reduction of biomechanical properties [28]. Doses as low as 100 Gy have been shown to cause direct killing of the cells of the skin in interphase [29].

Gamma irradiation has been shown to be a highly effective method of sterilisation of sclera allowing for extended storage and convenient use in the operating room [30].

Irradiation effects on the mechanical properties of meniscal grafts have also been determined. The elastic and viscoelastic properties of rabbit meniscal grafts were measured using a tensile testing machine and an indentation test. Doses commonly used for sterilising grafts had a significant adverse influence on both the elastic and viscous response of the grafts leading to the conclusion that irradiation sterilisation is unsuitable for clinical use [31].

TENDONS

Allograft patellar tendons present one option for reconstruction of the torn anterior cruciate ligament (ACL) of the knee [32,33]. Clinically, allografts have several advantages over autografts including decreased surgical time, decreased surgical morbidity, unaltered patellar-femoral tracking and a wider graft selection. One major concern for use of allografts has been the risk of infectious diseases such as AIDS, hepatitis or spore forming organisms which have been shown to be transmitted by allograft tendons [2,34-36].

Sterilisation can be achieved, depending on the bioburden and sensitivity of the infectious agent, by ethylene oxide or irradiation, using either gamma or e-beam sources, but each can have detrimental effects on the allograft. In the case of ethylene oxide, chemical derivatives such as ethylene glycol remain on the tissue after treatment despite extensive aeration techniques. These residuals can be retrieved at surgery in reconstructed knees and this method has been abandoned for allograft sterilisation [37,38].

Depending on dose, irradiation can have adverse effects as well. It has been shown that 10-50 kGy can induce minor cleavage of alpha chains of the collagen triple helix molecule [39,40]. Intra molecular and intermolecular hydrogen bonds are broken at levels as low as 10 kGy as indicated by the disorganization of previously tightly packed macromolecules in the collagen fibril [41,42]. Cross-linking, even at 50-100 kGy which cross-links at 0°C, can be suppressed at below 0°C leading to the practice of irradiating frozen tissue [43]. Ionising radiation has been shown to cause scission and structural breakdown of collagens [44]. After 20 kGy of gamma irradiation, Haut and Powlison [45] demonstrated an induced crimp pattern in tendon collagen. The extent of pretensioning needed to allow ACL allografts to provide normal knee stiffness depended on structural properties [46]. They conjectured that irradiated grafts could be made to respond more like controls simply by pre-straining.

DeDeyne and Haut [47] also used 20 kGy on human bone patellar tendon grafts and showed a decreased shrinkage temperature and increased solubility of collagen to pepsin suggesting cleavage of polypeptide chains. Haut and Powlison [48] also found significant reductions in material properties of human tendons including strength (-27%) and modulus (-56%) following 20 kGy. Bone attachments were more adversely effected. This report was supported by a similar finding after 17.5 kGy [49]. Alternatively, Gibbons *et al.* reported no changes after 20 kGy of irradiation of goat tendons. Goat patellar bone units were exposed to 20 or 30 kGy. The latter dose significantly reduced the maximum force and strain energy to the maximum force of the composite unit by 27% and 40% respectively, but 20 kGy exposure did not result in significant alterations [50]. This group did demonstrate, however that 30 kGy produced significant reduction in the maximum stress-and-strain energy density of human bone-patellar tendon-bone specimens [51]. Effects were not as large as reported by Paulos *et.al.*, who exposed freeze-dried human patellar tendon-bone units to the same dose [52].

In a study in which patellar-tendon bone exposed to 40 kGy was tested with lower, higher and functional loads to failure, irradiation significantly shortened the tendon but did not alter static or cyclic creep at low loads. The static creep test results indicate that irradiated allografts respond to constant, lower functional loading similarly to frozen grafts. The greatest changes were produced during failure testing with reduced linear stiffness (-12%) and maximum force (-26%), both statistically significant [53]. These authors did suggest that these grafts would have the required mechanical properties to replace the anterior cruciate ligament at the time of implantation.

The effects of freeze-drying and gamma sterilisation on the mechanical properties of human Achilles tendons have been studied [54]. A significant decrease of tensile strength (33%) for the freeze-dried grafts for the whole tendon and 43% for the bisected tendon were observed. Gamma sterilisation showed a smaller effect. Considering only the primary mechanical strength, the authors concluded that complete freeze-dried or complete and bisected irradiated Achilles tendons showed sufficient tensile strength for ACL replacement.

The order of freeze-drying and irradiation was reported as being significant with irradiation prior to freeze-drying resulting in improved mechanical and material properties than the reverse order [45]. Using a rat patellar tendon model, Toritsuka *et al.* [55] showed that freeze-drying, freeze-drying followed by irradiation and fresh freezing followed by irradiation temporarily accelerated graft remodelling as measured by loss of radiolabelled collagen. These studies are supported by a separate study in which freeze-dried medial collateral ligaments from sheep, irradiated with 26 kGy, were measured for mechanical properties compared to controls. Properties changed in respect to maximal load (-29.9%, $p \leq 0.05$); stress (-20.1%); strain (-0.64%); stiffness (-10.2%); energy (-31.4%) and elastic modulus (-1.3%) [56]. The tensile mechanical properties of porcine toe extensor tendons were measured following freezing, freeze-drying, freezing then irradiation (25 kGy) or freeze-drying then irradiation [57]. The ultimate tensile stress of the freeze-dried irradiated group was significantly different from the fresh and frozen irradiated groups being reduced by 90%.

Higher doses of irradiation have also been tested. Human bone-patellar tendon-bone allografts from humans were exposed to 20, 30 and 40 kGy. Maximum force, strain energy, modulus and maximum stress demonstrated a statistically significant reduction after 20 kGy. Stiffness, elongation and strain were reduced but not significantly. A 10% to 24% and 19% to 46% reduction in all parameters were found after 30 kGy ($p < 0.005$) and 40 kGy ($p < 0.0005$) of irradiation respectively. There was a dose dependent effect on all biomechanical parameters studied [58].

A similar dose dependent effect was seen after exposure of goat bone-patellar tendon-bone to 40, 60 and 80 kGy of gamma irradiation [59]. Both mechanical and material properties were reduced following exposure. They also found significant decreases in hydroxypyridinum cross-link density with 60 kGy of irradiation ($p < 0.05$). These findings were consistent with an earlier report using 20 and 30 kGy [60].

Attempts have been made to protect grafts from sterilising doses. Goertzen *et al.* [61] irradiated (25 kGy) canine bone-ACL-bone allografts under argon gas protection. Grafts were implanted and after 12 months the irradiated grafts failed at a mean load of 718.3 N, 63.8% of the strength of the normal canine ACL. The non-irradiated allografts failed at 780.1 N, 69.1% of normal. All grafts showed well-oriented collagen structure and there were no differences between treated and control groups. Canine tendons were treated with solvents (sodium hydroxide and hydrogen peroxide) prior to or following 28 kGy gamma irradiation [62]. The tensile strengths were 85% for the solvent group, 39% for the solvent/gamma group and 86% for the gamma/solvent group. The tendons in the solvent/gamma group underwent the most severe changes in material properties. The tendons in the gamma/solvent group changed the least.

In conclusion, a variety of studies seem to indicate that significant changes in tendons occur following irradiation at doses as low as 17 kGy. Nevertheless, there are less significant changes in the doses of 17 kGy to 25 kGy than with higher doses. Differences can be explained on the basis of the methods used for testing tendons, such as material or mechanical properties, and the type of tissue, e.g. goat or human, that have led to a difference in results. It seems evident that doses above 25 kGy have deleterious effects on mechanical and material properties and these effects increase with increasing doses (Figure 1). There is also a physical change to the tendons following irradiation that includes changes in colour and odour that are considered to be undesirable (Figure 2). Nevertheless, irradiated tendons continue to be used in clinical practice albeit at a lesser level than for non-irradiated grafts.

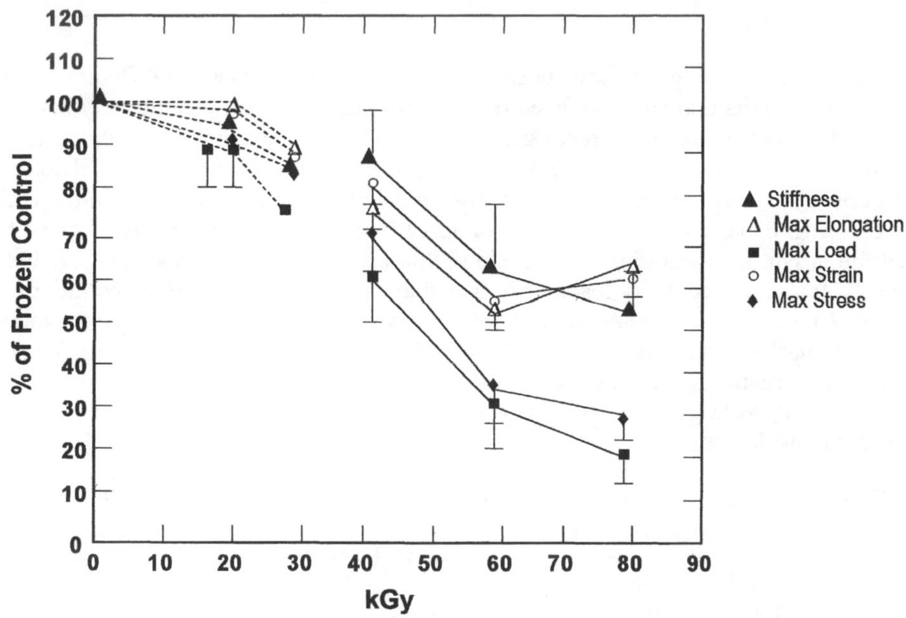

Figure 1. Average dose-dependent response for mechanical and material properties (Adapted from Salehpour *et al.*, *J. Orthop. Res.*, **1995**, 13, 898-906).

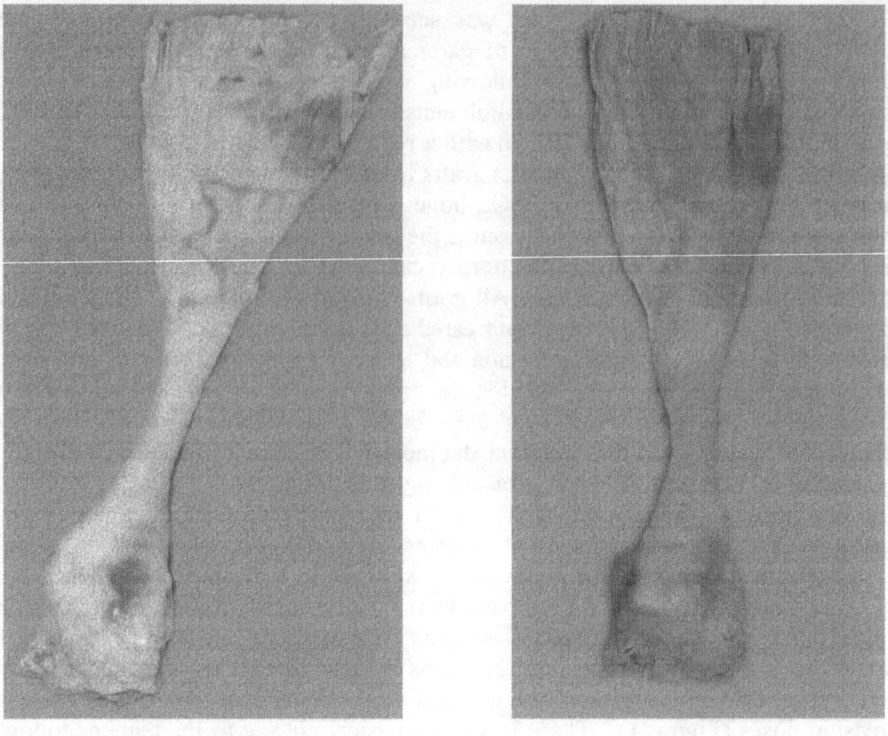

Pre-irradiation **17 kGy**

Figure 2. Achilles tendons pre- and post-irradiation (17 kGy).

CONCLUSIONS

Various soft tissue grafts have been studied following exposure to different doses of irradiation. Grafts that have not been required to maintain structural integrity but rather are used as coverings or in reconstruction such as cartilage have had more clinical success. Sterilising doses of irradiation that exceed 25 kGy have the most deleterious effect on all of these tissues. Recent studies in the blood banking industry in the United States using nucleic acid technology for identifying potentially infected donors have been important in calculating risks in blood donors [63]. These studies have also helped identify the expected viral loads that one may encounter in window period donors (Table 2.). Such viral loads significantly exceed the capability of current radiation doses to guarantee sterility of tissues. Nevertheless, irradiation in sterilising doses can significantly reduce the viral load and in combination with appropriate donor screening and laboratory testing will significantly enhance and improve the safety of tissues being used for transplantation.

Table 2. Window Period Blood Donors, USA, 1999-2003, Viral Yield.

	Range (copies/ml)
HIV	$3 \times 10^2 - 4.9 \times 10^6$
HCV	$1 \times 10^2 - 2.2 \times 10^8$
HBV	$1 \times 10^2 - 4.1 \times 10^8$
WNV	$1 \times 10^2 - 16 \times 10^4$

REFERENCES

1. T. Eastlund and D. M. Strong, Infectious disease transmission through tissue transplantation, In: *Advances in Tissue Banking, Volume 7,* G. O. Phillips, D. M. Strong, R. Von Versen and A. Nather (eds.), World Scientific, Singapore, **2004**, (in press).
2. CDC, Update: Allograft-associated bacterial infections - United States, 2002, *MMWR,* **2002**, 51, 207-210.
3. CDC, Hepatitis C virus transmission from an antibody-negative organ and tissue donor - United States, 2000-2002, *MMWR,* **2003,** 52, 273-276.
4. M. Cornelissen, H. Thierens and L. De Ridder, Effects of ionizing radiation on cartilage: emphasis on effects on the extracellular matrix, *Scanning Microsc.,* **1996,** 10, 833-840.
5. D. L. Schulte, C. L. Driscoll, T. J. McDonald, G. W. Facer and C. W. Beatty, Irradiated rib cartilage graft for reconstruction of the tympanic membrane: preliminary results, *Am. J. Otol.,* **1998,** 19, 141-144.
6. J. Komender, H. Malczewska, D. Sladowski and A. Komender, Allogenic preserved costal cartilage in reconstructive surgery, *Ann. Transplant.,* **1997**, 2, 49-51.
7. D. E. Schuller, J. Bardach and C. J. Krause, Irradiated homologous costal cartilage for facial contour restoration, *Arch. Otolaryngol.,* **1977**, 103, 12-15.
8. P. J. Donald, Cartilage grafting in facial reconstruction with special consideration of irradiated grafts, *Laryngoscope,* **1986**, 96, 786-807.
9. D.B. Welling, M. D. Maves, D. E. Schuller and J. Bardach, Irradiated homologous cartilage grafts. Long-term results, *Arch. Otolaryngol. Head Neck Surg.,* **1988**, 114, 291-295.
10. G. Lefkovits, Irradiated homologous costal cartilage for augmentation rhinoplasty, *Ann. Plast. Surg.,* **1990**, 25, 317-327.
11. W. P. Adams, R. J. Rorich, J. P. Gunter, C. P. Clark and J. B. Robinson, The rate of warping in irradiated and nonirradiated homograft rib cartilage: a controlled comparison and clinical implications, *Plast. Reconstr. Surg,.* **1999**, 103, 265-270.
12. J. V. Linberg, R. L. Anderson, J. J. Edwards, W. R. Panje and J. Bardach, Preserved irradiated homologous cartilage for orbital reconstruction, *Ophthalmic Surg.,* **1980**, 11, 457-462.
13. J. R. Bevivino, P. N. Nguyen and L. J. Yen, Reconstruction of traumatic orbital floor defects using irradiated cartilage homografts, *Ann. Plast. Surg.,* **1994**, 33, 32-37.
14. D. G. Heatley, R. A. Clary, F. T. Garner and R. P. Lusk, Auricular cartilage versus costal cartilage as a grafting material in experimental laryngotracheal reconstruction, *Laryngoscope,* **1995**, 105, 983-987.
15. L. Kangesu, T. E. Goodacre and P. R. Stanley, Survival of irradiated glutaraldehyde preserved bovine cartilage in nasal reconstruction: a retrospective study, *Br. J. Plast. Surg.,* **1991**, 44, 483-485.
16. G. Murray, Homologous aortic valve segment transplants as surgical treatment for aortic and mitral insufficiency, *Angiology,* **1956**, 7, 446.
17. B. G. Barratt-Boyes, A method for preparing and inserting a homograft aortic valve, *Brit. J. Surg.,* **1965**, 52, 847-56.
18. H. Davies, M. H. Lessof, C. I. Roberts and D. N. Ross, Homograft replacement of the aortic valve, *Lancet,* **1965**, 1, 926.

19. I. A. Meeker and R. E. Gross, Sterilization of frozen arterial grafts by high voltage cathode-ray irradiation, *Surgery,* **1951**, 30, 19.
20. J. D. Malm, F. O. Bowman, P. D. Harris and A. T. W. Kovalik, An evaluation of aortic valve homografts sterilized by electron beam energy, *J. Thorac. Cardiovasc. Surg.*, **1967,** 54, 471-477.
21. J. R. Malm, F. O. Bowen, P. D. Harris, G. A. Kaiser and A. T. Kovalik, Results of aortic valve replacement utilizing irradiated valve homografts, *Ann. N. Y. Acad. Sci.*, **1969**, 147, 740-747.
22. P. M. Beach, F.O. Bowman, G.A. Kaiser and J. R. Malm, Frozen aortic valve homografts. Long-term evaluation, *N.Y. State J. Med.*, **1973**, 73, 651-655.
23. P. M. Beach, F. O. Bowman, G. A. Kaiser, E. Parodi and J. R. Malm, Aortic valve replacement with frozen irradiated homografts. Long term evaluation, *Circulation*, **1972**, 45/46, I29-I35.
24. G. Cantore, B. Guidetti and R. Delfini, Neurosurgical use of human dura mater sterilized by gamma rays and stored in alcohol: long-term results, *J. Neurosurg.*, **1987**, 66, 93-95.
25. CDC, Creutzfeld-Jacob disease associated with cadaveric dura mater grafts – Japan, January 1979 – May 1996, *MMWR,* **1997**, 46, 1066-1069.
26. R. Klen and J. Pacal, Influence of ionizing sterilization on the permeability of human chorio-amniotic, dermo-epidermal and facial grafts, *Res. Exp. Med (Berl).*, **1976**, 167, 15-21.
27. M. M. Bari and R. Begum, Use of radiation-sterilized amniotic membrane grafts as temporary biologic dressings for treatment of leprotic ulcers, In: *Advances in Tissue Banking, Volume 3,* G. O. Phillips, J. N. Kearney, D. M. Strong, R. Von Versen and A. Nather (eds.), World Scientific, Singapore, **1999**, pp. 477-483.
28. S. C. Bourroul, M. R. Herson, E. Pino and M. B. Matho, Sterilization of skin allografts by ionizing radiation, *Cell Mol. Biol. (Noisy-le-grand),* **2002**, 48, 803-807.
29. J. W. Hopewell, The skin: its structure and response to ionizing radiation, *Int. J. Radiat. Biol.*, **1990**, 57, 751-773.
30. D. M. Colvard, R. R. Waller, R. J. Campbell and R. Friedt, Sterilization of sclera homografts with ionizing irradiation, *Am. J. Ophthalmol.*, **1979**, 87, 494-496.
31. L. H. Yahia, G. Drouin and D. Zukor, The irradiation effect on the initial mechanical properties of meniscal grafts, *Biomed. Mater. Eng.*, **1993,** 3, 211-221.
32. G. C. Bessette and R. E. Hunter, The anterior cruciate ligament, *Orthopaedics,* **1990**, 13, 551-562.
33. F. R. Noyes and S. D. Barber-Westin, Reconstruction of the anterior cruciate ligament with human allograft. Comparison of early and later results, *J. Bone Joint Surg. Am.*, **1990**, 78, 524-537.
34. R. J. Simonds, S. D. Holmberg, R. L. Hurwitz, T. R. Colemand, S. Bottenfield, L. J. Conley and S. H. Kohlenberg, Transmission of human immunodeficiency virus type 1 from a seronegative organ and tissue donor, *N. Eng. J. Med.*, **1992,** 12, 726-732.
35. C. M. Li, Y. R. Ho and Y. C. Liu, Transmission of human immunodeficiency virus through bone transplantation: a case report, *J. Formos. Med. Assoc.*, **2001**, 100, 350-351.

36. E. U. Conrad, D. R. Gretch, K. R. Obermeyer, M. S. Moogk, M. Sayers, J. L. Wilson and D. M. Strong, Transmission of the hepatitis C virus by tissue transplantation, *J. Bone Joint Surg.*, **1995**, 77, 214-224.
37. D. W. Jackson, G. E. Windler and T. M. Simon, The intra-articular reactions associated with the use of freeze-dried, ethylene oxide-sterilized bone-patella tendon-bone allografts in the reconstruction of the anterior cruciate ligament, *Am. J. Sports Med.*, **1990**, 18, 1-11.
38. F. H. Fu, C. H. Bennett, C. Lattermann and C. B. Ma, Current trends in anterior cruciate ligament reconstruction. Part 1: Biology and Biomechanics of reconstruction, *Am. J. Sports Med.*, **1999**, 27, 821-830.
39. A. J. Bailey, Effect of ionizing radiation on connective tissue components, *Int. Rev. Connect. Tissue Res.*, **1968**, 4, 233-281.
40. D. T. Cheung, N. Perelman, D. Tong and M. E. Nimni, The effect of gamma-irradiation on collagen molecules, isolated alpha-chains, and cross linked native fibers, *J. Biomed. Mater. Res.*, **1990**, 24, 581-589.
41. A. J. Bailey and W. J. Tromans, Effects of ionizing radiation on the ultrastructure of collagen fibrils, *Radiat. Res.*, **1964**, 23, 145-155.
42. R. A. Grant, R. W. Cox and C. M. Kent, The effects of gamma irradiation on the structure and reactivity of native and cross-linked collagen fibrils, *J. Anat.*, **1973**, 115, 29-43.
43. A. J. Bailey, D. N. Rhodes and C. W. Cater, Irradiation-induced cross-linking of collagen, *Radiat. Res.*, **1964**, 22, 606-621.
44. J. H. Cassel, Effects of gamma radiation on collagen, *J. Am. Leather Chemist Assoc.*, **1959**, 54, 432-449.
45. R. C. Haut and A. C. Powlison, Order of irradiation and lyophilization on the strength of patellar tendon grafts, In: *Trans. 35th Meeting of the Orthopaedic Research Society*, **1989**, p. 514.
46. S. M. Belkoff and R. C. Haut, Microstructurally based model analysis of gamma-irradiated tendon allografts, *J. Orthop., Res.*, **1992**, 10, 461-464.
47. P. De Deyne and R. C. Haut, Some effects of gamma irradiation on patellar tendon allografts, *Connect. Tissue Res.*, **1991**, 27, 51-62.
48. R. C. Haut and A. C. Powlison, The effects of test environment and cyclic stretching on failure properties of human patellar tendons, *J. Orth.. Res.*, **1990**, 8, 532-540.
49. S. Rezaiamiri, C. Davis, A. F. Tencer, E. U. Conrad and D. M. Strong, The effects of cryopreservation and irradiation on human patellar tendon allografts, In: *Advances in Tissue Banking, Volume 2*, G. O. Phillips, D. M. Strong, R. Von Versen and A. Nather, (eds.), World Scientific, Singapore, **1998**, pp. 191-206.
50. M. J. Gibbons, D. L. Butler, E. S. Grood, D. I. Bylski-Austrow, M. S. Levy and F. R. Noyes, Effects of gamma irradiation on the initial mechanical and material properties of goat bone-patellar tendon-bone allografts, *J. Orthop. Res.*, **1991**, 9, 209-218.
51. D. L. Butler, F. R. Noyes, K. A. Walz and M. J. Gibbons, Biomechanics of human knee ligament allograft treatment, *Trans. Orthop. Res. Soc.*, **1987**, 12, 128.
52. L. E. Paulos, E. P. France, T. D. Rosenberg, D. J. Drez, P. J. Abbott, C. B. Straight, D. J. Hammon and R. R. Oden, Comparative material properties of allograft tissues for ligament replacement: effects of type, age, sterilization and preservation, *Trans. Orthop. Res. Soc.*, **1987**, 12, 129.

53. T. J. Rasmussen, S. M. Feder, D. L. Butler and F. R. Noyes, The effects of 4 Mrad of gamma irradiation on the initial mechanical properties of bone-patellar, *Arthroscopy,* **1994,** 10, 188-197.
54. G. Rauch, M. Gerbersdorf, P. Dorner, M. Lengsfeld and P. Griss, Biomechanic study of the tensile strength of lyophilized and deep frozen human Achilles tendons following gamma and ethylene oxide sterilization, *Z. Orthop. Ihre Grenzgeb.,* **1991,** 129, 393-399.
55. Y. Toritsuka, K. Shino, S. Horibe, N. Kakamura, N. Matsumoto and T. Ochi, Effect of freeze-drying or gamma-radiation on remodeling a tendon allograft in a rat model, *J. Orthop. Res.,* **1997,** 15, 294-300.
56. D. Bettin, V. Rullkotter, J. Polster and S. Fuchs, Primary biomechanical influence of different sterilization methods on a freeze-dried bone-ligament transplant, *Arch. Orthop. Trauma Surg.,* **1999,** 119, 236-240.
57. C. W. Smith, I. S. Young and J. N. Kearney, Mechanical properties of tendons: changes with sterilization and preservation, *J. Biomech. Engin.,* **1996,** 118, 56-61.
58. B. M. Fideler, C. T. Vangsness, B. Lu, C. Orlando and T. Moore, Gamma irradiation: effects on biomechanical properties of human bone-patellar tendon-bone allografts, *Am. J. Sports Med.,* **1995,** 23, 643-646.
59. A. Salehpour, D. L. Butler, F. S. Proch, H. E. Schwartz, S. M. Feder, C. M. Doxey and A. Ratcliff, Dose-dependent response of gamma irradiation on mechanical properties and related biochemical composition of goat bone-patellar tendon-bone allografts, *J. Orthop. Res.,* **1995,** 13, 898-906.
60. A. Salehpour, D. L. Butler and F. S. Proch, Dose-dependent response of gamma irradiation on mechanical properties of goat bone-patellar tendon-bone allografts, *Trans. Orthop Res. Soc.,* **1994,** 19, 557.
61. M. J. Goertzen, H. Clahsen, K. F. Burrig and K. P. Schulitz, Sterilization of canine anterior cruciate allografts by gamma irradiation in argon. Mechanical and neurohistological properties retained one year after transplantation, *J. Bone Joint Surg. Br.,* **1995,** 77, 205-212.
62. A. Maeda, M. Inoue, K. Shino, K. Nakata, H. Nakamura, M. Tanaka, Y. Seguchi and K. Ono, Effects of solvent preservation with or without gamma irradiation on the material properties of canine tendon allografts, *J. Orthop. Res.,* **1993,** 11, 181-189.
63. S. L. Stramer, S. Caglioti and D. M. Strong, NAT of the United States and Canadian Blood Supply, *Transfusion,* **2000,** 50, 1165-1168.

THE EFFECT OF PRESERVATION PROCEDURES AND RADIATION STERILISATION CONDITIONS ON CONNECTIVE TISSUE GRAFTS AND THEIR CONSTITUENTS

Anna Dziedzic-Goclawska [1], Artur Kaminski [1], Izabela Uhrynowska-Tyszkiewicz [1], Jacek Michalik [2] and Waclaw Stachowicz [2]

[1] *Department of Transplantology and the Central Tissue Bank, Centre of Biostructure Research, The Medical University of Warsaw, ul. Chalubinskiego 5, 02-004 Warsaw, Poland {E-mail: agoclaw@ib.amwaw.edu.pl}*

[2] *Institute of Nuclear Chemistry & Technology, ul. Dorodna 16, 02-195 Warsaw, Poland*

ABSTRACT

The Central Tissue Bank, established in 1963, and two other multi-tissue banks operating in Poland provide connective tissue allografts such as bone, cartilage, tendons, sclera, skin, acellular dermis and amnion. All grafts are radiation sterilised with a dose of 35 kGy in a ^{60}Co source and/or with electron beam 10 MeV accelerator. Over 250,000 radiation-sterilised tissue grafts have been prepared and used in hospital departments throughout Poland and not one case of infectious disease transmission has been reported to date.

High doses of ionising radiation can evoke numerous chemical and physical changes that may affect biological quality of tissue allografts, such as osteoinductive potential of bone, the mechanical properties of bone and other connective tissue grafts as well as the rate of their resorption *in vivo*. The origin and stability of free radicals and other paramagnetic entities radiation-induced in bone will be discussed. The effect of various preservation procedures (e.g. lyophilisation, deep-freezing) and radiation sterilisation conditions (doses, temperature of irradiation) on osteoinductive potential of bone as well as on degradation of collagen, a major constituent of connective tissue grafts, will be presented.

The results of interdisciplinary research performed at the Central Tissue Bank in Warsaw, in collaboration with radiation chemists from the Institute of Nuclear Chemistry and Technology indicate, that radiation-induced changes can be diminished by modification of tissue preservation methods and that, to some extent it is possible to reduce undesired radiation-induced damage to the tissue grafts.

KEYWORDS

Tissue banking; tissue allografts; radiation sterilisation; bone allografts; free radicals

40-YEARS OF RADIATION STERILISATION & TISSUE BANKING IN POLAND

There are currently eight tissue banks active in Poland: three multi-tissue banks and five mono-tissue banks. Three mono-tissue banks prepare heart valves (in Warsaw, Krakow and Zabrze) and two of them provide corneas (in Warsaw and Lublin).

Multi-tissue banks

In 2003, we celebrated the 40th anniversary of radiation sterilisation and tissue banking in Poland. The Central Tissue Bank was created at the Medical University of Warsaw in 1963 and since then ionising radiation has been used routinely for sterilisation of connective tissue grafts. Under the supervision of the Central Tissue Bank, two multi-tissue banks were established at Blood Transfusion Centres in Katowice (1967) and Kielce/Morawica (1979). The scheme of their organisation is shown in Figure 1. Considering the extremely high resistance of prions and the high resistance of some viruses to many sterilising agents, including ionising radiation, careful donor screening and selection are performed to reduce the potential risk of infectious disease transmission of donor-origin. Monitoring adverse reactions and reverse information on early and long-term results after grafting (i.e. tissue graft traceability from procurement until transplantation) are carried out.

The multi-tissue banks provide allografts such as bone, rib cartilage, tendons, ligaments, sclera, skin, acellular dermis and amnion. Tissue procurement performed in morgues (dissecting rooms) is carried out under non-aseptic conditions (using sterile instruments and devices), because tissue grafts are subsequently radiation sterilised.

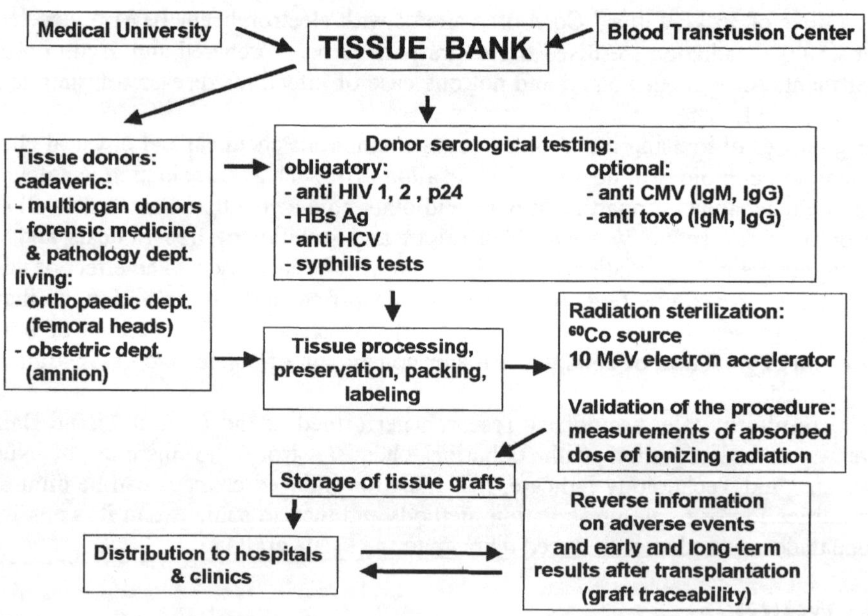

Figure 1. Scheme of organisation of multi-tissue banks in Poland

The main preservation procedures used in multi-tissue banks are shown in Figure 2. All tissue grafts are radiation sterilised with a dose of 35 kGy with gamma rays in a ^{60}Co source at the Institute of Applied Radiation Chemistry in Lodz and/or with electron beam 10 MeV accelerator at the Institute of Nuclear Chemistry & Technology in Warsaw. The history of application of ionising radiation to sterilise tissue grafts is shown in Figure 3. Between 1963 and 1966, bone allografts were sterilised in channels of a shutdown nuclear reactor [1].

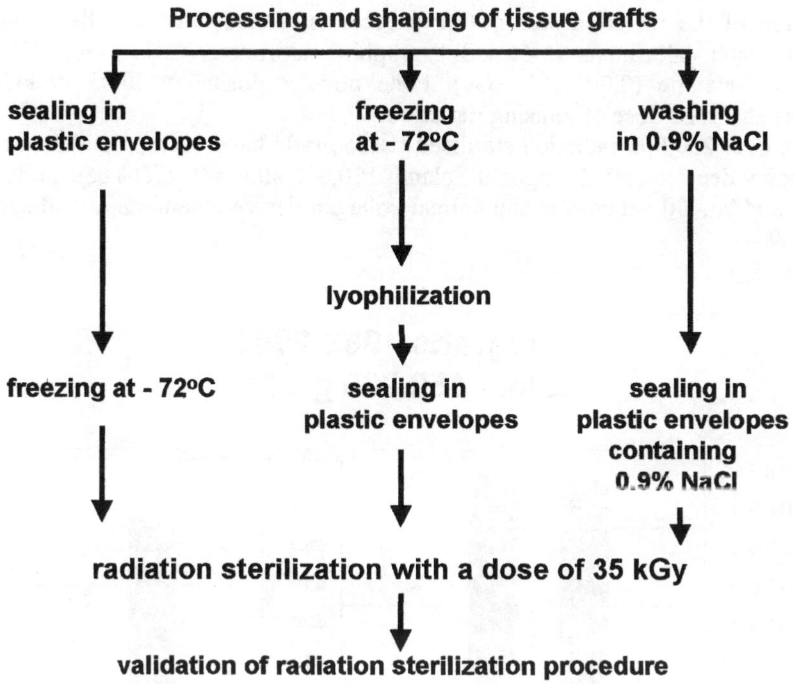

Figure 2. Main procedures used in multi-tissue banks for preservation of various types of connective tissue grafts.

1963 - 1966	**in channels of switched off nuclear reactor in cadmium screened aluminium containers in order to absorb remaining neutrons as present in a nuclear fuel; in the same containers ionization chambers were placed to measure the absorbed dose of ionizing radiation (Institute of Nuclear Research at Swierk/Warsaw)**
1967 - till now	**with gamma rays in an irradiation chamber loaded with 20,000 Ci of ^{60}Co (Institute of Applied Radiation Chemistry of the Polytechnic School in Lodz)**
1973 - till now	**linear electron accelerator LAE-13/9 with energy of 10 MeV and beam power of 6 kW (Institute of Nuclear Chemistry and Technology in Warsaw)**
1963 - 1997	**sterilization dose: 33 kGy**
1997 - till now	**sterilization dose: 35 kGy**

Figure 3. History of radiation sterilisation of tissue grafts in Poland.

Validation of the radiation sterilisation process is carried out using the following dosimeters: water calorimeter (5-50 kGy), graphite calorimeter (2-15 kGy), PVC foil (5-50 kGy), l-alanine (0.001-100 kGy), bone powder dosimeter (0.05-40 kGy) to measure the absorbed dose of ionising radiation.

To date, over 250,000 radiation sterilised tissue grafts have been prepared and used in 250 hospital departments throughout Poland: 150,000 allografts (75% constitute bone allografts) and 96,000 xenografts and animal collagen-derived membranes and sponges (Figures 4 & 5).

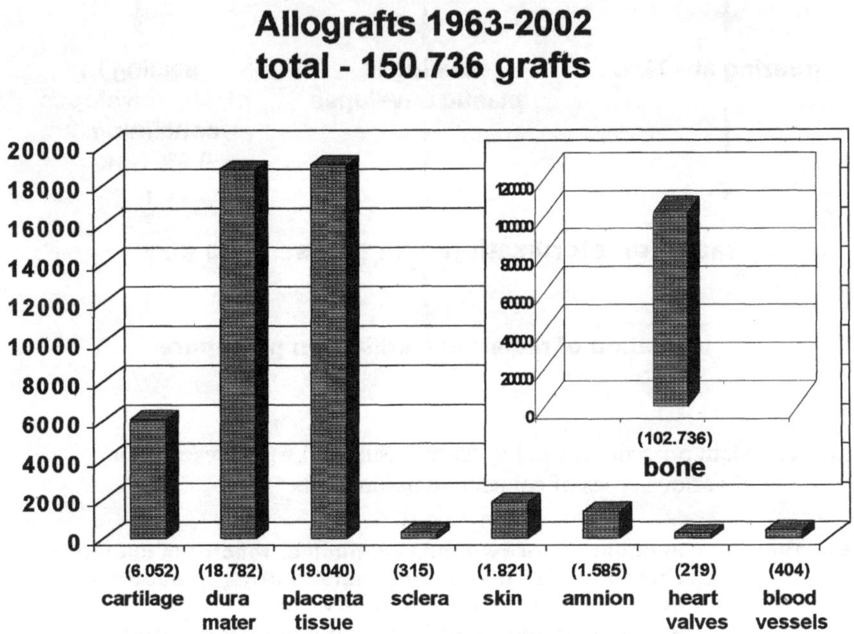

Figure 4. Tissue allografts prepared by 3 multi-tissue banks and used in hospitals.

Due to possible contamination of bovine tissues with prions that are very resistant to any type of sterilisation, including radiation sterilisation, and porcine tissues that may be infected with PERV viruses, we ceased preparation of xenografts to avoid transmission of zoonotic diseases to the human population and stopped preparing human dura mater grafts that might be prion-infected.

It should be stressed that no infectious disease transmission or other adverse reactions post-transplantation of tissue grafts sterilised with a dose of 35 kGy have been reported to date.

Mono tissue banks

Some cells of heart valve (e.g. endothelium) or cornea (epithelium, endothelium) should be living, and thus these allografts cannot be radiation sterilised. These tissues are collected and prepared under aseptic conditions and, if necessary, they are decontaminated in antibiotic solutions.

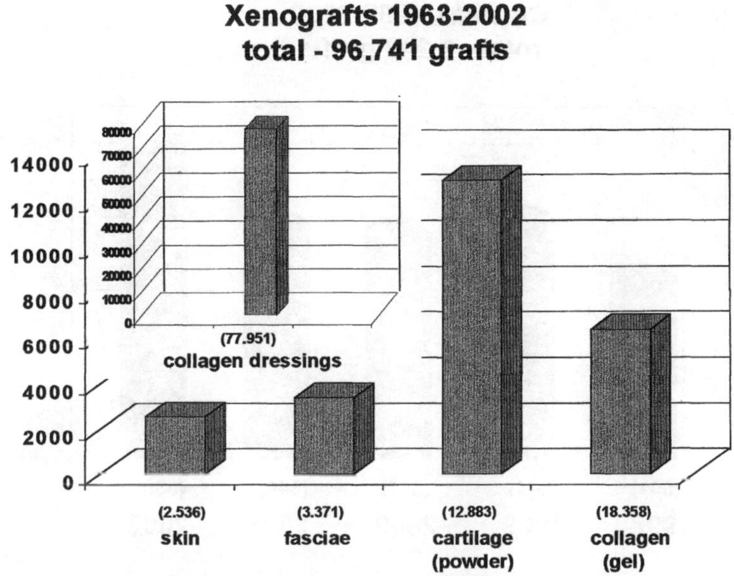

Figure 5. Tissue xenografts and animal-collagen derived materials prepared by three multi tissue banks and used in hospitals.

Figures. 6 & 7 illustrate the number of heart valves and corneas prepared by mono-tissue banks and used in hospitals in Poland over the last five years. It should be emphasised, however, that the number of procured heart valves and corneas were 2-3 times higher, due to the fact that many of them were disqualified, because contrary to non-viable, preserved other connective tissue grafts, the critical time of their preservation is short.

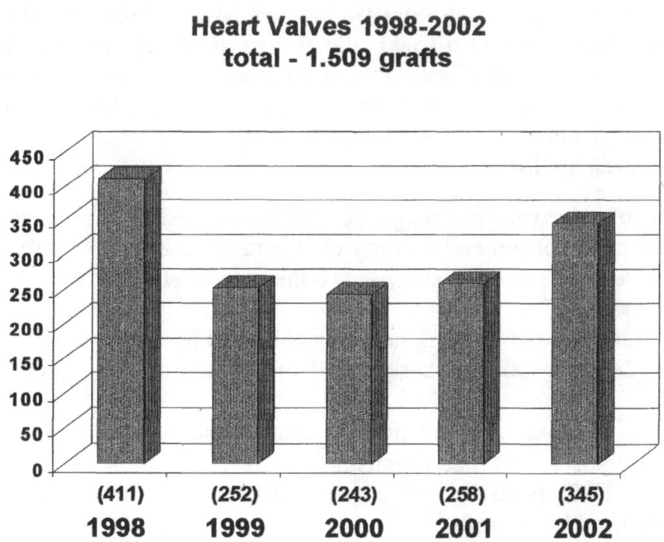

Figure 6. Number of heart valves prepared by 3 tissue banks and used in hospitals.

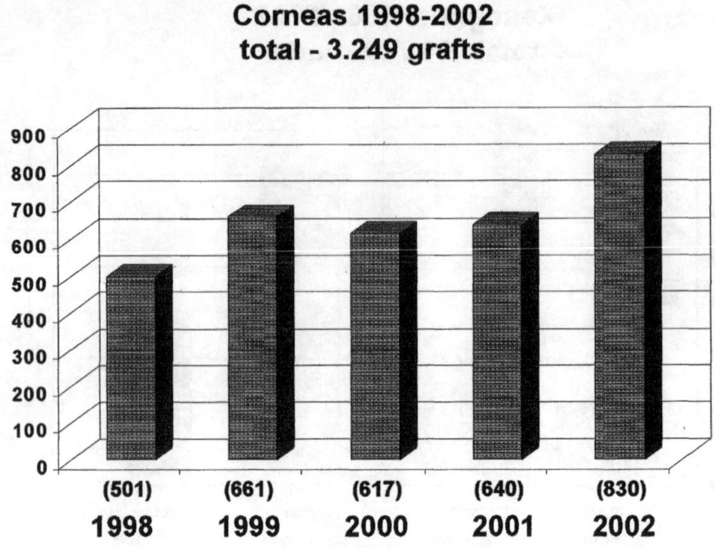

Figure 7. Number of corneas prepared by 2 tissue banks and used in hospitals.

INTERDISCIPLINARY RESEARCH CARRIED OUT AT THE CENTRAL TISSUE BANK IN WARSAW

The Central Tissue Bank at the Medical University of Warsaw is the leading institution for tissue banking in Poland. Its main duties are not only the preparation of connective tissue grafts, but also to carry out interdisciplinary research in this field, in co-operation with microbiologists, virologists, biochemists. Many of which were initiated by Professor Kazimierz Ostrowski, the former head of the Department of Histology and Embryology and Professor Janusz Komender, the former head of our Tissue Bank. Of particular importance is the collaboration with radiation chemists from the Institute of Nuclear Chemistry and Technology in Warsaw.

Main topics of research are listed below:

(1) Application of electron paramagnetic resonance (EPR) spectrometry for identification of the origin and stability of free radicals and other paramagnetic entities induced in radiation-sterilised bone and their use in research on mineralised tissues [2-9].

(2) Dosimetry of the dose of ionising radiation absorbed by living organisms in the case of accidental exposure, based on EPR spectrometry - skeleton as a dosimeter [10].

(3) Control of radiation-sterilisation process using bone powder as a dosimeter, based on quantitative EPR measurements [11].

(4) Application of EPR spectrometry for quantitative evaluation of remodelling of radiation-sterilised bone grafts [12,13].

(5) The effect of preservation procedures (lyophilisation and deep-freezing) with subsequent radiation sterilisation on immunogenicity of preserved tissue grafts.

(6) The effect of various preservation procedures (e.g. lyophilisation, deep-freezing) and radiation sterilisation conditions (doses, temperature) on osteoinductive potential of bone grafts [14-17].
(7) The effect of preservation procedures and radiation sterilisation conditions on *in vitro* solubility of bone collagen and on rate of bone resorption *in vivo* [16,18].
(8) The effect of preservation procedures and radiation sterilisation conditions on the mechanical properties of bone [19].
(9) The effect of various preservation procedures and radiation sterilisation conditions on degradation of collagen – a major constituent of connective tissue grafts [16,20].
(10) In *vitro* cytotoxicity testing of various polymers used for packing of radiation sterilised tissue grafts.
(11) Differences in cytotoxicity of medullary lipids of bone irradiated with a dose of 35 kGy with gamma rays and electron beam 10 MeV accelerator.
(12) Development of methods for heart valve preservation, which diminish their mineralisation *in vivo*.
(13) Long-term evaluation of the clinical results after transplantation of radiation-sterilised tissue allografts (bone, cartilage) [21-25].
(14) Evaluation of the interaction of artificial materials (e.g. ceramics and carbon composites) with tissues, *in vitro* biocompatibility testing based on cell culture.
(15) *In vitro* culture of human keratinocytes, chondrocytes, osteoblasts and stem cells.
(16) Preparation of xenogenic and allogenic collagen-derived scaffolds for tissue engineering [20].

The results of some of these experimental works will be presented below.

Application of electron paramagnetic resonance (EPR) spectrometry for identification of the origin and stability of free radicals and other paramagnetic entities induced in radiation-sterilised bone

The exposure of mineralised tissues of bone and teeth to ionising radiation results in the induction of free radicals and other paramagnetic entities that can be detected by electron paramagnetic resonance (EPR, ESR) spectrometry [26-29].

Since we started radiation sterilisation of tissue grafts in 1963 (at the beginning with gamma rays of the shutdown nuclear rector core, Figure 3), we became interested in studying the origin and stability of free radicals and other paramagnetic entities induced in radiation sterilised bone tissue [2]. We were particularly interested if whether radiation-induced paramagnetic entities are neutral or active in bone grafts transplanted into living organism. In degassed bone irradiated *in vacuo* at room temperature a complex EPR doublet was observed which decayed after admission of air oxygen and then a stable asymmetric EPR singlet can be seen (Figure 8).

Comparative EPR measurements performed on bone, decalcified bone collagen deproteinised bone mineral (Figure 9) showed that two distinct paramagnetic entities of different stability are observed in the EPR spectrum of irradiated bone. The EPR doublet present in bone samples irradiated *in vacuo* and decayed after admission of air oxygen is connected with collagen radicals. These radicals disappear completely within 10-20 days in the presence of atmospheric oxygen, and thus the storage of bone grafts for 3-4 weeks following irradiation completely eliminates organic radicals.

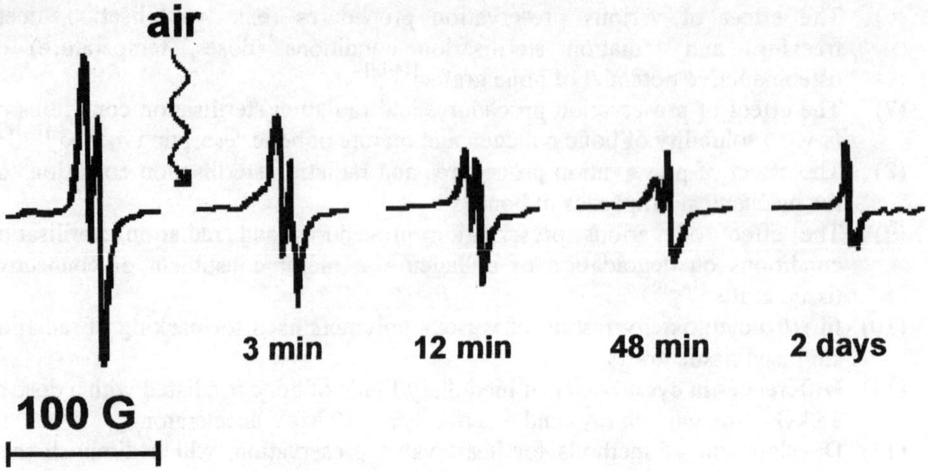

Figure 8. EPR spectra of human compact bone irradiated *in vacuo* and changes occurring upon contact with air oxygen.

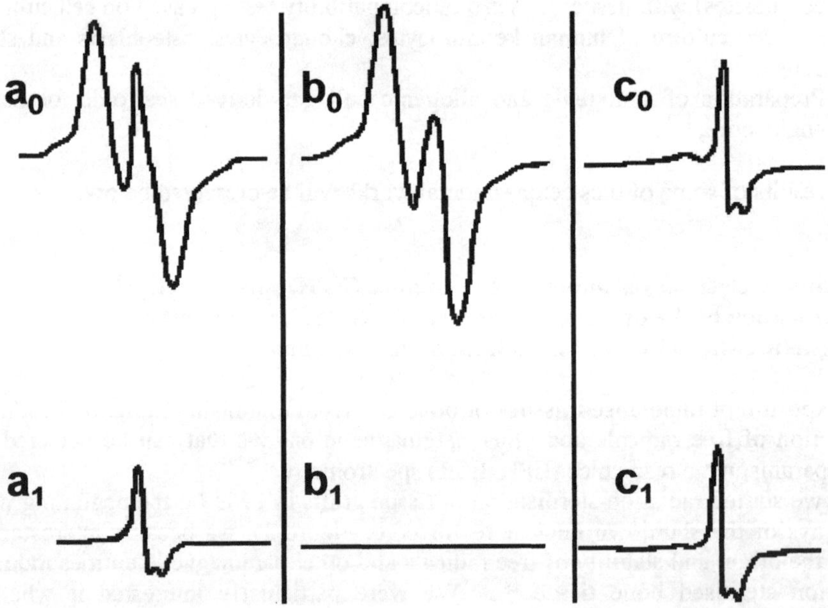

Figure 9. EPR spectra of compact bone (a_o), decalcified bone collagen (b_o) and mineral of deproteinised bone (c_o) recorded in samples irradiated at room temperature *in vacuo* and two weeks after admission of air oxygen (lower row: a_1, b_1, c_1).

The asymmetric EPR singlet is derived from paramagnetic centres radiation-induced in bone mineral (Figures 8 and 9) [2]. It has been calculated that the lifespan of these centres is extremely high (9×10^7 years at 15°C and 1.9×10^5 years at 37°C) [30]. The concentration of these centres was also determined to be proportional to the EPR signal intensity in a dose dependent relationship (Figures 10-12).

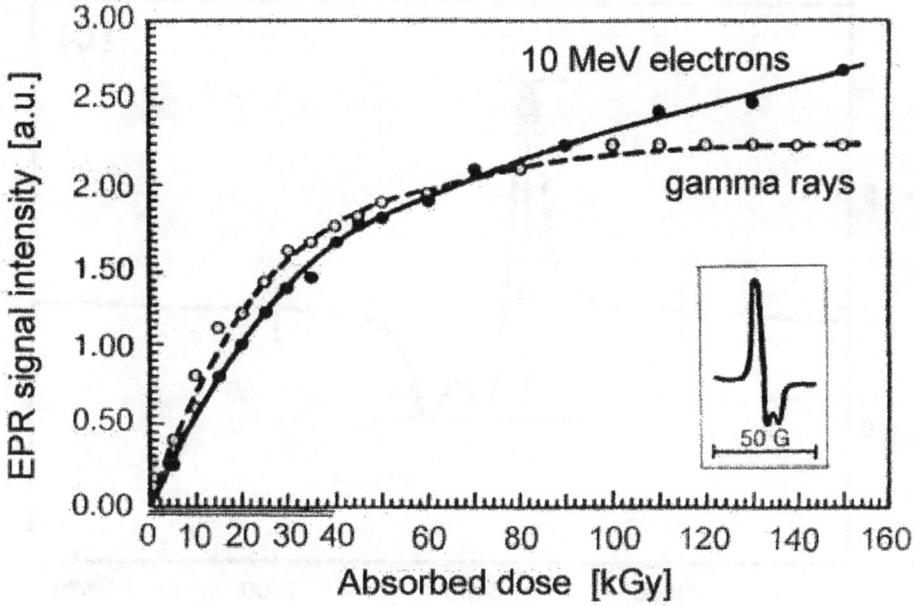

Figure 10. Relationship between the dose of ionising radiation and the intensities of the EPR singlet recorded with bone dosimeter.

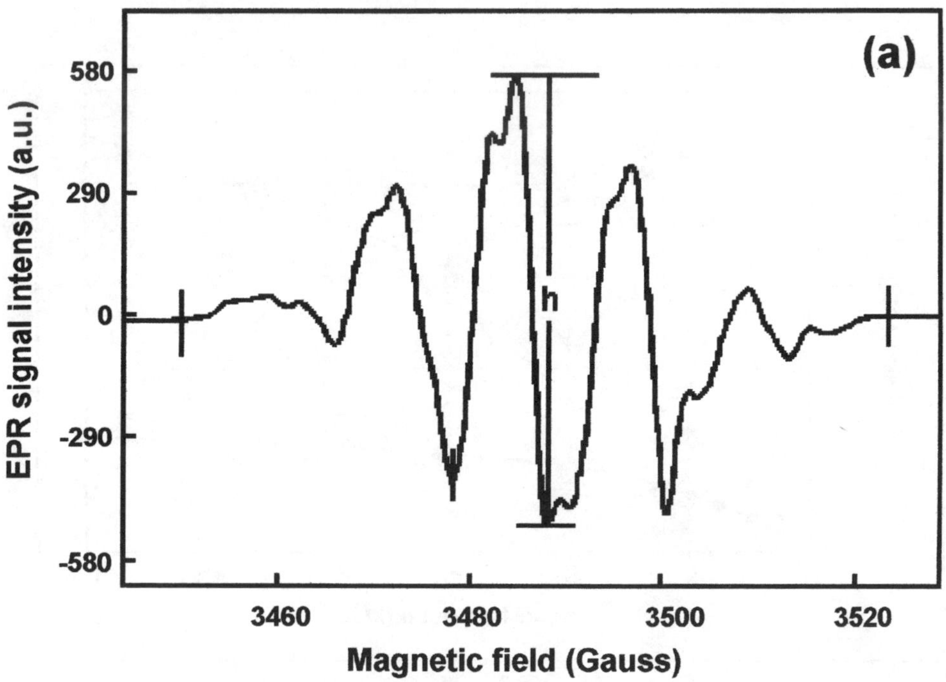

Figure 11 (a). EPR signals recorded with l-alanine dosimeter. (dose estimated from proportion of the peaks heights (h)).

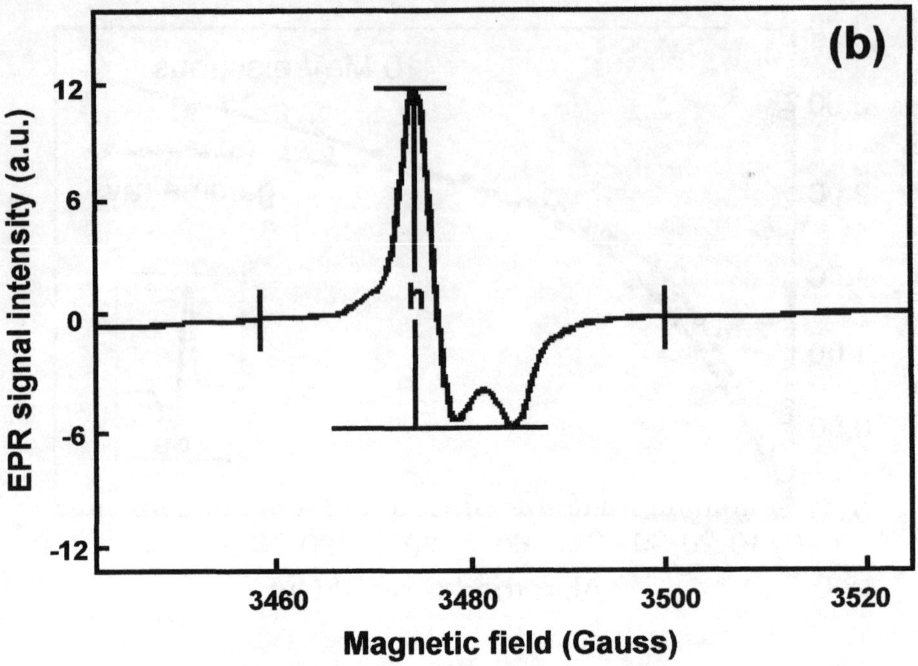

Figure 11 (b). EPR signals recorded bone dosimeter. (dose estimated from proportion of the peaks heights (h)).

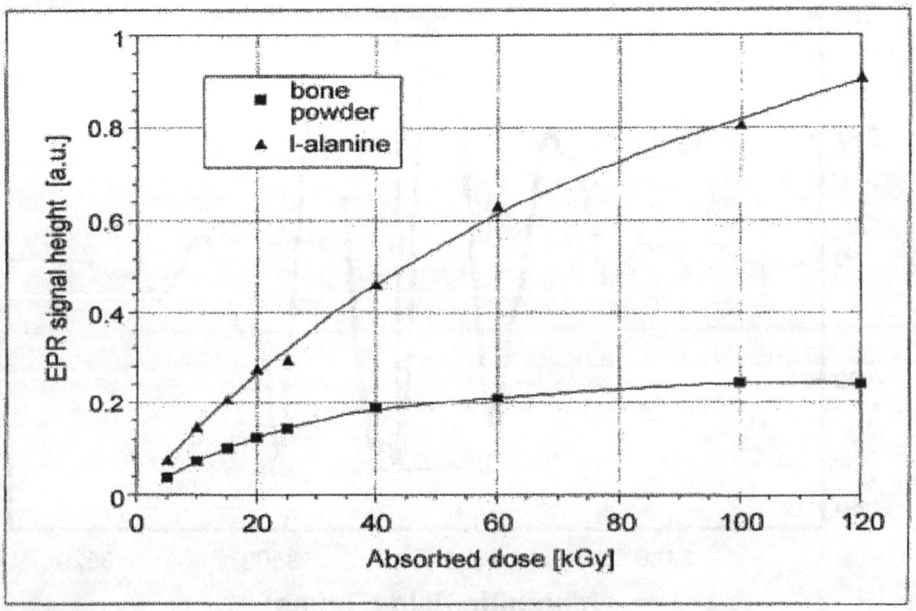

Figure 12. Relationship between the heights of the EPR signals (recorded with l-alanine and bone powder) and the dose of gamma radiation.

These observations were the basis for further studies in which radiation-induced paramagnetic centres were used as a kind of marker in research on mineralised tissues for evaluation of the crystallinity of bone mineral [31-35], remodelling process of radiation-sterilised bone grafts [12-14], estimation of the dose of ionising radiation absorbed by the living organisms in the case of accidental exposure [6,10] as well as for evaluation the absorbed dose in the course of radiation sterilisation of tissue grafts [11].

Bone powder as well as l-alanine is used as dosimeters to measure the absorbed dose and to validate the radiation sterilisation procedure of tissue grafts (Figures 11 and12). The advantage of l-alanine is the broad range of doses to be measured, from about 0.001 kGy up to 100 kGy. The amplitude of the central line (h) is proportional to the absorbed dose of gamma or electron beam irradiation (Figures 11 (a) and 12).

The advantage of bone powder dosimeter lies in the fact that its composition resembles that of bone grafts most frequently used for transplantation. The amplitude (h) of EPR signal is shown in Figure 11 (b). This dosimeter can be used in the range of 0.05 kGy to 40 kGy, but the curvature in the dose-dependence relationship around 25 kGy makes the dose estimation at higher doses less accurate (Figure 12).

Effect of various preservation procedures (lyophilisation, deep-freezing) & radiation sterilisation conditions (doses, temperature) on osteoinductive properties of bone grafts

The value of bone allografts depends greatly on their ability to induce new bone formation at the site of transplantation. Since the osteoinductive capacity of bone, caused by bone morphogenetic proteins (BMPs) present in the organic bone matrix, is of great clinical importance, an attempt should be undertaken to protect the osteoinductive properties of bone in the course of bone allograft processing, preservation and sterilisation.

The classic model of bone induction in heterotopic places (mainly in muscles) after transplantation of nonviable, decalcified bone matrix described by Urist in 1965 [36] is very useful in tissue banking practice to evaluate the effect of various processing, preservation and sterilisation procedures on osteoinductive potential of bone grafts. Controversial results concerning the effect of radiation sterilisation on osteoinductive potential of bone allografts have been published [14-17,37-40] as well as on osteoinductive potential of BMPs alone or combined with collagen carrier [41]. This is probably due to the fact bone samples were irradiated with different doses at various temperatures (ambient or low), in dry or wet states. These factors dramatically influence radiation-induced collagen damage [42] – a major constituent of bone matrix and BMP carrier.

In our Tissue Bank, a model of heterotopically induced osteogenesis has been used to study the effect of various preservation procedures (fresh, deep-frozen, lyophilised bone samples) and radiation sterilisation conditions (doses, irradiation at room temperature and at -72°C) on osteoinductive potential of allogenic rat bone matrices. It has been found that allogenic deep-frozen bone matrices irradiated with doses of 35 kGy and 50 kGy at -72°C induced *de novo* bone formation in an amount comparable with that of non-irradiated controls, while matrices preserved by lyophilisation and irradiated at room temperature with the same doses were completely resorbed and did not induced osteogenesis [14-17] (Figures 13 and 14). It thus seems that radiation-induced damage of bone allografts depends on two factors: (i) conditions of irradiation (dose, temperature), and (ii) physical state of samples, particularly the presence or absence of water. Radiation-induced damage of frozen large-molecule biological specimen may be as much as tenfold lower, if related to the damage rate at room temperature [43].

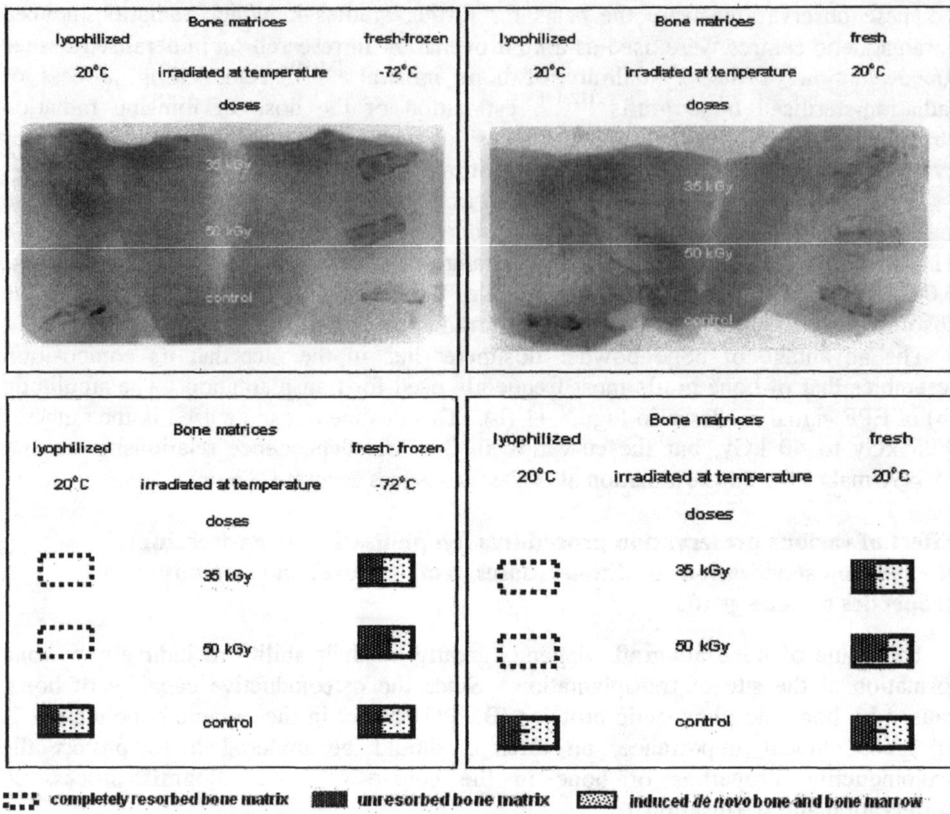

Figure 13. Upper row: X-ray examination of rat abdominal wall muscles five weeks after transplantation of lyophilised and frozen allogenic bone matrices irradiated with 35 kGy and 50 kGy at 20°C and at –72°C, respectively (left roentgenogram) and after transplantation of lyophilised and fresh matrices irradiated with the same doses, but at 20°C (right roentgenogram). Lower row: scheme illustrating the results of computerised morphometric analysis done on serial histological sections of the above-mentioned irradiated lyophilised, frozen and fresh matrices and on non-irradiated controls.

It has also been observed that fresh bone matrices irradiated at room temperature, even with a dose of 50 kGy, induced osteogenesis after transplantation (Figures 13 and 14). This might be due to the fact that the presence of water in fresh bone matrices irradiated at room temperature also strongly influences the nature of chemical reactions involved [42]. To elicit this problem, *in vitro* solubility of rat bone collagen has been studied (*vide infra*).

The effect of preservation procedures and radiation sterilisation conditions on *in vitro* solubility of rat bone collagen

As in above mentioned experiments, lyophilised, frozen and fresh rat bone matrices were irradiated with doses of 35 kGy and 50 kGy at 20°C or –72°C (on dry ice) respectively. Non-irradiated matrices served as controls.

Samples were pulverised in the SPEX freezer mill and extracted with 0.5 M NaCl (pH 7.0) at 4°C for 48 hrs, centrifuged to determine neutral soluble collagen (NSC) in extracts, then the residues were extracted with citric buffer (pH 3.6) at 4°C for 48 hrs and centrifuged to determine acid soluble collagen (ASC) in extracts. The amount of hydroxyproline (Pro-OH) in extracts was measured and calculated as mg Pro-OH per g of dry tissue mass. The total soluble collagen (TSC) was calculated as a sum of NSC and ASC. The total hydroxyproline content was also measured in dry non-irradiated bone matrices.

Figure 14 illustrates *in vitro* collagen solubility of rat bone matrices preserved by different methods and irradiated at various conditions (doses, temperature of irradiation). Dose-dependent, dramatic increase of collagen solubility was observed when lyophilised matrices were irradiated at room temperature with doses of 35 kGy and 50 kGy, respectively. In lyophilised samples irradiated with the same doses, but at -72°C (on dry ice) the increase of collagen solubility was lower. Solubility of collagen was low in frozen samples irradiated at -72°C, and, unexpectedly, also in fresh bone samples irradiated at room temperature.

It should be stressed that lyophilised matrices irradiated at room temperature with doses of 35 kGy or 50 kGy were quickly resorbed and did not induce osteogenesis, while frozen irradiated at -72°C, as well as fresh matrices irradiated at room temperature induced *de novo* bone formation even after irradiation with a dose of 50 kGy (Figure 13).

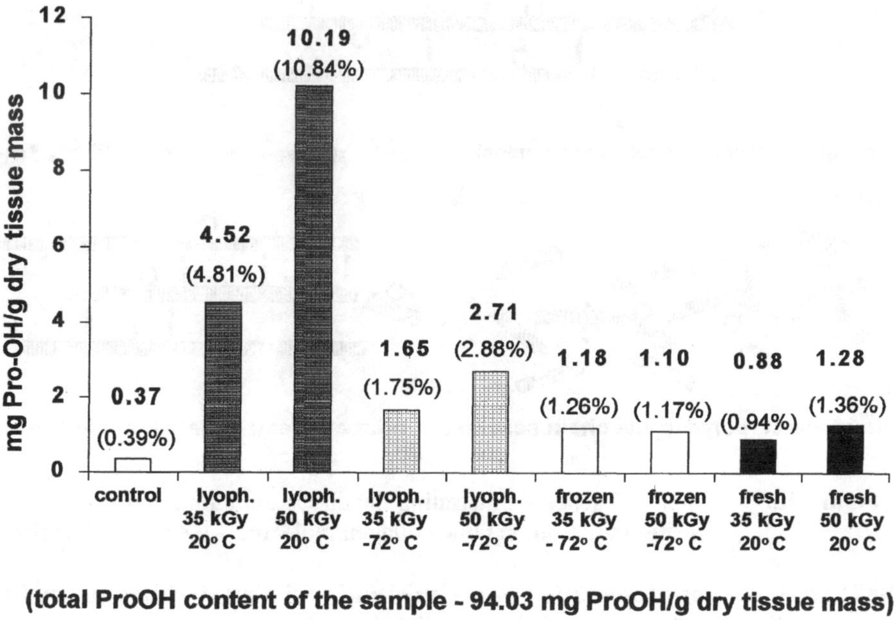

Figure 14. *In vitro* collagen solubility (TSC – total soluble collagen) of rat bone matrices preserved by different methods and irradiated at various conditions.

Numerous studies have been carried out on native tendon collagen irradiated both in the absence and the presence of water [42]. It has been postulated that polypeptide chain scissions predominate when collagen is irradiated at dry state due to the direct effect of ionising radiation, and this, in turn, dramatically increases collagen solubility *in vitro* (Figure 14) and the rate of bone matrix resorption *in vivo* (Figure 13). It has been found, however, that cross-linking reactions appear during irradiation of collagen in the presence of water (indirect effect), probably due to the action of highly reactive, short-lived hydroxyl radicals (*OH) resulting from water radiolysis (Figure 15).

Dramatic differences in the solubility of bone collagen between lyophilised and fresh (water-containing) matrices irradiated at room temperature (Figure 14) as well as osteogenesis observed after transplantation of fresh matrices irradiated with a dose as high as 50 kGy and lack of new bone formation after transplantation of lyophilised matrices irradiated with a dose of 35 kGy at room temperature (Figure 13), indicate that small BMP molecules (MW about 30 kDa) are not affected by irradiation, but degradation of large molecule of bone collagen, a carrier of BMPs, occurs in the samples irradiated at dry state (Figure 15).

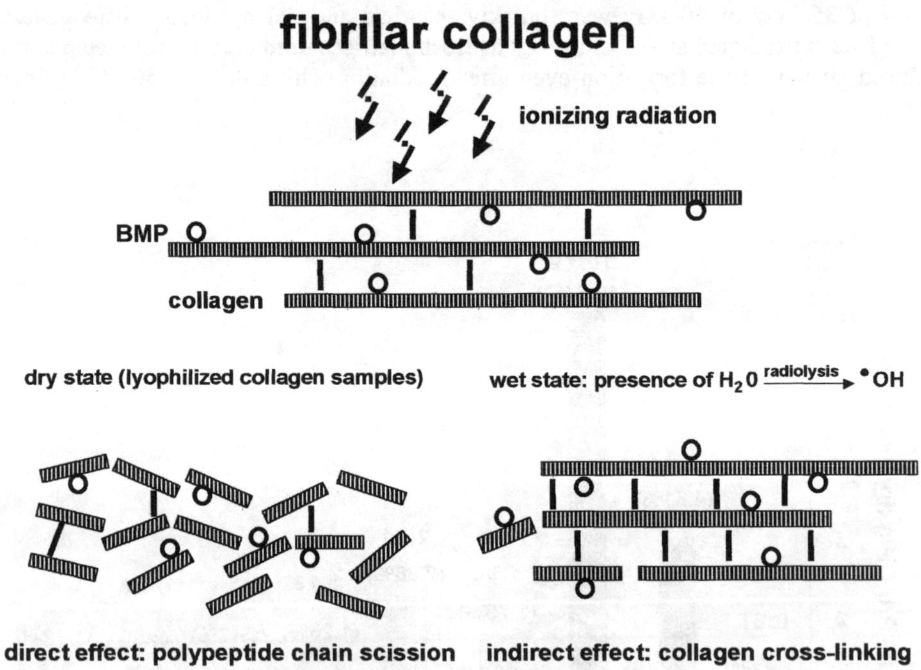

Figure 15. Simplified scheme illustrating the direct and indirect effects of ionising radiation on bone collagen molecule – a carrier of BMPs.

The effect of various preservation procedures and radiation sterilisation conditions on degradation of collagen – a major constituent of connective tissue grafts

Fresh and lyophilised samples of human compact bone, human rib cartilage and calf Achilles tendon were gamma irradiated in a ^{60}Co source at room temperature with doses of 25-100 kGy, then *in vitro* solubility of collagen was measured according to the above described method (Figures 16-18).

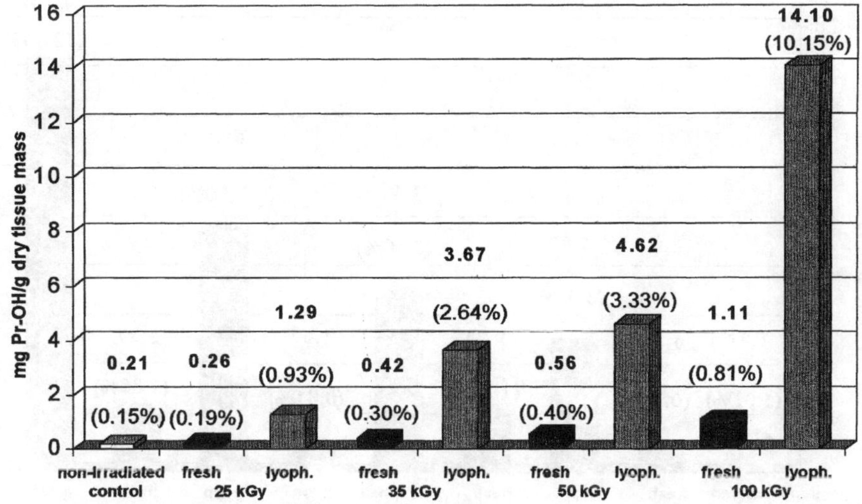

(total ProOH content of the sample - 138.9 mgProOH/g dry tissue mass)

Figure 16. *In vitro* collagen solubility (TSC – total soluble collagen) of fresh and lyophilised human compact bone samples gamma irradiated in a ^{60}Co source at room temperature with doses of 25, 35, 50 and 100 kGy.

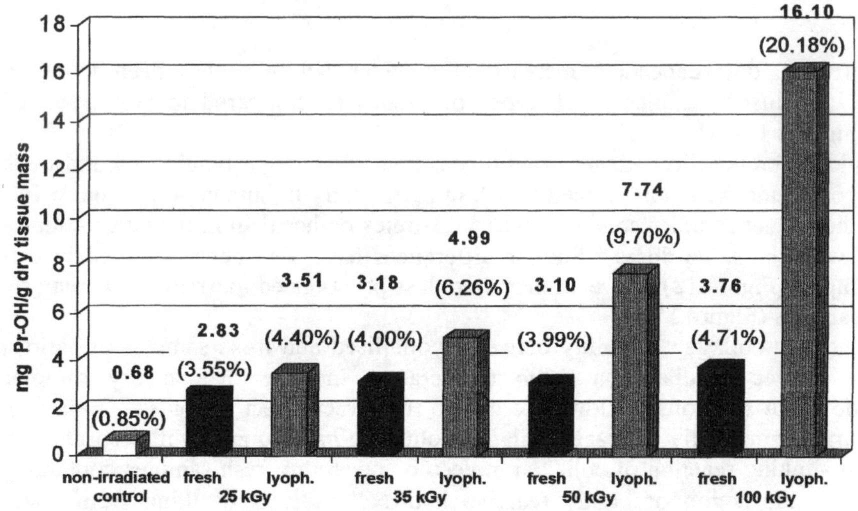

(total ProOH content of the sample - 79.78 mg ProOH/g dry tissue mass)

Figure 17. *In vitro* collagen solubility (TSC – total soluble collagen) of fresh and lyophilised human rib cartilage samples gamma irradiated in a ^{60}Co source at room temperature with doses of 25, 35, 50 and 100 kGy.

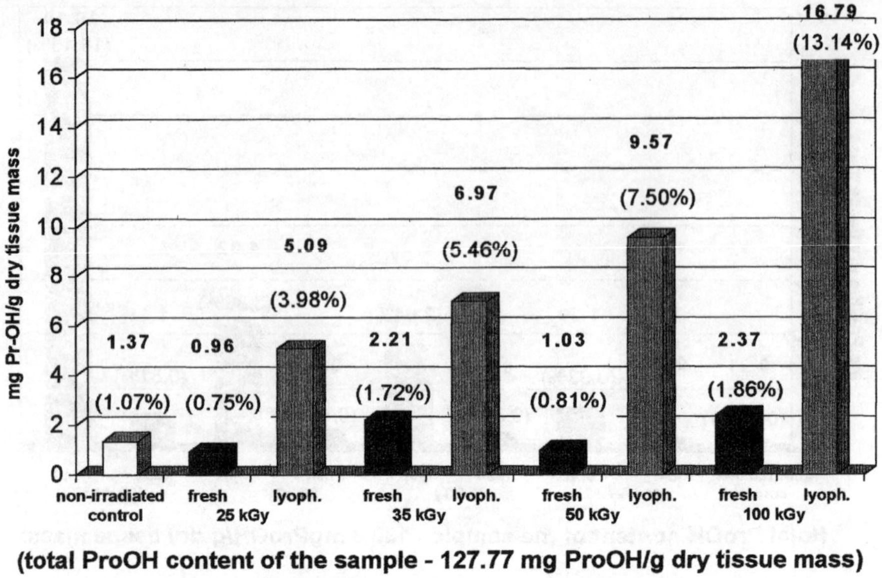

Figure 18. *In vitro* collagen solubility (TSC – total soluble collagen) of fresh and lyophilised calf Achilles tendon samples gamma irradiated in a ^{60}Co source at room temperature with doses of 25, 35, 50 and 100 kGy.

A significant, dose-dependent increase of collagen solubility has been found in irradiated lyophilised samples of all types of tissues as compared to non-irradiated controls (Figures 16-18).

Dramatic differences in collagen solubility between irradiated lyophilised and fresh samples (containing water) have been noted, in particularly in human bone (Figure 16), where in the former samples solubility was 5-13 times higher than in the latter samples irradiated with the same doses. Similar differences have also been observed in calf tendon samples (Figure 18), however they were less pronounced in irradiated human rib cartilage samples (Figure 17).

Differences in collagen solubility between lyophilised and fresh samples of various connective tissues irradiated at room temperature indicate that in dry samples polypeptide chain scissions predominate due to the direct effect of ionising radiation, and this turn, dramatically increases collagen solubility *in vitro* and graft resorption *in vivo*. Cross-linking reaction of collagen molecule appears in fresh samples containing water due to the action of highly reactive hydroxyl radicals resulting from water radiolysis (indirect effect of ionising radiation).

Determination of *in vitro* collagen solubility of various connective tissue grafts is a simple method, which allows one to predict the rate of their resorption *in vivo*, and is very useful in tissue banking practice when testing various processing, preservation and sterilisation procedures.

Evaluation of *in vitro* susceptibility to pepsin digestion of fresh and lyophilised human bone and calf Achilles tendon samples irradiated at room temperatures with doses of 25-100 kGy

Fresh and lyophilised human bone and calf Achilles tendon samples were gamma irradiated in a ^{60}Co source at room temperature with doses of 25, 50 and 100 kGy. The samples were digested with 20% (w/w) pepsin in 0.5 M acetic acid at 4°C for 12 hrs with constant stirring. The solution was centrifuged and the amount of hydroxyproline was measured in supernatant and in the residue (pellet) separately.

Figures 19 and 20 illustrate the susceptibility to pepsin digestion of collagen of human bone and calf Achilles tendon, respectively. In non-irradiated human bone samples, both fresh and lyophilised, solubility of collagen was low and did not exceed 5%. Dramatic differences between irradiated fresh and lyophilised bone samples have been observed. In irradiated lyophilised bone samples the solubility of collagen increased significantly in a dose-dependent manner, while in irradiated fresh samples the increase was very moderate (Figure 19).

The solubility of collagen of non-irradiated calf tendon samples, both fresh and lyophilised, was about nine-time higher than that of human bone samples and constitutes over 40%. In fresh tendon samples irradiated with doses of 25, 50 and 100 kGy solubility was even lower than in non-irradiated controls. This might be due to the fact that in the presence of water cross-linking reaction of collagen appears making the collagen less susceptible to pepsin digestion. I n irradiated lyophilised samples a significant, dose-dependent increase of solubility has been observed (Figure 20).

In Figure 21 simplified scheme summarising some of the effects of ionising radiation on collagen - a major constituent of connective tissue graft is shown.

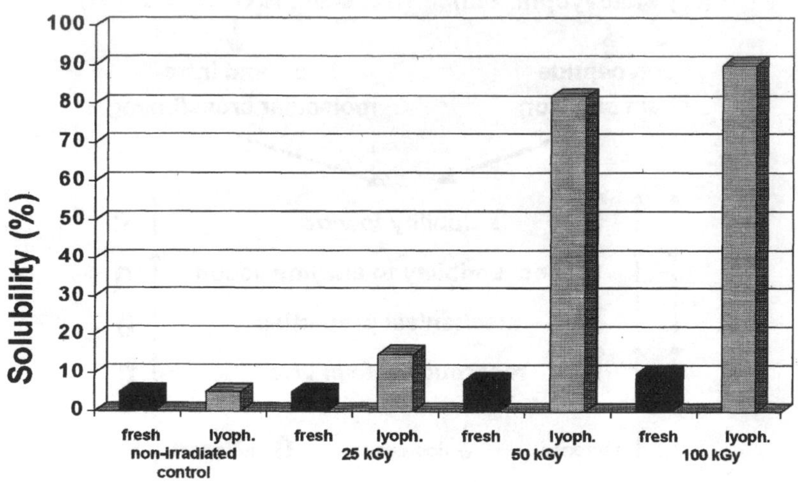

Figure 19. Susceptibility to pepsin digestion of fresh and lyophilised human bone samples gamma irradiated at room temperature with doses of 25, 50 and 100 kGy.

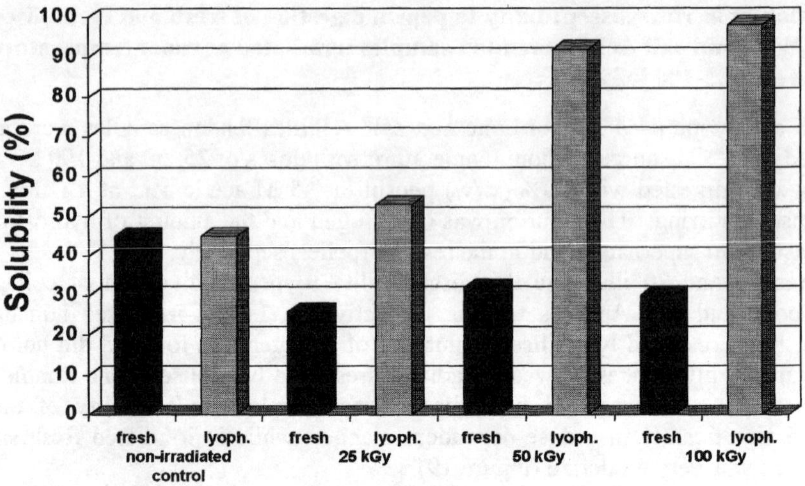

Figure 20. Susceptibility to pepsin digestion of fresh and lyophilised calf Achilles tendon samples gamma irradiated at room temperature with doses of 25, 50 and 100 kGy.

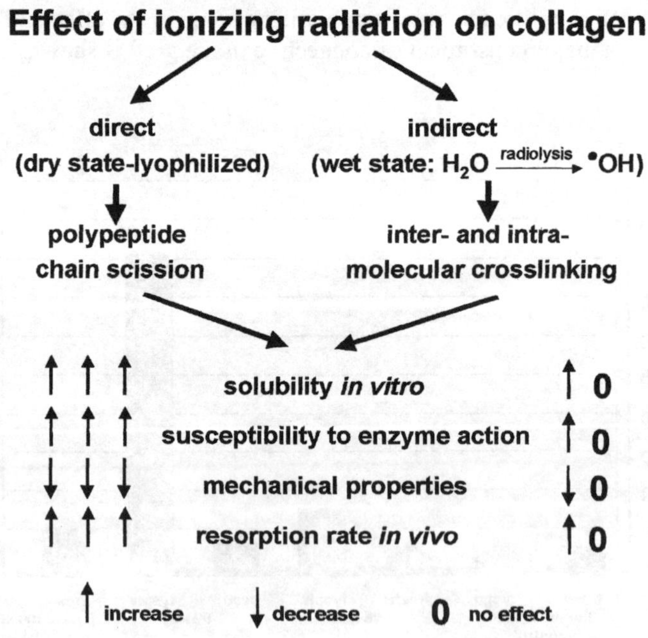

Figure 21. Simplified scheme summarising some of the effects of ionising radiation on collagen - a major constituent of connective tissue grafts.

Differences in cytotoxicity of medullary lipids of bone irradiated with 35 kGy with gamma rays in a ^{60}Co source and with electron beam 10 MeV accelerator

Large amount of lipids are present in bone tissue, particularly in medullary spaces of cancellous bone. It has been found that gamma irradiation of human bone allografts alters medullary lipids that become toxic for osteoblast-like cells [44], therefore three years ago we introduced bone defatting procedure in our tissue bank. Bone tissue is defatted either in the mixture of chloroform (20%, v/v) and ethanol containing 4% of ether (80%, v/v) or only in ethanol-ether. Our studies aimed to establish if non-defatted and defatted bone samples irradiated with a dose of 35 kGy in a ^{60}Co source at room temperature and with electron beam 10 MeV accelerator at room temperature and at -72°C (on dry ice) are toxic to human osteoblastic-like cells (SAOS-2) and murine fibroblast-like cells (3T3), using a MTT reduction test. The assay is based on reduction of tetrazolium salt (MTT) in living cells by the mitochondrial enzyme - succinate-dehydrogenase. Water-insoluble formazan crystals are formed and can be solubilised using either DMSO or another organic solvent. The optical density of the dissolved material is measured spectrophotometrically. Yielding absorbance directly correlates to the number of metabolically active cells in the culture. The preliminary results are shown in Figures 22 and 23.

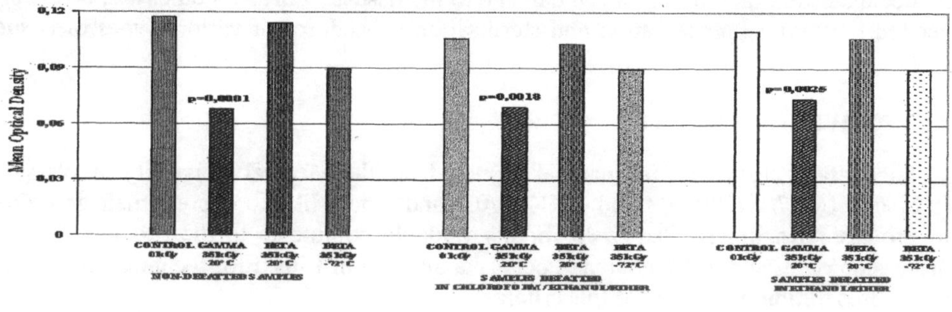

Figure 22. Results of MTT reduction test performed on fibroblast-like 3T3 cells.

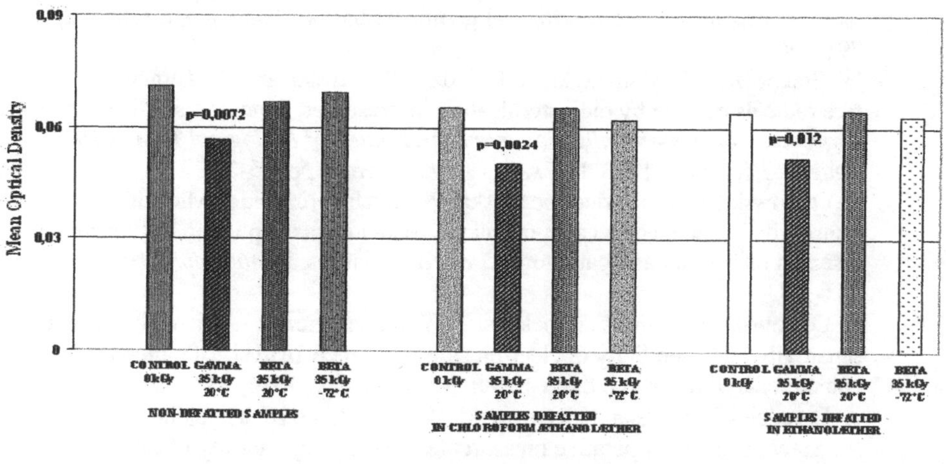

Figure 23. Results of MTT reduction test performed on human osteoblast-like cell line SAOS-2.

Unexpectedly, in all experiments performed, no statistically significant differences between samples irradiated with electron beam 10 MeV accelerator (both at 20°C and -72°C) and non-irradiated controls have been found. On the other hand, highly statistically significant differences between samples of bone irradiated with gamma rays at room temperature and non-irradiated controls have been found for two cell lines examined.

Other experiments on the effect of irradiated medullary lipids on cell viability and proliferation are currently being carried out, and from our preliminary results it appears that defatting procedure should be introduced for bone allografts sterilised with gamma rays. Further biochemical studies on medullary lipids irradiated with gamma and electron beam are planned so as to explain the differences observed in the studies of their cytotoxity mentioned above.

CONCLUSIONS

The results of interdisciplinary research performed at the Central Tissue Bank in Warsaw, in collaboration with radiation chemists from the Institute of Nuclear Chemistry and Technology indicate, that radiation-induced changes can be diminished by modification of tissue preservation methods and that, to some extent, it is possible to reduce undesired radiation-induced damage to the tissues. Further studies are, however, needed to optimise preservation and sterilisation procedures of various types of tissue grafts.

ACKNOWLEDGEMENTS

The authors acknowledge financial support from the Medical University of Warsaw (projects: 1M17/W/2001-03 and 1M17/N/03) and would like to express their gratitude to Professor Kazimierz Ostrowski for his valuable comments to the text and fruitful discussion on it, and to Magdalena Puchalska and Kamil Lipski for technical assistance in the preparation of the text of this article.

REFERENCES

1. M. Czerniewski, P. Panta, M. Zielczynski, W. Zak and K. Zarnowiecki, Bone tissue sterilization using reactor fuel gamma radiation, *Nukleonika*, **1965**, 12, 791-806.
2. W. Stachowicz, K. Ostrowski, A. Dziedzic-Goclawska and A. Komender, On free radicals evoked by radiosterilization in preserved bone grafts, *IAEA Panel Proceeding Series - Sterilization and Preservation of Biological Tissues by Ionizing Radiation*, **1970**, IAEA-PL-333/3, Vienna, pp. 15-27.
3. K. Ostrowski, A. Dziedzic-Goclawska, W. Stachowicz and J. Michalik, Sensitivity of the electron spin resonance technique as applied in histochemical research on normal and pathological calcified tissues, *Histochemie*, **1972**, 32, 343-351.
4. K. Ostrowski, A. Dziedzic-Goclawska, W. Stachowicz & J. Michalik, Accuracy, sensitivity and specificity of electron spin resonance analysis of mineral constituents of irradiated tissues, *Ann. NY Acad. Sci.*, **1974**, 238, 186-201.
5. A. Dziedzic-Goclawska, K. Ostrowski, W. Zasacki and J. Michalik, W. Stachowicz, Comparative measurements of the crystallinity of callus and compact bone by electron spin resonance spectrometry, *Bull. Polish Acad. Sci.*, **1974**, 22, 545-548.

6. K. Ostrowski, A. Dziedzic-Goclawska and W. Stachowicz, Stable radiation-induced paramagnetic entities in tissue mineral and their use in calcified tissue research, In: *Free Radicals in Biology, Volume IV*, W. Pryor (ed.), Academic Press, New York, **1980**, pp. 321-344.
7. K. Ostrowski, A. Dziedzic-Goclawska, W. Stachowicz and J. Michalik, Radiation-induced paramagnetic centers in research in bone physiopathology, *Clin. Orthop.*, **1991**, 272, 21-29.
8. A. Dziedzic-Goclawska, Effect of radiation sterilization on biostatic tissue grafts and their constituents, In: *Sterilization by Ionizing Radiation*, E. R. L. Gaughran and J. Goudie (eds.), Multiscience, Montreal, **1978**, pp. 156-187.
9. W. Stachowicz, G. Burlinska, J. Michalik, A. Dziedzic-Goclawska and K. Ostrowski, Application of EPR spectroscopy to radiation treated materials in medicine, dosimetry and agriculture, *Appl. Radiat. Isot.*, **1993**, 44, 423-437.
10. W. Stachowicz, J. Michalik, A. Dziedzic-Goclawska and K. Ostrowski, Evaluation of absorbed dose of gamma and X-ray radiation using bone tissue as a dosimeter, *Nukleonika*, **1974**, 19, 845-850.
11. W. Stachowicz, J. Michalik, A. Dziedzic-Goclawska and K. Ostrowski, Deproteinized bone powder as a dosimeter for radiosterilization of biostatic grafts, *Nukleonika*, **1972**, 18, 425-431.
12. A. Dziedzic-Goclawska, *Application of the electron paramagnetic resonance spectroscopy for quantitative evaluation of remodelling process of radiation-sterilized bone grafts*, Doctoral Thesis, The Medical University of Warsaw, **1971**.
13. K. Ostrowski, A. Dziedzic-Goclawska, W. Stachowicz, J. Michalik, E. Torsoly and A. Komender, Application of the ESR technique for quantitative evaluation of the resorption rate of irradiated bone grafts, *Calc. Tiss. Res.*, **1971**, 7, 58-66.
14. A. Dziedzic-Goclawska, K. Ostrowski, W. Stachowicz, J. Michalik and W. Grzesik, Effect of radiation sterilization on the osteoinductive properties and the rate of remodelling of bone implants preserved by lyophilization and deep-freezing, *Clin. Orthop.*, **1991**, 272, 30-37.
15. A. Dziedzic-Goclawska and W. Stachowicz, Sterilisation of tissue allografts, In: *Advances of Tissue Banking, Volume 1*, G. O. Phillips, R. von Versen, D. M. Strong and A. Nather (eds.), World Scientific, Singapore, **1997**, pp. 261-321.
16. A. Dziedzic-Goclawska, Application of ionising radiation to sterilise connective tissue allografts, In: *Radiation and Tissue Banking*, G. O. Phillips (ed.), World Scientific, Singapore, **2000**, pp.57-99.
17. A. Dziedzic-Goclawska, Evaluation of the osteoinductive potential of preserved bone grafts, In: *Research Methodology in Orthopaedics and Reconstructive Surgery*, A. Nather (ed.), World Scientific, Singapore, **2002**, pp.263-281.
18. A. Dziedzic-Goclawska, M. Wasilewska, A. Kaminski and J. Marowska, The effect of radiation-sterilization conditions on solubility *in vitro* of bone matrix collagen as a carrier of bone morphogenetic proteins, In: *Proceedings of the 5th International Conference on Tissue Banking*, EATB, Berlin, **1996**.
19. A. Komender, Influence of preservation on some mechanical properties of human haversian bone, *Materia Medica Polona*, **1976**, 8, 13-17.
20. A. Sliwowski and A. Dziedzic-Goclawska, Influence of gamma radiation on the solubility of collagen-derived membranes, *Materia Medica Polona*, **1976**, 4, 1-3.
21. J. Komender, H. Malczewska and A. Komender, Therapeutic effects of transplantation of lyophilized and radiation-sterilized allogenetic bone, *Clin. Orthop.*, **1991**, 272, 38-49.

22. W. Zasacki, The efficacy of application of lyophilized, radiation-sterilized bone grafts in orthopaedic surgery, *Clin. Orthop.*, **1991**, 272, 82-87.
23. J. Komender, H. Malczewska, D. Sladowski and A. Komender, Allogenic preserved costal cartilage in reconstructive, **1997**.
24. W. Marczynski, D. Tylman and J. Komender, Long term follow up after transplantation of frozen and radiation sterilized bone grafts, *Annals of Transplantation*, **1997**, 2, 64-66.
25. J. Komender, W. Marczynski, D. Tylman, H. Malczewska, A Komender and D. Sladowski, Preserved tissue allografts in reconstructive surgery, *Cell and Tissue Banking*, **2001**, 2, 103-112.
26. U. T. Slager and M. J. Zucker, The occurrence of electron spin resonance signals in bone grafts sterilized with high voltage electron beams, *Transplant. Bull.*, **1962**, 30, 536-537.
27. U. T. Slager, M. J. Zucker and E. B. Reilly, The persistence of electron spin resonance in bone grafts sterilized by ionizing radiation, *Radiat. Res.*, **1964**, 22, 556-563.
28. H. M. Swartz, Long-lived electron spin resonance in rats irradiated at room temperature, *Radiat. Res.*, **1965**, 4, 579-586.
29. J. L. Houben, Free radicals produced by ionizing radiation in bone and its constituents, *Int. J. Radiat. Biol.*, **1971**, 20, 373-389.
30. G. H. Henning, W. Herr, E. Weber and N. I. Xirotiris, ESR dating of the fossil huminid cranium from Petralona Cave, Greece, *Nature*, **1980**, 292, 533-536.
31. A. Dziedzic-Goclawska, J. Michalik, W. Stachowicz, W. Zasacki and K. Ostrowski, Theoretical basis for evaluation of the kinetics of mineralization of newly formed bone and the process of ageing of calcified tissues by electron spin resonance spectroscopy, *Nova Acta Leopoldina*, **1976**, 44, 217.
32. A. Dziedzic-Goclawska, K. Ostrowski, W. Stachowicz and R. Moutier, Decrease of crystallinity of bone mineral in osteopetrotic rats, *Metab. Bone Dis. Relat. Res.*, **1979**, 2, 33.
33. A. Dziedzic-Goclawska, U. Fuchs, K. Ostrowski, W. Stachowicz and J. Michalik, Crystallinity of mineral deposited in arterial walls in the course of arteriosclerosis in diabetics and in patients with normal carbohydrate metabolism, *Basic Appl. Histochem.*, **1984**, 28, 21.
34. A. Dziedzic-Goclawska, *Crystallinity of mineral and spatial distribution of collagen fibers in normal and pathologically changed bone tissue*, Doctoral Thesis, Medical Academy, Warsaw, **1984**.
35. A. Dziedzic-Goclawska, J. Emerich W. Grzesik, W. Stachowicz, J. Michalik and K. Ostrowski, Differences in the kinetics of the mineralization process in enchondral and intramembranous osteogenesis in human fetal development, *J. Bone Miner. Res.*, **1988**, 3, 533.
36. M. R. Urist, Bone: formation by autoinduction, *Science*, **1965**, 159, 893-899.
37. K. Buring and M.R. Urist, Effects of ionizing radiation on induction principle in the matrix of bone implants, *Clin. Orthop.*, **1967**, 55, 225-234.
38. E. Munting, J. F. Wilmart, A. Wijne, P. Hennebert and Ch. Delloye, Effect of sterilization on osteoinduction. Comparison of five methods in demineralised rat bone, *Acta Orthop. Scand.*, **1988**, 59, 34.
39. N. Schwarz, H. Reddi, A. Schiesser, G. Schlag, M. Thurnher, F. Lintner and H. P. Dingers, Irradiation-sterilization of rat bone matrix gelatin, *Acta Orthop. Scand.*, **1988**, 59, 165.

40. S. Wientroub and A. H. Reddi, Influence of irradiation on the osteoinductive potential of demineralised bone matrix, *Calcif. Tissue Int.*, **1988**, 42, 255.
41. S. Ijiri, T. Yamamuro, T. Nakamura, S. Kotani and K. Notoya, Effect of sterilization on bone morphogenetic protein, *J. Orthop. Res.*, **1994**, 12, 628-636.
42. A.J. Bailey, Effect of ionizing radiation on connective tissue components, In: *International Review of Connective Tissue Research, Volume 4*, D. A. Hall (ed.), Academic Press, New York, **1968**, pp.233-281.
43. R. M. Glaeser and K. A. Taylor, Radiation damage relative to transmission electron microscopy of biological specimens at low temperature, *A. Review J Microsc.*, **1978**, 112, 127.
44. M. F. Moreau, Y. Gallois, M. F. Basle and D. Chappard, Gamma irradiation of human bone allografts after medullary lipids and releases toxic compounds for osteoblast-like cells, *Biomaterials*, **2000**, 21, 369-376.

EFFECTS OF RADIATION ON THE INTEGRITY AND FUNCTIONALITY OF AMNION AND SKIN GRAFTS

J. Koller

Teaching Department of Burns and Reconstructive Surgery, Central Tissue Bank University Hospital Ruzinov, Ruzinovksa 6, 826 06 Bratislava, Slovakia {E-mail: koller@nspr.sk}

ABSTRACT

Amnion and skin grafts represent a very effective treatment of extensive skin and soft tissue defects of various origins such as burns, large soft tissue injuries, granulating wounds and chronic wounds. For temporary skin replacement (grafting) they are used mostly as allografts. As in any other kinds of tissue allotransplantation, the most important requirements for the grafts are quality, safety and effectivity. Both amnion and skin are well differentiated tissues with very unique anatomical structures and physiologic functions [1-5]. Skin is, as a matter of fact, an organ with many functions, which are important for the integrity and functionality of the organism. When skin is used as a temporary graft, it can replace just a few of its original functions, mainly the protective and barrier ones. The same can be applied for amnion, although its effectiveness is enhanced by the content of different very potent healing agents like cytokines and growth factors.

Anatomically, skin is composed of two principle layers – the epidermis and dermis [1]. Epidermis is responsible for the major part of the barrier function of the skin. It is relatively thin, composed of several cell layers, and very few extracellular matrix. The most important chemical substances of the cells are DNA, RNA, and proteins. The basal cell layer of the epidermis is attached to the basement membrane system representing the borderline between epidermis and dermis. The basement membrane contains special proteins like fibronectin, laminin, collagen IV, and collagen VII. Compared to epidermis, dermis is composed mostly of connective tissue fibres and extracellular matrix, with very few cells. The main chemical compounds of the extracellular matrix include collagen, elastin, and glycosaminoglycans (mostly hyaluronic acid and heparan sulphate). The principal functions of the dermis include mechanical strength, carrier of blood vessels, nerves, and skin appendages.

Skin thickness is variable and depends on the anatomical region. The thinnest skin is in the area of upper eyelids and genitalia (0.5 mm), and the thickest is on the back (2 mm). Amnion, compared to skin, is much thinner (0.05-0.2 mm) and it is composed of five layers. It contains a single layer of cuboid cells attached to a basement membrane; the other three layers include compact layer, fibroblastic layer, and spongy layer. It does not contain any structures of blood vessels, lymphatic vessels, or nervous tissue. It is of embryonic origin.

Radiation sterilisation is used to increase the safety of the biological tissue grafts in order to prevent transmission of microorganisms causing diseases from the donor tissue to the recipient [6-9]. The irradiation doses used to sterilise amnion and skin grafts do not merely devitalise the biological tissue, but can also cause structural changes of the main anatomical components and structures of the grafts.

Structures containing large molecules like collagen, some other proteins, and hyaluronic acid, are most vulnerable to irradiation. The most important issue is that these changes will not adversely affect to a large extent the structural integrity, mechanical strength, and adherence of the grafts, which properties are the major determinants of the most substantial temporary skin substitute functions.

INTRODUCTION

Amnion and skin grafts represent a very effective treatment of extensive skin and soft tissues defects of various origins such as burns, large soft tissue injuries, granulating wounds, and chronic wounds [10-11]. As in any other kinds of tissue allotransplantation, the most important requirements for the grafts are quality, safety and effectivity [12]. Both amnion and skin are well differentiated tissues with very unique anatomical structures and physiological functions.

Amnion anatomy

Amnion is one of the two foetal membranes (FMs), which engulf the embryo during intrauterine development. They are also often called amniotic membranes. From the anatomical point of view, they represent two loosely connected membranes, the *amnion* and the *chorion*. They rupture at birth, and are delivered together with the placenta. Amnion is usually a 0.05 to 0.2 mm thick, shiny and tough membrane. It is composed of five layers [4,5]:

- Inner layer of cuboidal and flattened epithelial cells
- Basement membrane
- Compact layer
- Fibroblast layer
- Spongy layer

The *epithelium* is made up of a single layer of cuboidal to polygonal epithelial cells. Transmission electron microscopy reveals numerous microvilli on the free surface of the cells, which face into the amniotic fluid. The cytoplasm of the cells contains filaments composed by actin, cytokeratin and vimentin. The cells generally have a single nucleus with one or two nucleoli. Amniotic epithelial cells are specially adapted to perform three major functions: covering, secretion and transcellular transport.

The *basement membrane* is a thin acellular structure adhered to the base of the amniotic epithelium. It consists of two distinctive layers – lamina lucida, and lamina densa. The first one is adherent to the base of the epithelial cells and extends itself to intercellular space. The second one is interposed between the lamina lucida and the deeper compact layer. The basement membrane contains collagens of type III, IV, V, fibronectin and laminin.

The *compact layer* is a thin, acellular structure comprising a network of reticular fibres, which create a uniform woven mesh. The spaces within the mesh are filled with mucus. The fibres contain collagen types I, III, and V. This layer provides remarkable tensile strength to the entire amniotic membrane.

The *fibroblast layer* is responsible for the main thickness and variations in diameter of the amnion. The fibroblasts are usually stellate or fusiform in shape, and they vary in size, depending upon their age and physiological state. This layer also contains so called Hoffbauer cells, which are histiocytes morphologically similar to fibroblasts, and macrophages.

The *spongy layer* is compressed between the chorion and amniotic sac and adheres better to the amnion than chorion. It consists of a complex network of fibrils surrounded by mucus. The fibrils are composed of reticulin and collagens of type I and III. This layer contains two types of cells – fibroblasts, and Hoffbauer cells. The spongy layer acts as a viscoelastic pad between the two membranes protecting the amnion against trauma and rupture.

Chorion

The chorion is connected by means of its mesenchymal tissue to the amnion. The outer layer of chorion is composed of the fairly thick transitional epithelium. Its clinical use is very limited; sometimes it can be used together with the amnion.

From the practical point of view, amnion is thinner, stronger and shiny, in contrast to the less homogeneous, weaker and dull chorion.

Amnion – properties as a graft

Amnion is used as a temporary skin replacement/cover, or as a healing agent, due to its content of various substances and its beneficial properties as a membrane. It is pliable and, following application adheres tightly to the wound bed. Burleson and Eisenman characterized this adherence as 'fibrin-elastin biological-bond mechanism' [13]. It was quantified in the FMs and, more recently, also compared with other skin substitutes [14]. Amnion is translucent, thus enabling visual control of the wound healing through the membrane. The antibacterial properties of the FMs were proved both in animal and clinical experiments [15,16]. It was shown, that the content of alantoin, immunoglobulins, lysozyme, progesterone or other specific substances in FMs alone is not responsible purely for the beneficial effects found. They need to be potentiated by the excellent adherence of the FMs to the wound site. The antigenicity of amnion is very low, or absent [17,18]. Amnion contains a variety of active substances, which play a role in wound healing and scar formation. The presence of angiogenetic factors in amnion was verified histochemically [19] and by isolation and purification [20]. Findings of Longaker and Adzick about scarless healing of foetal wounds hint at a new field where the usefulness of the FMs application could still pay off [21].

Amnion – procurement and processing

Processing of human foetal membranes for their use as biological skin substitutes or dressings was first practiced more than 90 years ago [22,23]. Since then, the donor selection, testing, processing, and preservation methods have improved substantially to assure good quality, safe, and efficient grafts for the end-users.

FMs are collected after obtaining informed consent from healthy mothers following regular term physiological deliveries, or Caesarean sections. The donor exclusion criteria are identical to any other tissue donations [24]. Samples of the mother's blood and cord blood are collected for serological screening. The procured FMs are placed into sterile transport containers filled with sterile saline solution, and transported to the tissue bank for processing. Processing can be performed under aseptic, or clean non-sterile conditions. Where terminal sterilisation by a validated sterilisation method will be performed, clean non- sterile processing is sufficient. In all the other cases aseptic processing is mandatory. The processing itself starts in the tissue bank with thorough cleaning of the FM grafts from blood remnants by rinsing them in saline solution.

Processing details are described elsewhere [25,26]. In cases, where viability of the amnion is required, the only possible preservation method is cryopreservation [27,28]. For non-viable grafts several methods can be used, including freeze-drying [29,30], glycerolisation [31,32], glutaraldehyde cross-linking [33], etc. According to Ward [34] it is advisable, however, to pay attention to the risk of possible elimination of the active substances through the processing techniques used.

Skin

Skin is the largest and most visible organ of the human body. The average adult human skin area is approximately 2 m^2 while its average weight is about 5 kg. Skin represents a very unique interface between the organism and its environment with many functions. It is adapted to withstand several physical, chemical and biological stresses.

Skin anatomy

Anatomically, the skin is composed of two major layers: the *epidermis and dermis.* The tissue lying deeper to the dermis is called *hypodermis* (subcutis) (Figure 1). Both skin layers contain cells and extra cellular structures [1,3].

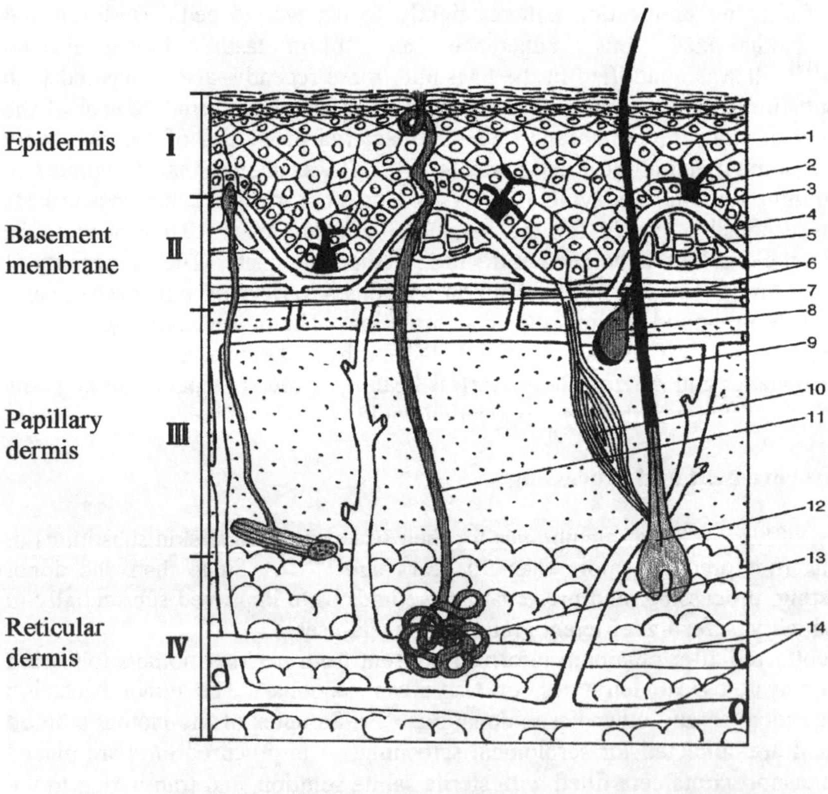

Figure 1. Skin anatomy: (1) keratinocyte; (2) melanocyte; (3) basal layer; (4) basement membrane; (5) capillaries; (6) dermal papillae; (7) subpapillary plexus; (8) sebaceous gland; (9) hair shaft; (10) arrector pilli; (11) sweat duct; (12) hair bulb; (13) sweat gland; (14) subdermal plexus.

In the epidermis there is a prevalence of cells with very scarce or almost no extracellular matrix. The epidermal cells are permanently renewed from the basal layer. The viability of the cells decreases towards the surface of the epidermis. Dead cells completely filled by keratohyaline granules are desquamated from the surface of the skin (stratum corneum, Figure 2.).

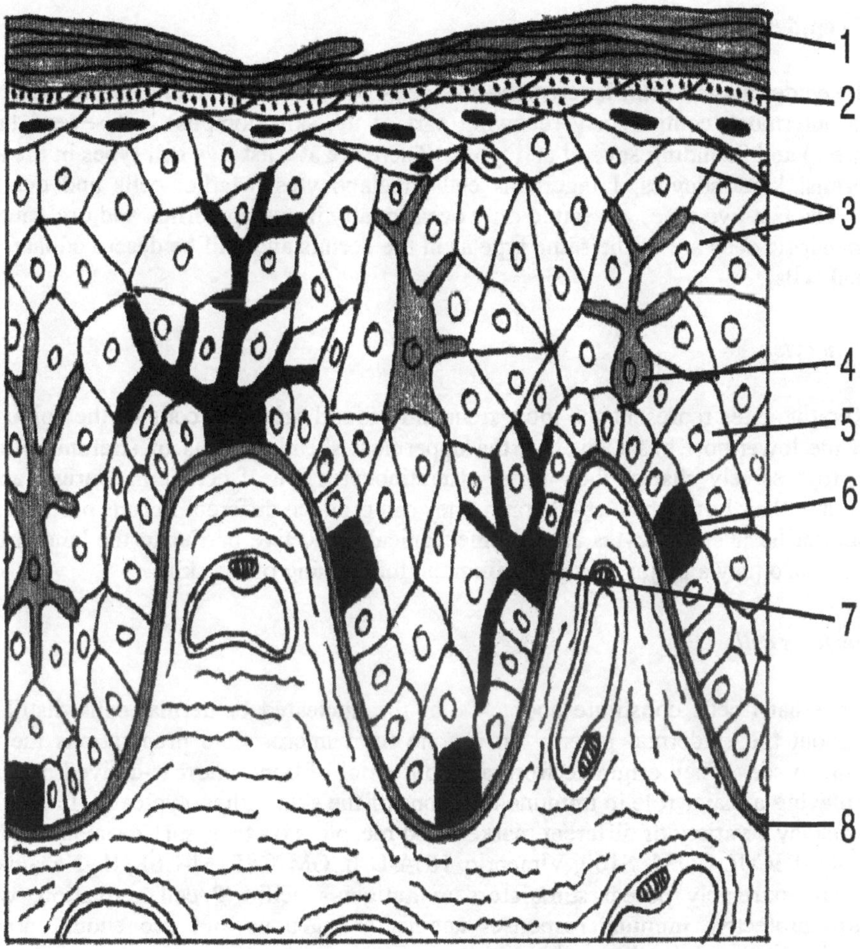

Figure 2. Epidermis: (1) stratum corneum; (2) stratum granulosum; (3) stratum lucidum; (4) Langerhans cell; (5) stratum spinosum; (6) Merkel cell; (7) melanocyte; (8) basement membrane.

In the dermis there are more extracellular components than cells. They include an integrated system of fibrous, filamentous and amorphous connective tissue that facilitates vascular, nerve and cutaneous immune system networks. The organisation of the collagen and elastic tissue of the dermis is a distinctive feature of human skin.

The uppermost part of the dermis adjacent to epidermis is called *papillary dermis* because of the dermal papillae interdigitating with the deeper epithelial layers. The junction itself between the epidermis and dermis is represented by the *basement membrane zone system* (Figure 3).

Deeper layers of the dermis are called ***reticular dermis*** according to the arrangement of the fibres. Blood vessels and nerve fibres are included in the dermis only, whereas epidermis is avascular. Other structures situated mostly in the dermis are skin appendages including sweat glands and ducts, hair follicles, sebaceous glands, arrectores pillorum smooth muscle fibers, special nerve receptors, and nail beds / nails.

Skin - epidermis

The epidermis is a multilayered sheet of cells with very little extracellular matrix. It is the outermost, continuously renewing part of the skin composed of several layers (Figure 2) and including several cell types. There are at least five cell types in the adult epidermis: keratinocytes, Langerhans cells, melanocytes, Merkel cells and dendritic cells. The last two types are found only occasionally in the epidermis and oral mucosa. The dendritic cells are of the same type as in the dermis and will be described later (see Dermal cells).

Keratinocytes

Keratinocytes represent the most frequently found principal cells of the epidermis. From the lowermost basal layer to the uppermost shedding cells of stratum corneum they progressively change their form. Keratinocytes are of ectodermal origin and in addition to their basic product – keratin, they can produce different fibrous proteins such as tonofibrils. Keratinocytes act as a mechanical protective barrier to the human body and they also play a major role in the immune functioning of the skin.

Langerhans cells

Langerhans cells constitute about 4% of the nucleated epidermal cells distributed throughout the epidermal layers. In routine light microscopic preparations they are difficult to see. They originate from a mobile pool of bone marrow-derived precursor cells playing a major role in immune functions of the skin. Phenotypically, Langerhans cells display a variety of different markers and receptors on their surface such as CD45, MHC-I, MHC-II, CD54, S100, Vimentin, HLA-D-li, GM-CSF, M-CSF, IL-2 chains etc. They are extremely potent stimulators of antigen-specific T cell activation, which initiates protective immune responses against endogenous and exogenous antigens. Other functions of the Langerhans cells include phagocytosis, antigen presentation, participation in cutaneous immune surveillance, and involvement in skin allograft rejection. Impairment of these cells can have deleterious consequences for the immunological defence of the host.

Melanocytes

Melanocytes are pigment-producing cells derived from the neural crest. They are evenly distributed in the basal layer of the epidermis with a frequency of 1 melanocyte for every 10 basal keratinocytes. Through their dendrites, one cell can acquire a relation with up to 36 keratinocytes. It is very interesting, that the number of melanocytes in epidermis is the same, regardless of race and skin colour. Racial differences in skin colour are determined by the density and size of the melonosomes (see later).

Melanin

The pigment produced by the melanocytes is synthesised in a complex organelle called melanosome. Chemically, there are two basic types of melanin – namely eumelanin, which is brown-black and insoluble, and phenomelanin, which is yellow-red and soluble in dilute alkali. Produced melanin granules move from melanocytes to other cells where they assume a static array. In keratinocytes they form a supranuclear cap that acts as a shield against UV radiation. It has been shown that sun exposure stimulates the melanocytes to produce larger melanosomes making the distribution of these proteins resembling the pattern found in dark-skinned individuals.

Merkel cells

They are mostly found in special regions such as the lips, oral cavity, hair follicles, the glabrous skin of the digits, or as a part of certain tactile discs. They are attached to adjacent keratinocytes by desmosomes (Figure 3). There are two prevailing hypotheses regarding the origin of the Merkel cells: the neural crest and cutaneous origin hypotheses. The *tactile Merkel cells* are opposed to small nerve plates connected by short, nonmyelinated axons to myelinated axons. These complex structures serve as tactile mechanoreceptors (Figure 4). *Epidermal Merkel* cells seem to stimulate local proliferation and differentiation of keratinocytes.

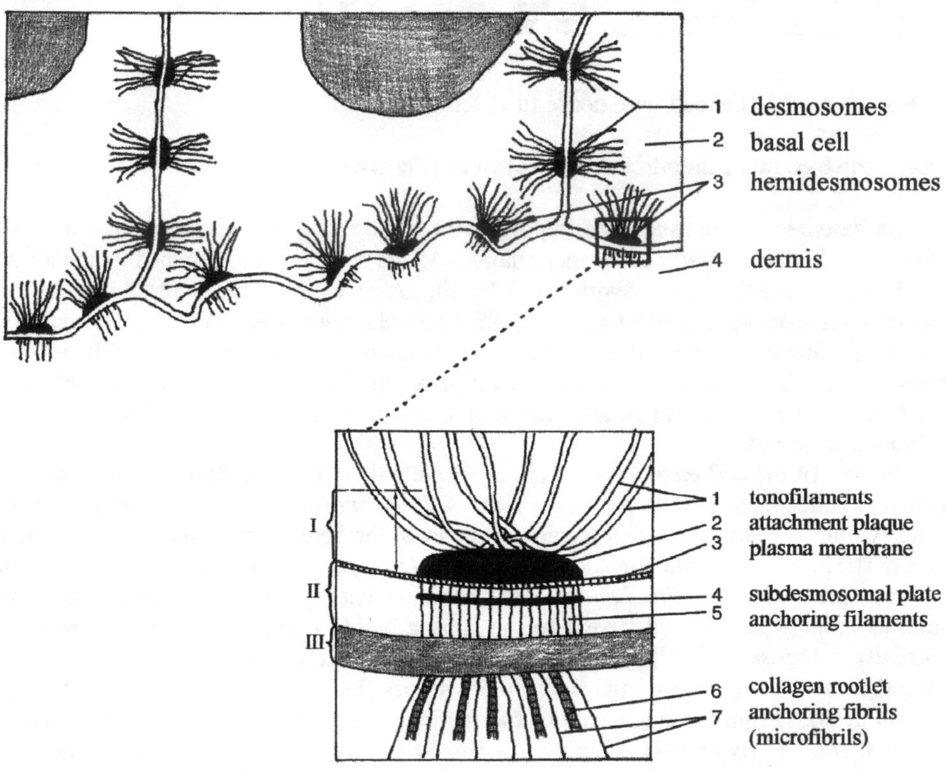

Figure 3. The basement membrane zone.
{I. Hemidesmosome; II. Lamina lucida; III. Lamina densa}

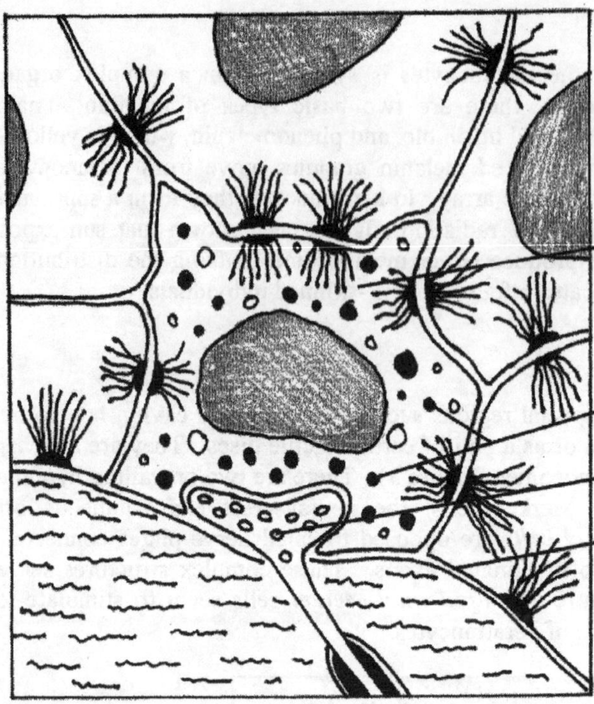

Figure 4. Merkel cell with nerve fibre terminal.

Skin - the basement membrane zone system (Figure3)

The basement membrane is an important interface that separates the epidermis from the dermis both physically and functionally. All its components, except the anchoring fibrils and microfibrils, are synthesised by the basal cells of the epidermis. As the epidermis is a nonvascularized structure, the basement membrane zone helps to regulate proper proliferating and differentiating mechanisms of the epidermis. It is also responsible for epidermal-dermal adherence, probably serves as a selective macromolecular filter, and is also the major site of immune reactant localisation in cutaneous diseases.

The part of the *cell membrane* of the basal cells that faces the dermis and includes a structure called *hemidesmosome*. Tiny fibres called *tonofilaments* are crossing the basal cells cytoplasma and attach to the epidermal part of the hemidesmosome. To the dermal part of the hemidesmosome the *subdesmosomal dense plate* is attached. Adjacent to the cell membrane is the *lamina lucida* – an electron lucent layer where adherence proteins such as laminin are located. Next to the lamina lucida is the *lamina densa, which* is basically composed of collagen type IV mesh-like scaffold. The *anchoring filaments* extend from the subdesmosomal dense plate across the lamina lucida and insert to the lamina densa. From the inner face of the lamina densa *anchoring fibrils* composed of collagen rootlets (type VII) extend for a short distance into the papillary dermis. In addition to the anchoring fibrils, which are of collagen, *microfibrils*, which are delicate, long, elastic fibrils, extend and blend with the underlying elastic fibrillary system of the dermis.

Skin - Dermis

The dermis is composed of cells, extracellular matrix, blood and lymphatic vessels, and skin appendages. The dermis contributes with its bulk, density, compliance, elasticity and tensile strength to the skin properties. This is due to the presence of dermal matrix containing fibrous and no fibrous connective tissues. Among the fibrous molecules the most important are collagen and elastin. Non-fibrous molecules are represented mainly by proteins and glycosaminoglycans (ground substance).

Collagen is the ultimate product of fibroblasts and its presence results in tensile strength of the skin. At least seven types of mature collagen are currently recognised.

Elastic fibres consist of two protein components - the more common *elastin* (amorphous appearance at electron microscopy, around 90%), and the *elastic microfibrils* composed of a specialised glycoprotein. Elastin is the second major fibrous protein in connective tissue. Elastin contributes to a great extent to skin elasticity.

Ground substance constituents are the glycosaminoglycans (GAGs), glycoproteins, and mucoproteins, in addition to water and electrolytes. Current interest in some ground substance components relates to their hypothetical capacity for *actively directing tissue repair.*

Other Components

Laminin is a large glycoprotein, an *essential component of basement membranes,* adjacent to the cell membranes. *Fibronectin* is a ubiquitous, high-molecular weight glycoprotein. It is found in plasma and can be associated with cell surfaces, basement membranes, and pericellular matrices. Fibronectin can bind some macromolecules, including collagen, fibrin, heparin, and proteoglycans. Its role is important in wound repair as a functional and structural component.

The two main anatomical portions of the dermis are the *papillary dermis* and *reticular dermis* [35]. The *papillary dermis* is situated immediately deep to the epidermis and basal membrane zone. It is relatively thin and has little structure when viewed with the light microscope. The papillary dermis contains different protein forms with a high proportion of type III collagen. The boundary between the papillary and reticular dermis is defined by a horizontal subpapillary vascular plexus. Papillary dermis is populated by dermal cells, more densely than the reticular one. The *reticular dermis* represent the bulk of the dermis containing the majority of dermal collagen organised into coarse bundles. It is composed primarily of type I collagen. Each collagen bundle is associated with elastic fibres that can be demonstrated microscopically only by special stains.

The blood flow required for the nutrition of the skin is very small. In normal conditions at ordinary skin temperature the amount of blood flowing through the skin is 10 times more than is needed for nutrition.

Dermis is a highly vascularised structure with a special kind of vascular network (plexus). The superficial vascular plexus, called *subpapillary plexus,* is situated on the boundary between papillary and reticular dermis and is composed of arterioles and postcapillary venules. From this plexus a terminal arteriole extends into each dermal papilla, where an arterial capillary is formed. The arterial capillary makes a U-turn and becomes a venous capillary and a postcapillary venule coming back to the subpapillary plexus. A second, larger and deeper vascular network is situated in the subcutaneous tissue immediately deeper to the dermal layer and is called *subdermal plexus.* The vascular connections between these two networks are realized by arterioles and venules running perpendicularly to the skin surface through the reticular dermis. This means that the vascular supply of the reticular dermis is less abundant than that of the papillary dermis, which can play an important role in wound healing, particularly in deep dermal burns. The blood flow through the two plexuses is involved in the regulation of body temperature as well as in the metabolic supply of the whole skin. There are also some direct vascular communications between the arterial and venous plexuses which are present in some skin areas exposed to maximal cooling, such as the volar surfaces of hands and feet, the lips, the nose and the ear.

The lymphatics of the skin form a complex and random network beginning as lymphatic capillaries near the epidermis. A superficial lymphatic plexus is formed from which lymphatic channels drain to regional lymph nodes. The lymphatic channels are important for the clearance of fluids, macromolecules, and cells from the dermis.

Dermal cells

The majority of dermal cells are of mesodermal origin such as fibroblasts, mast cells, macrophages, dendritic cells, and T-lymphocytes.

Fibroblasts

Fibroblasts are the most frequent connective tissue cells. In the papillary dermis they are located mostly in the papillary region and around vessels, in the reticular part in the interstices between collagen fiber bundles. Fibroblasts play an important role in wound healing processes. In newly formed tissue they can migrate along capillaries and produce matrix components. They produce many different extracellular matrix and structural proteins. Among the extracellular matrix proteins the most important are all types of collagens, further elastin, fibronectin and proteoglycans; the structural proteins include enzymes, enzyme inhibitors, integrins, actin, vimentin, and tubuline.

Mast cells

They are derived from the bone marrow. Mast cells are present in all regions of the dermis. They are more frequent in the upper dermis around vessels and epidermal appendages and in the subcutaneous fat. Dermal mast cells are surrounded by fibronectin, which helps them to anchor to the extracellular matrix in inflammatory sites where they proliferate and release different mediators. They are involved in a variety of physiological and pathological events. They store active proteins and respond to a variety of immunologic and non-immunologic stimuli. They release a variety of vasoactive mediators, chemotactic mediators and enzymes.

Macrophages

Macrophages are large, mobile, phagocytic cells. They look very similar to neutrophils, from which they differ by unlobed nucleus and absence of specific granules. They are derived from the bone marrow precursor cells, which differentiate into monocytes in the blood and macrophages in the tissue. They play an active role in cell-mediated immune mechanisms. They are capable of phagocytosis of foreign particles such as cellular debris and bacteria. Their number increases after local inflammatory stress. Macrophages carry high amounts of major histocompatibility complex (MHC) class-II antigens and bear different receptors. They are also active antigen-presenting cells. Macrophages play a key role in wound healing mechanisms.

Dendritic cells

Dermal dendritic cells are located in perivascular areas. They are different from the epidermal Langerhans cells. Their cytoplasm contains organelles involved in active cellular metabolism. They can be located also in the basal layer of epidermis. Dermal dendritic cells bear several receptors such as CD 36, CD54 etc., and carry large amounts of MHC class-II antigens. Their functional role in the skin's immune system is still not clear.

T-lymphocytes

T-lymphocytes migrate from the blood predominantly to the dermis and are located mostly around postcapillary venules and the skin appendages. They contribute to the immune surveillance and homeostasis of the skin.

Skin appendages

Skin appendages include hair follicles, sebaceous glands, sweat glands with sweat ducts, nails and special nerve receptors.

Skin functions

The skin is structured to provide mechanical strength and protection to the delicate body components, to prevent loss of essential body fluids, and to protect the body against the entry of toxic environmental chemicals. This important function of the skin is called 'protective and barrier function'. The stratum corneum, which is the outermost part of the epidermis, with its content of overlapping cells and intercellular lipid, makes diffusion of water into the environment very difficult, and vice versa. Collagen and elastic fibres contained in the dermis assure the mechanical strength and elasticity of this barrier. The immune function of the skin is a part of the innate immunity of the body against invasion by microorganisms. Many factors do play an active role in this protection, including the normal bacterial flora of the skin, the fatty acids of sebum and lactic acid of sweat; they all represent natural defence mechanisms against invasion of microorganisms. Langerhans cells of the epidermis have an antigen-presenting capacity and might play an important role in delayed hypersensitivity reactions. They also play a role in immunosurveillance against viral infections. Langerhans cells interact with neighbouring keratinocytes, which secrete a number of immunoregulating cytokines, and epidermotropic T-cells forming the skin immune system: SALT (skin associated lymphoid tissue).

Skin pigmentation by melanin pigment of the skin protects the nuclear structures against damage from ultraviolet radiation. Sensory functions of the skin are provided by special receptors for heat, cold, pain, touch, and tickle. Parts of the skin are considered as erogenous zones. Thermoregulatory function - the skin contributes to a great extent to the body's temperature regulation system, protecting us against hypothermia and hyperthermia. This is assured by regulation of skin blood circulation, by sweat evaporation, and partly by insulation properties of the subcutaneous fat. Exposure to extreme cold reduces the rate of cutaneous blood flow to very low values to prevent the loss of heat. On the other hand, in hot environment the rate of cutaneous blood flow can increase up to 7 times the normal value to assure maximal heat loss from the body. The loss of heat is further enhanced by sweat excretion and evaporation.

The excretory functions of the skin include excretion of sweat and sebum. *Sweating* is the normal response to exercise or thermal stress by which the human organism controls its body temperature through evaporative heat loss. Under extreme conditions the amount of perspiration can reach several litres a day. In addition to the secretion of water and electrolytes the sweat glands serve as excretory organ for heavy metals, organic compounds, and macromolecules. The sweat is composed of 99% water, electrolytes, lactate, urea, ammonia, proteolytic enzymes, and other substances. *Sebum* has a protective and nutritive effect to the skin and hair. Sebaceous glands are found on all areas of the skin with the exception of the palms, soles, and dorsa of the feet. They are holocrine glands, which means their secretion is formed by complete destruction of the cells. The sebum is composed of triglycerides and free fatty acids, wax esters, squalene and cholesterol. The sebum controls moisture loss from the epidermis. It also protects against fungal and bacterial infections of the skin due to its contents of free fatty acids.

Skin plays an important role in calcium homeostasis by contributing to the body's supply of vitamin D. Vitamin D3 (cholecalciferol) is produced in the skin by the action of ultraviolet light on dehydrocholesterol. It is then hydroxylated in the liver and kidneys to the active form of vitamin D. This anti-rachitic vitamin acts on the intestine increasing calcium absorption, as well as on the kidneys promoting calcium reabsorption.

Skin also provides the cosmetic packaging of the individual organism. The fingers and toes, the palms of the hands and soles of the feet, are covered with a system of ridges that form certain patterns. The term 'dermatoglyphics' is applied to both the configurations of the ridges, and also to the study of fingerprints. Fingerprints are unique to each individual, which has a substantial medico-legal importance.

The skin has great psychological importance at all ages. It is an organ of emotional expression and a site for the discharge of anxiety. Caressing favours emotional development, learning and growth of newborn infants.

Epidermal regeneration

The cells in the basal layer (stratum basale) renew by cell division and as they ascend towards the surface they undergo a process known as keratinisation which involves the synthesis of the fibrous protein – keratin [36]. The cells on the surface of the skin forming the horny layer (stratum corneum) are fully keratinised dead cells that are gradually peeled off. The rate of cell production must be balanced by the rate of cell loss at the surface. The control mechanism of epidermopoiesis consists of a balance of stimulatory and inhibitory signals mediated by diffusible factors including cytokines and growth factors. Some of them such as epidermal growth factor, transforming growth factor-alpha, interleukins, basic fibroblast growth factor, are stimulating the epidermal cells; others like chalones, transforming growth factor-beta, interferons, and tumour necrosis factor have inhibitory action. The epidermis renewal time under normal conditions varies between 50 and 75 days. Damage to, and/or loss to large areas of the skin such as in extensive burns or other affections can cause severe systemic alterations, and even death of the individual.

Skin - procurement, processing, preservation, sterilization

All the tissue procurement and/or banking organisations use approved general and specific standards, protocols, and standard operating procedures in their practices of procurement, testing, preparation, processing, preservation, storage and distribution of skin allografts [24,37-39]. The majority of these organisations use sterile techniques for the skin banking without terminal sterilisation methods in order to retain the vitality of the prepared grafts. Some of the methods use techniques, where the grafts are devitalised, such as preservation by 85% glycerol [40]. Chemical sterilisation methods by glutaraldehyde, ethylene-oxide, etc., did not become very popular. Several countries, especially those where tissue banking was supported by IAEA projects (like Argentina, Brazil, Mexico, Peru, etc.), use radiation sterilisation for skin grafts.

SKIN GRAFTS

Skin grafts are portions of skin, which are detached from their original positions and transferred to a host bed. The process of transferring grafts is called *grafting, or transplantation.* Skin grafts consist of epidermis and dermis. The dermal component is important because epidermis is avascular and the healing is assured by establishing connection between the host bed and graft vasculatures. These grafts are commonly referred to as *dermo-epidermal grafts.*

Types of skin grafts

<u>*Skin autograft (isograft)*</u> is a graft transferred from a donor to a recipient site in the same individual.

<u>*Skin allograft (homograft)*</u> is a graft transplanted between genetically disparate individuals of the same species.

<u>*Skin xenografts (heterografts)*</u> are grafts transplanted between individuals of different species.

Autografts

Autologous skin grafts can be partial (split) or full thickness. In *split thickness grafts* only a part of the dermis is included, whereas in *full thickness grafts* the entire dermis is included. Both require a recipient bed that is well vascularised and free of devitalised tissue and bacterial contamination (< 10^5 microorganisms per gram of tissue). Haemostasis at the recipient site is important, as haematoma beneath the graft is a common cause of graft failure. The transplanted skin derives its initial nutrition via serum from the recipient site; the graft then gains blood supply from the recipient bed by in-growth of blood vessels. At this stage the graft is susceptible to mechanical shearing and should be protected by immobilisation.

The *full-thickness skin graft* was the first skin graft described. It gives an excellent cosmetic result with limited graft contraction but has the disadvantage of less reliable graft 'take'. The amount of full-thickness skin graft available is also limited if primary closure of donor site is to be achieved. In cases in which large areas are to be covered with a full-thickness graft, as in resurfacing a face after burns, the donor area can be increased by preoperative tissue expansion, or the donor area can be covered with a split-thickness skin graft.

Autologous *split-thickness skin grafting* is the most commonly practiced form of tissue transplantation in plastic surgery today. The graft can be taken at different thickness depending on the level at which it is harvested through the dermis. It has the advantages of large available donor areas and better graft 'take', but is prone to increased graft contraction and hypertrophic scarring, especially in children. Expansion of the split-thickness skin graft by meshing with expansion ratios from 1:1.5 to 1:9 can be useful and sometimes essential in extensive burns.

Donor sites for split-thickness skin graft may be limited in patients with extensive burns. This lack of available tissue has spurred the development of alternatives to conventional skin grafts. One method involves growing autologous keratinocytes in culture with the ability to expand the available tissue 10,000-fold. This technique has been applied in the treatment of large thermal injuries [41]. Cultured autologous keratinocytes have also been used to treat leg ulcers and other benign conditions. There are reported disadvantages with the use of cultured keratinocytes. This technique is more susceptible to bacterial contamination than split-thickness grafts and its 'take' has been reported to be less reliable than meshed graft. After healing, cultured autograft has been found to blister spontaneously, to be more prone to minor trauma, and to contract more in comparison to split-thickness skin graft. These effects are purported to be related to a poorly developed dermoepidermal junction. Increased 'take' has been reported in recipient beds of early granulation tissue and/or allogenic dermal support rather than chronic granulating wounds [42].

The lack of a dermal component in these autografts was overcome by a combination of cultured autologous keratinocytes and allogeneic dermis (after removal of the more antigenic epidermis) [43]. The technique has had favourable reports in patients with extensive burns but the problem of dermal antigenicity remains. An acellular or 'artificial skin' consisting of dermal components, collagen, and a glycosaminoglycan overlaid with a Silastic sheet was developed to combat this antigenic problem [44]. A disadvantage of this approach is the need to skin graft the 'dermis' after removal of the outer Silastic dressing. Development of a skin substitute containing allogenic or xenogenic structural proteins and ground substance seeded with autologous cells has also been described; this is comprised of cultured autologous fibroblasts populating the 'dermis' and cultured autologous keratinocytes covering the 'dermis' [45].

Mode of survival ('take') of skin grafts

The survival of skin autografts is permanent, whereas the survival of skin allografts is only temporary until rejection occurs. The fate of amnion grafts and skin xenografts is different and will be described later.

The phases of skin grafts survival [35,46]

At the time of surgical excision (removal, harvesting) of the skin graft from its donor site the grafts are completely detached from the surrounding skin and subjacent tissue layer, which means that the circulation, lymphatic drainage, and nerve continuity are abruptly terminated. The survival of the graft is dependent on how rapidly it can acquire new blood supply for nutrition and disposal of metabolic products.

The Phase of Serum Imbibition

In the meantime between transplantation and revascularisation the survival of the graft appears to be ensured by the absorption of fluids very similar to plasma from the host bed. This early process of fluid nourishment was termed *'plasmatic circulation'*. The sequence of events can be described as follows:

The blood vessels of a freshly cut skin graft are collapsed and empty as a result of the spasms of vessels after their separation from the donor site vasculature. As early as eight hours after grafting a faint pink tint in the graft can be noted. Within 24 hours after transplantation the graft vessels are again dilated, although they continue only a few haemic elements. By 48 hours the vessels are more distended and contain large numbers of erythrocytes. The exudate which accumulates between the graft and the host tissues consists of plasma, erythrocytes, and polymorphonuclear leucocytes. This can explain the rapid colour change observed within hours after transplantation. The fibrinogen from the plasma precipitates and forms fibrin on the surface of the host bed. In addition to fibrin the initial bond (so called fibro-elastic bond) of the graft with the host bed is assured by 'binding proteins' such as fibronectin and integrins. In summary, the 'phase of serum imbibition' is a period during which the graft vessels fill with a fibrinogen-free fluid and cells from the host bed [35,46]. There is no real 'plasmatic circulation', because the fluid absorbed by the graft is passively trapped within the graft. However, this mechanism can assure the nutrition of the graft only for a short time until the establishment of vascular connection occurs. It is not capable of long-term maintenance of graft survival when the graft fails to become successfully vascularised.

Revascularisation of the skin grafts

Revascularisation can occur by one or combination of three mechanisms:

1. Direct connection of the graft and host vessels referred to as 'inosculation' [35],
2. In-growth of vessels from host bed into endothelial channels of the graft itself,
3. In-growth of host vessels into graft dermis creating new endothelial channels.

Immediately after application the blood vessels of the graft are less filled with the host bed fluid described above. On the day after grafting many vessels show distension and rapid filling with static blood. On the second day the vessels distension continues but blood circulation has not commenced. A sluggish flow of blood occurs in the graft vasculature on the third and fourth day and continues to improve until the fifth or sixth day. During the next days the blood vessels return to normal calibre and circulation in all *autografts*. The process of graft visualization is completed on the sixth or seventh day (Table 1).

Allografts

Skin allograft was the first 'organ' transplant achieved and constituted the foundation of modern transplant immunology. However, skin is strongly antigenic and is subject to rejection even in the presence of surviving organ allografts in the same experimental animal. Rejection of allogeneic tissue occurs through cellular and humoral immunologic responses. These responses are generated when the host defence system detects certain antigens expressed on the donor cell surface. These antigens are referred to as major histocompatibility complex (MHC) antigens.

The revascularisation process in *allografts* is identical to autografts only until the onset of the *allograft rejection.* Early symptoms of rejection are increased distension of the vascular system, followed by sluggish circulation and clumped elements. Complete obstruction of the blood flow and vascular disruption in most of the skin allografts occur usually between seven to ten postoperative days in immunologically non-compromised individuals. The immunocompromised state of patients after a major burn usually delays rejection of allografts for several weeks.

Table 1. Skin grafts vascularisation.

Time	Event	Clinical appearance
24 hours	Absorption of wound bed fluids, dilation of vessels, fibroelastic bond of graft to graft bed (fibrin and fibronectin)	Pink hue, can be lifted easily
48 hours	So called 'plasmatic circulation', vessels are more distended	Pale, fixed by fibroelastic bond
Day 2-5	Starting vascular in-growth by connection of venous site of capillaries, sluggish blood flow, start of fibrotisation	Livid appearance, sluggish refill phenomenon
Day 6-7	Completion of vascularisation, collagen fibres proliferation	Cherry-red hue, intensive, good capillary refill
Day 7 <	Firmer bond, partly by collagen, gap closure, increased blood flow	Cherry-red hue, more pale, detaching more difficult, bleeding if lifted
Day 14 <	Firm fibrous bond, cessation of blood flow	Firm, almost impossible to lift, normal appearance

The mechanism of healing in combined grafting techniques (such as intermingled, and sandwich grafting, see later) in humans was studied histologically by Omi and co-workers [47]. They found that at 5 days after transplantation variable numbers of lymphocytes and neutrophils were scattered throughout the graft with fibrin strands at the borders between grafted and recipient tissues. At 2 weeks after transplantation the allografted epidermis was completely sequestrated and gradually replaced by autologous epidermal cells. At 3 weeks after transplantation the new epidermis of recipient origin became acanthotic and covered the remaining allografted dermis, which appeared basophilic and contained increasing number of fibroblasts and capillaries. At 4 weeks after transplantation the capillaries tended to be arranged perpendicularly to the epidermal surface. Fibroblasts migrated through the gaps in the basement membrane where they appeared to participate in the formation of new connective tissue elements. In this phase the connective tissue elements from the allograft skin became indistinguishable in areas subjacent to the new epidermis of recipient origin.

Xenografts

It has been generally held that the survival time of xenografts is too restricted to permit a reestablishment of blood circulation. In animal xenografting experiments (mouse to rat, rabbit to rat, pig to rat or rabbit) it was shown, that blood flow in the xenografts was initiated usually on the fourth day after transplantation, attained the maximal rate soon thereafter, and suddenly ceased on or around the sixth day [48].

The vasculature of the xenograft was not newly formed, but the original graft vascular system established direct connection with that of the host. In the human host it was unable to distinguish between xenograft vascularisation and invasion of the graft-host interface by granulation tissue formation. At 14 days there was no evidence of vascularisation or viability of any xenografts. In human burn recipients of skin xenografts from pigs [49] an initial non-immune, inflammatory cellular response during the first week was followed by an increasingly immunocompetent cellular reaction that peaked at 30 days. Anti-pigskin humoral factors could not be detected. There was no clinical manifestation of any host sensitisation.

The most commonly used xenografts are grafts from the skin of domestic pigs. Porcine xenograft has been used as a temporary dressing in both superficial burns without excision, and in deep excised burns as a temporary skin substitute. Xenografts were used also in the sandwich grafting technique for covering of microskin grafts, or largely meshed autografts in large burns [50,51]. The application of xenogeneic dermis has also been found valuable in preparing a wound for subsequent grafting by stimulation of granulation tissue formation. Porcine skin xenografts are not suitable for temporary biological dressing of skin graft donor sites because the porcine collagen can be incorporated in the subepithelial area of the donor sites leading to donor site inflammation and delay in repair. The acellular artificial skin described by Yannas and associates [52] uses a bovine collagen 'dermis', which recipient fibroblasts repopulate. Other animal skin used for grafting was from calves, frogs, and sheep.

CLINICAL USE OF AMNION AND SKIN GRAFTS

Amnion

Since the first reported applications [22,23], the utilisation of FMs, particularly amnion, became quite popular in many indications like in flame burns [53]; in paediatric burns [54-56], for coverage of split-thickness skin donor sites [27,57-59]; for coverage of clean superficial 2nd degree burns [53,56,58,60-62]; for coverage of freshly excised deep dermal burns [27,63]; for post-burn sequelae; in granulating chronic wounds [15,22,34,64-66], for ulcerated surfaces of different aetiology [67]; in leg ulcers [64,66], for prevention of postoperative adhesions [28,68].

The beneficial effects observed following application of the FMs included alleviation of pain, prevention of wound desiccation, cleaning of the wounds by repeated graft changes, enhancement of the growth of new granulations, improvement of the vascularity of the wounds, enhancement of epithelialisation from the wound margins, and preparation of the wound for better 'take' of the autografts. In ophthalmology, amnion is used mostly for treatment of corneal defects, conjunctival reconstruction, and adhesion prevention [69-72].

Although application of amnion grafts is referred to as grafting, in fact amnion, which is avascular, regardless of its application form (fresh, fresh-frozen, freeze-dried, glycerol-preserved), can never establish vascular connections between the host and the graft. Amnion on the wound surface behaves as a biological dressing with all its beneficial properties [25]. Due to its thinness, pliability and elasticity, amnion adheres very well to the wound surface initially. Several hours following the application the adherence is reinforced by fibroelastic bonding mechanism as in other types of grafts. It has been found, that amnion application can contribute to the control of bacterial proliferation in the healthy wound bed. The beneficial effects observed after application of amnion were prevention of desiccation of the wounds, cleaning the wounds by repeated graft changes, enhancing the growth of new granulation tissue, improvement of the vascularity of the wounds, enhancement of epithelisation from the wound margins, and, finally, preparation of the wounds for better 'take' of autografts. In painful wounds the pain disappeared very soon after the application of grafts, and the comfort of the patients was much improved. In covering the post-dermabrasion defects the postoperative pain was almost completely absent during the whole healing time, and the need of several painful postoperative dressing changes was eliminated [26]. Amnion proved to be safe and very beneficial in treating both clean and problem wounds, due to its low antigenicity, availability, low cost, and potentials of enhancing the wound healing process. Another beneficial effects of amnion included diminution of pain, prevention of wound desiccation, and shortening of the healing time. However, in full-thickness and infected wounds it disintegrates rapidly, requiring frequent reapplications. The advantages of amnion compared to other biological skin substitutes include availability, simple harvesting and processing, low antigenicity, and relatively low costs. Another advantage is the possibility of performing confirmatory serological testing of donors after 6 months, which is not available in cadaveric skin donors.

Skin grafts

For temporary skin replacement (grafting) they are used mostly as allografts. When used as temporary grafts, they can replace just some of the original skin functions, mainly its protective and barrier functions. Skin allografts are used mostly in situations, where there are extensive skin losses, and sources for autograft harvesting are very limited, such as in extensive burns. The use of skin allografts has been found to be beneficial, and/or life saving in large burns with or without concurrent skin autografts. In cases when the allografts are used alone, it is not advisable to leave them in place until they will become rejected, but to remove, or change them before rejection occurs. *Modified allografts rejection* was observed in some of the so-called 'combined' grafting techniques. The Mowlem-Jackson technique [73] used alternate placement of narrow auto- and allograft stripes on the wound. As both the grafts healed, the rejection was starting in allograft epidermis, which was gradually replaced by the neighbouring autograft epidermis with the result of a healed wound. The same basic principle was modified later in the seventies by the Chinese [74], who reported to use large sheets of allografts with chess-like fashion made tiny holes, where small pieces of autografts were placed. The allografts provided favourable wound conditions for the spread and growth of the autograft skin islands, with the autograft epidermis gradually replacing the rejected allograft epidermis while the allograft dermis stayed in place and was not rejected. The mechanism described above was called 'sandwich phenomenon'.

In the USA the technique was modified by using a double meshgraft technique, which was called 'sandwich grafting' [10]. Widely meshed (1:6 - 1:9) autografts were placed on the wound surfaces and covered by allografts meshed 2:1 with similar results after healing. Cultured allogenic keratinocytes have also been used as temporary covering. Such grafts can be grown in culture pre-emptively for burn treatment but are susceptible to rejection in addition to the problems associated with cultured autografts. They can be used also in combined grafting techniques.

RADIATION STERILISATION OF AMNION AND SKIN GRAFTS.

Radiation sterilisation is used to increase the safety of the biological tissue grafts in order to prevent transmission of microorganisms causing diseases from the donor tissue to the recipient.

The immune response in rats to gamma-irradiated human amnion and human skin collagen was characterised through histological and immunological methods. Pepsin-extracted human amnion collagen and skin collagen were purified and reconstituted. Implants of amnion collagen demonstrated greater persistence than skin collagen. For amnion collagen implants, no significant inflammatory response was found. Fibroblast and adipocyte in-growth and neovascularisation were present. Conversely, obvious inflammatory infiltration was evident in the skin collagen implants.

Enzyme-linked immunosorbent assay results showed that anti-amnion collagen antibody levels were significantly lower than anti-skin collagen antibody levels against their respective implant materials. The ratios of type I to type III collagen are 56:44 and 95:5 for amnion collagen and skin collagen, respectively. These findings suggest that in this heterologous type system, type III collagen-rich amnion collagen preparations appear superior to skin collagen.

The effects of radiation to amnion collagen compared to the effect of glutaraldehyde (GA) was investigated by Fujisato and co-workers [75]. Radiation cross-linking was performed with gamma ray and electron beam while chemical cross linking was with GA. Both gamma ray and electron beam irradiation decreased the tensile strength and elongation at break of the amniotic membrane with an increase in the irradiation dose, whereas GA cross-linking had no effect on the tensile properties. This is probably due to the scission of collagen chains through irradiation.

No significant change was observed on the water content of cross-linked amniotic membranes for any of the crosslinking methods and in marked contrast with cross-linking of a gelatin membrane. A permeation study revealed that protein permeation through the amniotic membrane was not influenced by the GA concentration at cross-linking. These findings are ascribed to the structure characteristic of the amniotic membrane.

It is possible that cross-linking takes place in the interior of the fibre assembly without impairing the mesh structure, resulting in no change of the water content and protein permeability. In vitro degradation of cross-linked amniotic membranes revealed that radiation cross-linking appeared to be much less effective than GA cross-linking in retarding the degradation, probably because of low cross-linking densities. GA-cross-linked amniotic membranes were degraded more slowly as the GA concentration at cross-linking increased. When the GA-cross-linked amniotic membrane was subcutaneously implanted in the rat, the tissue response was mild, similar to that of the non-cross-linked native membrane.

Tyszkiewicz described the method of preparation of both frozen and freeze-dried amnion grafts sterilised by an irradiation dose of 35 kGy [76]. It has been observed, that lyophilised irradiated allografts were resorbed within a few days, while frozen irradiated ones better adhered to wound and persisted even 3 weeks after grafting, therefore, it has been decided to preserve amnion by deep-freezing.

SUMMARY & CONCLUSIONS

The use of radiation-sterilised skin grafts was described in the 1980's by Rudolf Klen and his co-workers [77]. In this time problems with relatively poor resistance of irradiated skin to infection led to decrease and cessation of its use in clinical practice. With the progress in surgical techniques and burn wound care, and following publication of reports of disease transmission by non-irradiated skin grafts [78,79], increasing the safety of the grafts became a more and more serious concern. Irradiation was used also in grafts preserved in 85% glycerol with good clinical results [9,80].

Amnion and skin are very delicate structural and functional entities used as biological dressings and/or grafts in many clinical situations. The irradiation doses used to sterilise amnion and skin grafts do not merely devitalise the biological tissue, but can also cause structural changes of their main anatomical components and structures. Structures containing large molecules like collagen, proteins, and hyaluronic acid, are most vulnerable to damages caused by irradiation. The most important issue is, that these changes will not affect adversely to large extent the anatomical integrity and functional efficacy of the grafts. It seems to be of utmost importance to use appropriate techniques to decrease the initial and in-process contamination of the tissue grafts, which will enable to reduce the radiation dose for achieving the optimal sterility assurance level. In irradiated amnion, the safety and efficacy was already well proved in clinical practice. In irradiated skin grafts more research is needed to prove their effectiveness by controlled clinical studies.

REFERENCES

[1] Champion, R. H., Burton, J. L. and Ebling, F. J. G. (eds.), *Textbook of Dermatology, 5th Edition*, Blackwell, Oxford, **1992**.

[2] Fitzpatrick, T. B., Eisen, A. Z., Wolff, K., Freedberg, I. M. and Austen, K. F. (eds.), *Dermatology in General Medicine, 3rd Edition*, McGraw-Hill, New York, **1987**.

[3] Graham-Brown, R. and Burns, T., *Lecture Notes on Dermatology, 6th Edition*, Blackwell, Oxford, **1990**.

[4] Farazdaghi, M., Adler, J. and Farazdaghi, S. M., Electron microscopy of human amniotic membranes, In: *Advances in Tissue Banking, Volume 5*, Nather, A. (ed.), World Scientific, Singapore, **2001**, pp.149-171.

[5] Mohamad, H., Anatomy and embryology of human placenta, amnion and chorion, In: *Advances in Tissue Banking, Volume 5*, Nather, A. (ed.), World Scientific, Singapore, **2001**, pp.139-148.

[6] Korlof, B., Simoni, E., Baryd, I., Lamke, L. O. and Eriksson, G., Radiation-sterilized split skin: a new type of biological wound dressing. A preliminary report, *Scand. J. Plast. Reconstr. Surg.*, **1972**, 6, 126-31.

[7] Klen, R. and Pacal, J., Influence of ionizing sterilization on the permeability of human chorio-amniotic, dermo-epidermal and fascial grafts, *Res. Exp. Med. (Berl.)*, **1976**, 167, 15-21.

[8] Tang, Z. Y., Irradiated porcine skin in the treatment of second degree burn, *Zhonghua Zheng Xing Shao Shang Wai Ke Za Zhi* (China), **1990**, 6, 187-188.

[9] Bourroul, S. C., Herson, M. R., Pino, E. and Matho, M. B., Sterilization of skin allografts by ionizing radiation, *Cell Mol. Biol. (Noisy-le-grand)*, **2002**, 48, 803-807.

[10] Alexander, J. W., Macmillan, B. G., Law, E. and Kittur, D. S.. Treatment of severe burns with widely meshed skin autograft and meshed skin allograft overlay, *J. Trauma*, **1981**, 21, 433-438.

[11] Herndon, D. N., Perspectives in the use of allograft, *J. Burn Care Rehab.*, **1992**, 18, S6-S9.

[12] Kearney, J., Quality issues in skin banking: a review, *Burns*, **1998**, 24, 299-305.

[13] Burleson, R. and Eisemann, B., Mechanism of antibacterial effect of biologic dressing, *Ann. Surg.*, **1973**, 177, 181.

[14] Morykwas, M. J., Thornton, J. N. and Barlett, R. H., Zeta potential of synthetic and biological skin substitutes: effects on initial adherence, *Plast. Rec. Surg.*, **1987**, 79, 732-739.

[15] Salisburry, R. E., Caines, R. and McCarthy, R., Comparison of the bacterial clearing effects of different biologic dressings on granulating wounds following thermal injury, *Plast. Rec. Surg.*, **1980**, 66, 596.

[16] Talmi, Y. P., Sigler, L., Inge, E., *et al.*, Antibacterial properties of human amniotic membranes, *Placenta*, **1991**, 11, 285-288.

[17] Reilly, B. D., Goldberg, E. H. and Tomasi, T. B., Prolongation of male skin graft survival in female mice treated with amniotic fluid, *Transplantation*, **1991**, 45, 996-998.

[18] Matsui, S., Yoshina, N. and Oka, T., In vitro and in vivo immunosuppressive activity of amniotic fluid in the rat, *Transplant.*, **1989**, 21, 3307-3308.

[19] Faulk, W. P., Stevens, P. J., Burgos, H., *et al.*, Human amnion as an adjunct in wound healing, *Lancet*, **1980**, 1, 1156-1158.

[20] Burgos, H., Angiogenetic and growth factors in human amnio-chorion and placenta, *Eur. J. Clin. Invest.*, **1983**, 13, 289-96.

[21] Longaker, M. T. and Adzik, T., The biology of fetal wound healing: a review, *Plast. Rec. Surg.*, **1991**, 87, 788-798.

[22] Davis, J. S., Skin transplantation with a review of 550 cases at the Johns Hopkins Hospital, *JHH Report*, **1910**, 15, 307.

[23] Sabella, N., Use of fetal membranes in skin grafting, *Med. Rec. N.Y.*, **1913**, 83, 478.

[24] *EATB General Standards for Tissue Banking*, **1994**.

[25] Panáková, E. and Koller, J., Utilisation of foetal membranes in the treatment of burns and other skin defects. In: *Advances in Tissue Banking, Volume 1*, Phillips, G. O., Strong, D. M., von Versen, R. and Nather, A. (eds.), World Scientific, Singapore, **1987**, pp. 165-181.

[26] Koller, J. and Panáková E., Experience in the use of foetal membranes for the treatment of burns and other skin defects, In: *Advances in Tissue Banking, Volume 2*, Phillips, G. O., Strong, D. M., von Versen, R. and Nather, A. (eds.), World Scientific, Singapore, **1990**, pp.353-359.

[27] Colocho, G., Graham, W. P., Greene A. E., *et al.*, Human amniotic membrane as a physiologic wound dressing, *Arch. Surg.*, **1974**, 109,370-373.

[28] Rennekampf, H. O., Dohrmann, P., Foery, R. and Demertzis, A., In vitro and vivo evaluated viability of cryoconservated amniotic membrane, *1st Europ. Conf. on Tissue Banking and Clinical Application*, Berlin, **1991**, p 103.

[29] Klen, R., *Biological Principles of Tissue Banking*, Pergamon Press, Oxford, **1982**, p. 264.

[30] Visuthikosol, V., Somna, R., Nitiyanant, P. and Navikarn, T., The preparation of lyophilised fetal membrane for biological dressing, *J. Med. Assoc. Thai.*, **1992**, 75, 52-59.

[31] Hoekstra, M. J., Buitenwerf, J., Kreis, R. W., *et al.*, Virology of glycerol-preserved allografts, *3rd Congress of EBA*, Prague, Czechoslovakia, **1989**.

[32] Hoekstra, M. J., Van Ess, I., Ligtvoet, E., et al., Glycerinisation of human cadaver allograft donor skin. A study of the inhibition of growth microorganism, *Abstracts of 2nd Congress of EBA*, Aachen, **1987**.

[33] Xue-Xia, M., *et al.*, The use of glutaraldehyde preserved bovine amniotic membranes, *Chin. J. Plast. Burns*, **1987**, 3, 265.

[34] Ward, D. J., Bennett, J. P., Burgos, H. and Fabre, J., The healing of chronic venous ulcers with prepared human amnion, *Brit. J. Plast. Surg.*, **1989**, 42, 463-467.

[35] Converse, J. M., Šmahel, J., Ballantyne, D. L. and Harper A. D., Inosculation of vessels of skin graft and host bed. A fortuitous encounter, *Br. J. Plast. Surg.*, **1975**, 28, 282.

[36] Hunt, T. K., Basic principles of wound healing, *J. Trauma*, **1990**, 30, S123-S128.

[37] *AATB Standards for Tissue Banking*, **2002**.

[38] *Council of Europe Guide to safety and quality assurance for organs, tissues and cells*, Council of Europe Publishing, **2002**.

[39] *IAEA International Standards for Tissue Banks*, **2002**.

[40] van Baare, J., Buitenwerf, J., Hoekstra, M. J. and Dupont, J. S., Virucidal effect of glycerol as used in donor skin preservation, *Burns*, **1994**, 20, S77-S79.

[41] Rheinwald, J. G. and Green, H., Serial cultivation of strains of human keratinocytes: the formation of keratinizing colonies from single cells, *Cell*, **1975**, 6, 337-343.

[42] Heck, E. L., Sergstrasser, P. R. and Baxter, C. R., Composite skin graft: frozen dermal allograft support of the engraftment and expansion of autologous epidermis, *J. Trauma*, **1985**, 25, 106-111.

[43] Hickerson, W. L., Compton, C., Fletchall, S. and Smith, L. R., Cultured epidermal autografts and allodermis combination for permanent burn wound coverage, *Burns*, **1994**, 20, S52-S56.

[44] Burke, J. F., Yannas, I. V., Quinby, W. C., *et al.*, Successful use of a physiologically acceptable artificial skin in the treatment of extensive burn injury, *Ann. Surg.*, **1981**, 194, 413-428.

[45] Hansbrough, J. F., Cooper, M. L., Cohen, R., Spielvogel, R., Greenleaf, G. and Bartel, R. L., Evaluation of a biodegradable matrix containing cultured human fibroblasts as a dermal replacement beneath meshed skin grafts on athymic mice, *Surgery*, **1992**, 111, 438-446.

[46] Converse, J. M. and Rapaport, F. T., The vascularization of skin autografts and homografts: an experimental study in man, *Ann. Surg.*, **1956**, 143, 306.

[47] Omi, T., Kawanami, O., Matsuda K., et al., Histological characteristics of the healing process of frozen skin allograft used in the treatment of burns, *Burns*, **1996**, 22, 206.

[48] Toranto, I. R., Salyer, K. E. and Myers, M. B., Vascularization of porcine skin heterografts, *Plast. Reconstr. Surg.*, **1974**, 54, 195.

[49] McCabe, W. P., Rebuck, J. W., Kelly, A. P. and Ditmars, S. M., Cellular immune response of humans to pigskin, *Plast. Reconstr. Surg.*, **1973**, 55, 181.

[50] Sheng-de, G., Skin substitutes, In: *Modern Treatment of Burns*, Zhi-yang, F., Zhi-yong, S., Ngao, L. and Sheng-de, G. (eds.), Springer, Berlin, **1992**, pp. 253-260.

[51] Wang, H. J., Chen, T. M. and Cheng, T. Y., Use of a porcine dermis template to enhance widely expanded mesh autologous split-thickness skin graft growth: preliminary report, *J. Trauma*, **1997**, 42, 177-82.

[52] Yannas, I. V. and Burke, J. F., Design of artificial skin I. Basic design principles, *J. Biomed. Mat. Res.*, **1980**, 14, 65-81.

[53] Douglas, B., Homografts of fetal membranes as a covering for large wounds especially those from burns, *J. Tenness. Med. Assoc.*, **1952**, 45, 230.

[54] Maier, W., Eihautueberpflanzung bei Kindern mit ausgedehnten Verbrennungen, *Pädiat. Prax.*, **1965**, 4, 581-583.

[55] Bruck, J. C., Hormann, T. and Von Stetten, M., Indicators for the transplantation of amniotic membranes, In: 5th International Conference on Tissue Banking, Berlin, Germany, **1996**, p. 65.

[56] Goebel, P. and Schubert, W., Personal experiences with covering burns in children with amnion, *Brit. Orthop. Traumatol.*, **1990**, 37, 495-498.

[57] Unger, M. G. and Roberts, M., Lyophilized amniotic membranes on graft donor sites, *Brit. J. Plast. Surg.*, **1976**, 29, 99-101.

[58] Quinby, W. C., Hoover, H. C., Schefflan, M., *et al.*, Clinical trials of amniotic membranes in burn wound care, *Plast. Reconstr. Surg.*, **1982**, 70, 711 716.

[59] Waikakul, S., Chumniprasas, K., Setasubun, S., *et al.*, Application of freeze-dried amniotic membrane: a control trial at donor site of split thickness skin grafting, *Bull. Hosp. Joint Dis.*, **1990**.

[60] Pigeon, J., Treatment of second degree burns with amniotic membranes, *Can. Med. Assoc. J.*, **1960**, 83, 844.

[61] Bose, B., Burn wound dressing with human amniotic membrane, *Ann. Roy. Coll. Surg. Engl.*, **1979**, 61,444-7.

[62] Dahinterová, J. and Dobrkovsklá, M., Treatment of the burned surface grafts, *Suppl. Sborn. ved. Prací lék. Fak v Hradci Králové*, **1968**, 11, 513.

[63] Ding, Y.-L. and Han, Ch.-M., Recent advances in burn wound management in China, *Acta Chir. Plast.*, **1989**, 31, 84-91.

[64] Dlabalová, H., Klen, R. and Jandová, E., Treatment of leg ulcers with chorion grafts, *Čs. Dermatol.*, **1972**, 47, 42.

[65] Trelford, J. D., Hanson, F. W. and Anderson, D. G., Wound healing and the amniotic membrane, *J. Med. Clin. Exptl. Theoret.*, **1976**, 6, 383-386.

[66] Bennett, J. P., Matthews, R. and Faulk, W. P., Treatment of chronic ulceration of the legs with human amnion, *Lancet*, **1980**, 1, 1153.

[67] Stern, M., The grafting of preserved amniotic membranes to burned and ulcerated surfaces, substituting skin grafts, *JAMA*, **1913**, 60, 973.

[68] Young, R. L., Cota, J., Zund, G., *et al.*, The use of amniotic membrane graft to prevent postoperative adhesions. *Fertil. Steril.*, **1991**, 55, 624-628.

[69] Batmanov, L. E., Egorova, K.S. and Kolesnokova, L. N., Use of fresh amnion in the treatment of corneal diseases, *Vestn. Ottalmol.*, **1990**, 106, 17-19.

[70] Ti, S. E., Tow, S. L. and Chee, S. P., Amniotic membrane transplantation in entropion surgery, *Ophthalmology*, **2001**, 108, 1209-1217.

[71] Rama, P., Giannini, R., Bruni, A., Gatto, C. and Tiso, R., Further evaluation of amniotic membrane banking for transplantation in ocular surface diseases, *Cell Tissue Banking*, **2001**, 2, 155-163.

[72] Prabhasawat, P. and Tesavibul, N., Preserved amniotic membrane transplantation for conjunctival surface reconstruction, *Cell Tissue Banking*, **2001**, 2, 31-39.

[73] Jackson, D., A clinical study of the use of skin homografts for burns, *Brit. J. Plast. Surg.*, **1954**, 7, 26.

[74] Burns Unit, First Affiliated Hospital, Number 245 Unit, Chinese PLA, A review of the management of extensive third-degree burns in 14 successive years, *Chinese Med. J.*, **1973**, 11, 148.

[75] Fujisato, T., Tomihata, K., Tabata, Y., Iwamoto, Y., Burczak, K. and Ikada, Y., Cross-linking of amniotic membranes, *J. Biomater. Sci. Polym. Ed.*, **1999**, 10, 1171-1181.

[76] Tyszkiewicz, J. T., Uhrynowska-Tyszkiewicz, I. A., Kaminski, A. and Dziedzic-Goclawska, A., Amnion allografts prepared in the Central Tissue Bank in Warsaw, *Ann. Transplant.*, **1999**, 4, 85-90.

[77] Chylkova-Horakova, V., Buriankova, E. and Klen, R., Radiosterilization of dermoepidermal grafts, *Sb Ved Pr Lek Fak Karlovy Univerzity Hradci Kralove*, **1980**, 22, 87-96.

[78] Kealey, G. P., Aguiar, J., Lewis, R. W., Rosenquist, M. D., Strauss, R. G. and Bale, J. F., Cadaver skin allografts and transmission of human cytomegalovirus to burn patients, *J. Amer. Coll. Surg.*, **1996**, 182, 201-205.

[79] Kealey, G. P., Disease transmission by means of allograft, *J. Burn Care Rehab.*, **1997**, 18, S10.

[80] Bolgiani, A., The role of biosynthetic skin replacements, In: *Radiation and tissue banking*, Phillips, G. O. (ed.), World Scientific, Singapore, **2000**, pp. 173-216.

STRUCTURAL EFFECTS OF RADIATION STERILISATION ON SODIUM HYALURONATE

John F. Kennedy [1*], Maria P. C. da Silva [2], Linda L. Lloyd [1] and Charles J. Knill [1]

[1] *Chembiotech Laboratories, Institute of Research & Development, University of Birmingham Research Park, Vincent Drive, Edgbaston, Birmingham, West Midlands, B15 2SQ, UK.*
{E-mail: jfkennedy@chembiotech.co.uk}

[2] *Laboratório de Imununopatologia, Keizo Asami (LIKA) & Departamento de Bioquímica, Universidade Federal de Pernambuco (UFPE), Avenida Prof. Moraes Rêgo 1235, Cidade Universitária, 50670-420 PE, Recife, Pernambuco, Brazil.*

ABSTRACT

The molecular weight profiles, reducing and unsaturated carbohydrates in gamma and electron beam (E-beam) irradiated (0-100 kGys) sodium hyaluronate samples (NaHA, in solid form and 1% w/v aqueous solutions), were determined by gel permeation chromatography (GPC), 3,5-dintrosalicylic (DNS) acid assay, and Warren assay, respectively, in order to assess induced degradation, the latter via an anticipated free radical-induced β-elimination mechanism. NaHA was depolymerised by both gamma- and E-beam-irradiation, the extent of depolymerisation being dependent upon irradiation intensity. Gamma-irradiation generally resulted in higher degrees of depolymerisation than E-beam-irradiation, and NaHA solutions were depolymerised to a greater extent than NaHA solids by gamma-irradiation, at the same irradiation levels. Measured reducing sugars correlated with theoretical levels (calculated from determined peak molecular weights, M_P). Measured non-reducing unsaturated glucuronates were lower than measured / theoretical reducing sugars, indicating that not all non-reducing ends created by depolymerisation contained Warren assay active unsaturated sugars.

INTRODUCTION

Hyaluronan is a biocompatible / biodegradable, linear, water-soluble, glycosaminoglycan (GAG) composed of repeating (1→4) linked disaccharide units consisting of (1→3) linked β-D-glucopyranosyluronic acid (β-D-Glc*p*A) and 2-acetamido-2-β-D-glucopyranosyl (*N*-acetyl-D-glucosamine) (β-D-Glc*p*NAc) units (Figure 1) [1,2]. It has a high molecular weight (around 10^5-10^7 Da), depending on source, giving a DP range of ~ 250-25000 [3]. It is found in all vertebrates, being present in almost every tissue as a component of the extracellular matrix and is distributed throughout the mammalian body, especially in synovial fluid, connective tissue, umbilical cord and the vitreous body of the eye [4]. The largest amount of hyaluronan (7-8 g per average human, 50 % of the total in the body) is in the skin tissues (both the dermis and epidermis) [5(a)]. It is commercially available as the free acid (hyaluronic acid, HA) or in salt form (e.g. sodium hyaluronate, NaHA), and the main commercial sources are rooster combs, human umbilical cords and continuous bacterial fermentation (using *Streptococcus equi*). Using a microbial source in pharmaceutical and cosmetic applications is advantageous since it eliminates any risks associated viruses or prion proteins from avian / mammalian hyaluronan sources.

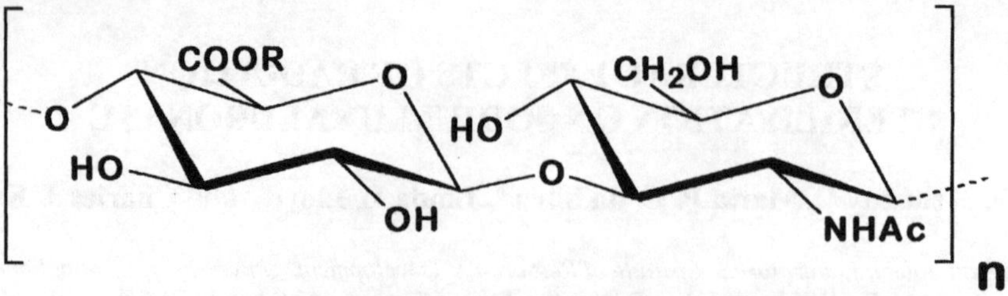

→4)-β-D-Glc*p*A-(1→3)-β-D-Glc*p*NAc-(1→

Figure 1. Disaccharide repeating unit structure in hyaluronan molecules (R = H = hyaluronic acid, HA; R = Na = sodium hyaluronate, NaHA)

In solution the hyaluronan backbone is stiffened by intermolecular hydrogen bonds, mutually repelling anionic groups, and solvent interactions, making it a rigid and highly hydrated molecule. It adopts an expanded random coil structure in physiological solutions, occupying a large domain. Small molecules, e.g. water and electrolytes, can freely diffuse through the domain, whilst large molecules are partially excluded due to their hydrodynamic size. At low concentrations, individual chains entangle forming a continuous network with viscoelastic and pseudoplastic properties. Entangled networks exhibiting elastic properties can be formed at higher concentrations, which can resist rapid, short duration fluid flow. However, fluid flow of longer duration can partially separate and align molecules, allowing movement and thus exhibit viscous properties [6].

The viscoelastic properties of hyaluronan solutions are ideal for use as a biological shock absorber and lubricant, which is why it is present in synovial fluid, where it lubricates cartilage between joints. Cartilage provides a cushion between bones allowing smooth joint movement. During the inflammatory stage of arthritis joint fluid elasticity / viscosity is reduced due to free radical depolymerisation of HA, diminishing shock absorbing and barrier properties [7-10]. As the joint is used, stress causes fibrillation and dislocation of cartilage and synovial tissue collagen networks. Viscoelastic hyaluronan-containing solutions can be injected into osteoarthritic joints (visco-supplementation) in order to restore the joint rheological environment and thus provide shock absorption and improved function by decreasing pain associated with mobility [5(b)]. Viscoelastic hyaluronan solutions are also used in ophthalmic surgery as vitreous supplement that ensures protection, lubrication and hydration of tissue surfaces, preventing post-operative adhesion and eye damage by induced shearing forces.

Biologically, hyaluronan is more than just a high viscosity space filler, since it is capable of interacting with many biomolecules. It forms aggregates with other GAGs, which along with a fibrous collagen matrix provide stability and elasticity to the extracellular environment [11,12]. In wounds it acts as a sacrificial free radical scavenger, modulating inflammation [5(c)]. It is recognised by receptors on cells associated with tissue repair and regeneration. Incorporation of hyaluronan into infected wounds, were the normal healing process is compromised, is reported to accelerate wound healing.

Hyaluronan is degraded by acid, alkali, enzymes, transition metals (ferrous, cuprous & stannous ions) [13-18], L-ascorbic acid [15(a),17], X-rays [18,19], uv light and electron flux [16,20-23], gamma irradiation [16,18,22,24-27], and ultrasonics [16,28]. Degradation induced by many of these factors is believed to involve free radicals, the glucuronate C5 radical being the most stable, where β-elimination results in production of unsaturated terminal non-reducing glucuronate residues (Figure 2) [16,29-35].

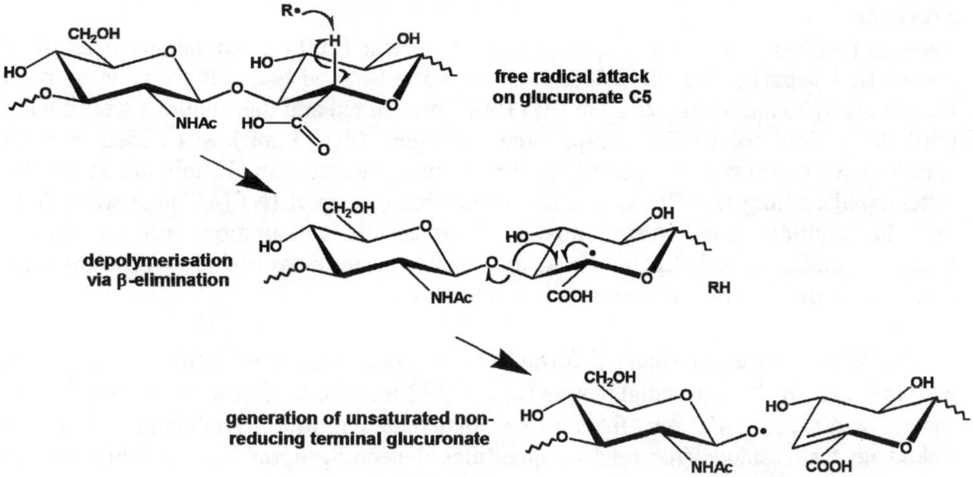

Figure 2. Depolymerisation of hyaluronan via free radical induced β-elimination resulting in unsaturated terminal non-reducing glucuronate residues.

In this paper NaHA (from *Streptococcus equi*) characterisation, pre- and post-sterilisation by gamma and electron beam irradiation (at 1-100 kGys), is presented. Elucidation of the degree of NaHA modification afforded by such sterilisation techniques (routinely used for biomedical products [36(a)&(b)]) is required since depolymerisation/modification can have a dramatic effect on activity and performance.

MATERIALS & METHODS

NaHA Samples

A range of sodium hyaluronate (NaHA) samples were produced by experimental pilot scale fermentations (using *Streptococcus equi*) and purified via formaldehyde treatment, alcohol precipitation, and chromatographic fractionation (protein contents < 1% w/w). NaHA samples were subjected to gamma and electron beam irradiation (at 1, 2, 5, 10, 25 & 100 kGys, in dried solid form and 1% w/v in 18 MΩ UHQ water) [16]. NaHA samples were dried to constant mass over several days in an Abderhalden drying pistol under reduced pressure using phosphorus pentoxide (desiccant) at 65°C (refluxing methanol jacket), and were stored in a vacuum desiccator (over silica gel) until required.

Determination of unsaturated sugars by Warren assay [14,16,37-39]

Reagents:
(i) Sodium metaperiodate (0.2 M) in phosphoric acid (9 M); (ii) sodium arsenite (6.5% w/v) and sodium sulphate (0.5 M) in sulphuric acid (0.05 M); (iii) 2-thio-barbituric acid (0.6% w/v) in sodium sulphate (0.5 M); (iv) cyclohexanone.

Calibration Standards:
N-acetylneuraminic acid, 2-deoxy-D-glucose and dextran (all 125 μg/mL in 18 MΩ UHQ water) were diluted accordingly with 18 MΩ UHQ water to produce a range of calibration standards.

Procedure:

Reagent (i) (100 μL) was added to samples/standards (200 μL, containing 0-25 μg of unsaturated sugars), and the solutions were vortex mixed and left to stand at room temperature (20 minutes). Reagent (ii) (1 mL) was added and the solutions were shaken until the yellow colouration disappeared. Reagent (iii) (3 mL) was added, and the solutions were vortex mixed, and heated in a boiling water bath (15 minutes at 100°C). After rapid cooling (25°C) using a cold-water bath, reagent (iv) (4.2 mL) was added, and the solutions were vortex mixed to extract the colouration into the organic (cyclohexanone) layer. Organic layers absorbances were measured (350-700 nm) using a Shimadzu uv/vis 240 scanning spectrophotometer.

The Warren assay produces β-formylpyruvic acid (λ_{max} ~ 549 nm) from N-acetyl-neuraminic acid [40], malondialdehyde (λ_{max} ~ 532 nm) from most deoxysugars [41], and formic acid (λ_{max} ~ 450 nm) from all terminal reducing, all non-reducing and (1→6)-linked neutral residues (the relative quantities depending upon the individual linkage configurations) [37] (Figure 3). In the case of the Warren assay performed on hyaluronan (and free radical degraded hyaluronan), formic acid is produced from terminal non-reducing glucuronate residues (C3), terminal reducing glucosamine residues (C5), and terminal unsaturated non-reducing glucuronate residues. β-Formylpyruvic acid is also produced from terminal unsaturated non-reducing glucuronate residues (Figure 4), which arise as a result of free radical-induced degradation (Figure 2). Non-terminal glucuronate and glucosamine residues in hyaluronan do not result in liberation of formic acid or β-formylpyruvic acid in the Warren assay as they do not possess the required hydroxyl group configurations.

The unsaturated uronic acid and formic acid producing end-group contents of all NaHA samples were determined from the N-acetylneuraminic acid and dextran calibration curves (mass *vs* absorbance) and the associated calibration equations (determined by linear regression analysis).

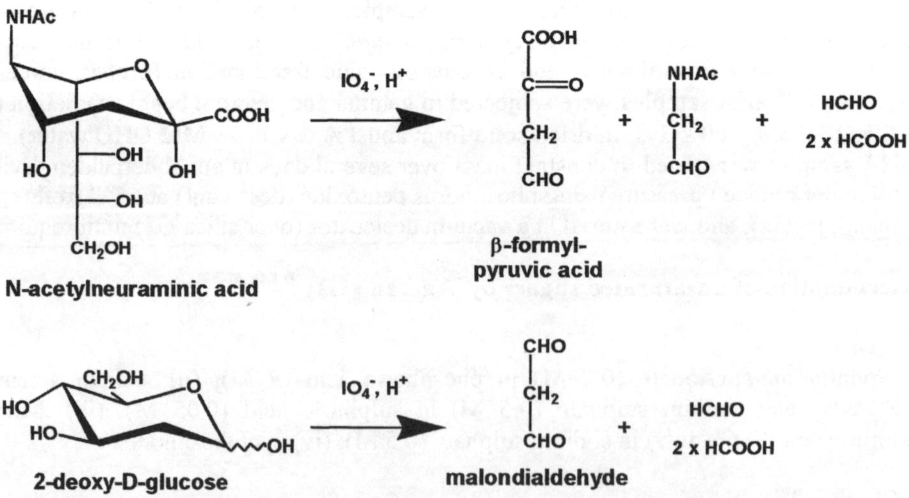

Figure 3. Production of β-formylpyruvic acid (λ_{max} 549 nm), malondialdehyde (λ_{max} 532 nm), & formic acid (λ_{max} 450 nm) in the Warren assay from N-acetylneuraminic acid and 2-deoxy-D-glucose.

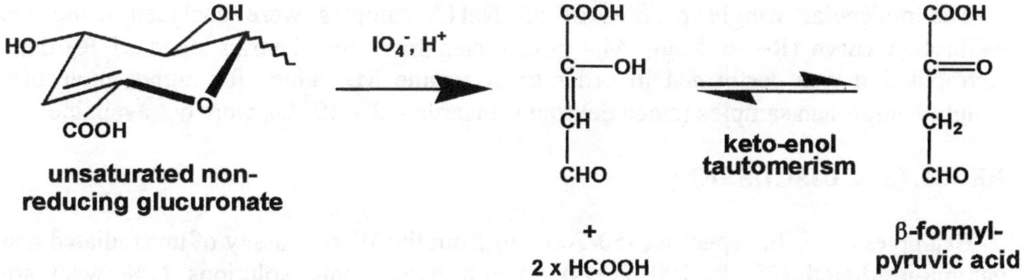

Figure 4. Production of β-formylpyruvic acid in the Warren assay from unsaturated terminal non-reducing glucuronate residues in degraded hyaluronan.

Determination of reducing sugars by 3,5-dinitrosalicylic acid assay [16,37,38,42]

Reagent:
3,5-Dinitrosalicylcic acid (0.25 g) and sodium potassium tartrate (75 g) dissolved in sodium hydroxide (2 M, 50 mL), diluted to 250 mL with 18 MΩ UHQ water.

Standard:
N-acetyl-D-glucosamine (D-GlcNAc, 5 mg/mL in 18 MΩ UHQ water) was diluted accordingly with 18 MΩ UHQ water to produce a range of calibration standards.

Procedure:
Reagent (1 mL) was added to samples/standards (100 μL, containing 0-500 μg of reducing sugars), and the solutions were vortex mixed, and heated in a boiling water bath (10 minutes at 100°C). After cooling (25°C) using a cold-water bath, the absorbances were measured (350-700 nm, $\lambda_{max} \sim 570$ nm) using a Shimadzu uv/vis 240 scanning spectrophotometer. Reducing sugar contents of all NaHA samples were determined from the calibration curve (mass *vs* absorbance) and the associated calibration curve equation (determined by linear regression analysis).

Determination of molecular weight profiles by GPC [16]

The molecular weight profiles of unirradiated and gamma/electron beam irradiated samples were determined by GPC analysis using the isocratic system detailed below:

Instrumentation:	LDC Analytical Constametric III pump Rheodyne 7125 injection valve (200 μL sample loop) LDC Analytical Refracto Monitor RI detector Knauer Twin Pen Model 42 chart recorder
GPC Columns:	2 x PL-GFC 4000 Å & PL-GFC 300 Å connected in series
Eluent:	0.05 M phosphate buffer containing 0.25 M NaCl, pH 7.0 (prepared using 18 MΩ UHQ water with helium sparging)
Flow rate:	0.5 mL/minute
Calibration:	Dextran standards (Pharmacia, Sweden) with M_P values of 2000, 500, 150, 110, 70, 40 & 10 kDa (1 mg/mL in eluent)
Injections:	100 μL for dextran standards & samples (NaHA solids, 0.5 mg/mL in 18 MΩ UHQ water; NaHA 1% w/v solutions diluted to 0.5 mg/mL using 18 MΩ UHQ water)

The molecular weight profiles of all NaHA samples were analysed using the calibration curve (R_T *vs* Log_{10} M_P) constructed from the dextran standard R_T data. Extrapolation was performed in order to determine M_P values for higher molecular weight hyaluronan samples (since dextran standards > 2×10^6 Da were not available).

RESULTS & DISCUSSION

Examples of visible spectra (350-700 nm) from the Warren assay of unirradiated and gamma-irradiated (25 & 100 kGys) sodium hyaluronate solutions (1% w/v) are displayed in Figure 5. The peak at $\lambda_{max} \sim 450$ nm corresponds to formic acid released from terminal non-reducing glucuronate residues (C3), terminal reducing glucosamine residues (C5), and terminal unsaturated non-reducing glucuronate residues. The peak at $\lambda_{max} \sim 549$ nm corresponds to β-formylpyruvic acid released from terminal unsaturated non-reducing glucuronate residues (Figure 4). There is a visible increase in both peaks as a function of increasing irradiation, which is indicative of increasing free radical-induced depolymerisation (generation of unsaturated glucuronate and other terminal residues, Figure 2). As expected, the unirradiated sample has essentially no β–formylpyruvic acid peak. There is also another peak ($\lambda_{max} \sim 510$ nm) in the gamma-irradiated samples, which is at too low a wavelength for malondialdehyde (from deoxysugars, $\lambda_{max} \sim 532$ nm, Figure 3), and may therefore correspond to periodate-released Warren assay active compounds from other free radical degradation products.

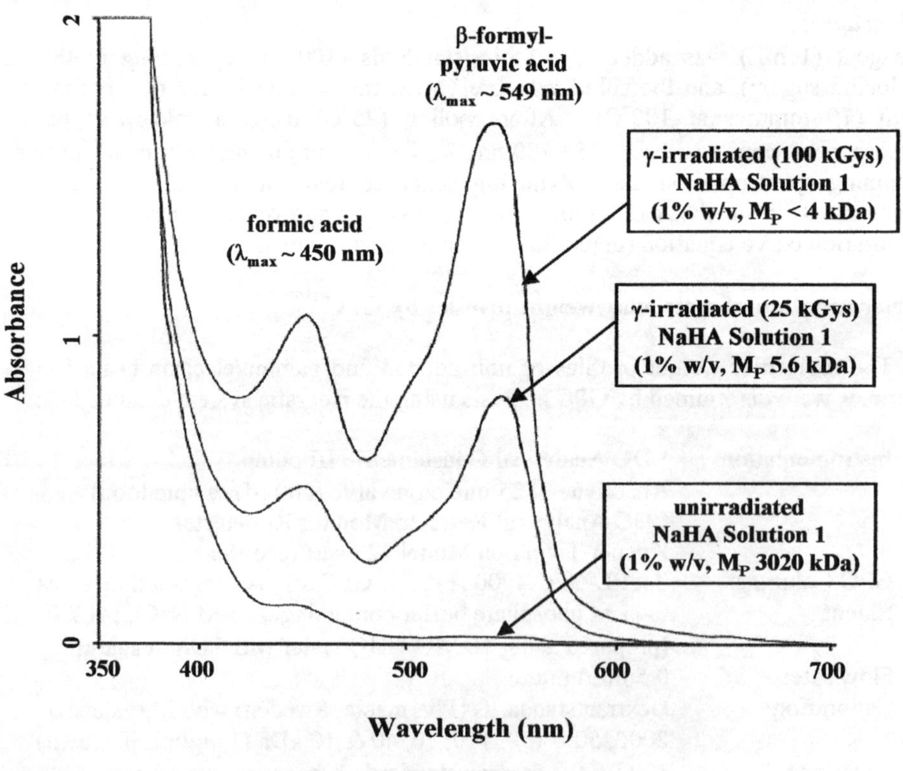

Figure 5. Warren assay spectra for unirradiated and gamma-irradiated (25 & 100 kGys) NaHA Solution 1 (1% w/v).

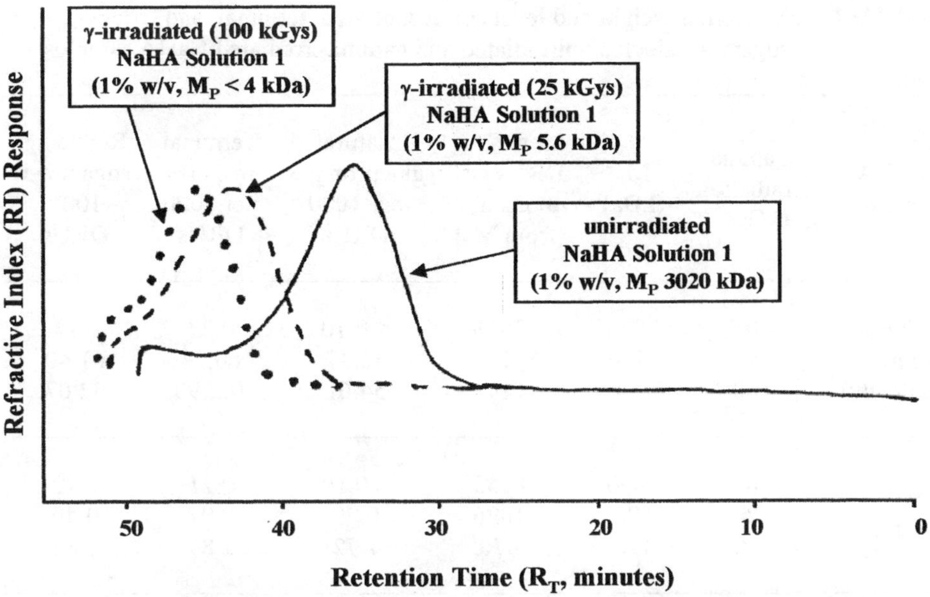

Figure 6. Molecular weight profiles of unirradiated and gamma-irradiated (25 & 100 kGys) NaHA Solution 1 (1% w/v).

Examples of molecular weight profiles (GPC chromatograms) of unirradiated and gamma-irradiated (25 & 100 kGys) sodium hyaluronate solutions (1% w/v) are displayed in Figure 6, which clearly shows the dramatic reduction in NaHA solution (1% w/v) peak molecular weight (M_P, from 3020 kDa to < 4 kDa) as a function of increased gamma-irradiation (0-100 kGys).

The peak molecular weight (M_P), estimated mean number of hyaluronan anhydro-disaccharide repeating units (DRUs, calculated from the M_P) and the levels of unsaturated, terminal, and reducing sugars in selected unirradiated and gamma-irradiated NaHA samples are displayed in Table 1. NaHA in solution is clearly more susceptible to free radical-induced depolymerisation than solid NaHA, since its M_P is more drastically reduced at similar irradiation levels, and its therefore has significantly higher levels of unsaturated, terminal and reducing sugars.

The fact that in all cases the unsaturated sugar levels are lower than the corresponding reducing sugar levels, indicates that not all depolymerisation has occurred via the β-elimination mechanism detailed in Figure 2. This means that there are either other non-reducing unsaturated residues arising from free radical-induced depolymerisation that do not yield β-formylpyruvic acid in the Warren assay (Figures 3 & 4), or there are some saturated non-reducing residues formed during the depolymerisation process.

The reducing sugar levels correlate reasonably well with the theoretical reducing sugar levels based upon the M_P data. Differences will arise as a result of the M_P data being significantly different from the number (M_n) and weight (M_W) average molecular weights (i.e. the samples having a significantly high degree of polydispersity (*d*). The sum of the unsaturated and terminal residue levels should theoretically equate to twice the reducing sugar levels, since the terminal levels arise from the formic acid released from the reducing and non-reducing ends of hyaluronan molecules.

Table 1. Molecular weight and levels of unsaturated, terminal, and reducing sugars in selected unirradiated and gamma-irradiated NaHA samples.

NaHA Sample	Gamma irradiation (kGys)	M_P (kDa)	Mean No. DRUs per molecule (from M_P)	Unsaturated glucuronates per 1000 DRUs	Terminal residues per 1000 DRUs	Reducing sugars per 1000 DRUs
NaHA Solution 1 (1% w/v)	0	3020	~ 7968	< 0.10	0.22	0.13
	25	5.6	~ 15	32.37	66.74	61.40
	100	< 4	< 11	54.61	142.90	92.07
NaHA Solid 1	0	3090	~ 8153	< 0.10	0.21	0.12
	25	692	~ 1826	0.29	0.97	0.40
	100	178	~ 470	1.72	2.83	1.84

{DRU = hyaluronan anhydro-disaccharide repeating unit, MW ~ 379 g/mole}

The peak molecular weight (M_P), estimated mean number of hyaluronan anhydro-disaccharide repeating units (DRUs, calculated from the M_P) and the levels of unsaturated sugars in selected unirradiated, gamma-irradiated and electron beam (E-beam)-irradiated NaHA solid samples are displayed in Tables 2 and 3, and represented graphically in Figure 7, respectively.

Table 2. Molecular weight and levels of unsaturated sugars in selected unirradiated and gamma-irradiated NaHA solid samples.

NaHA Sample	Gamma irradiation (kGys)	M_P (kDa)	Mean No. DRUs per molecule (from M_P)	Unsaturated glucuronates per 1000 DRUs
NaHA Solid 2	0	5010	~ 13219	< 0.10
	5	2090	~ 5515	0.14
	10	1000	~ 2639	0.32
	25	400	~ 1055	0.79
NaHA Solid 3	0	5890	~ 15541	< 0.10
	5	2820	~ 7441	0.10
	10	1740	~ 4591	0.21
	25	580	~ 1530	0.59

{DRU = hyaluronan anhydro-disaccharide repeating unit, MW ~ 379 g/mole}

Table 3. Molecular weight and levels of unsaturated sugars in selected unirradiated and electron beam-irradiated NaHA solid samples.

NaHA Sample	Gamma irradiation (kGys)	M_P (kDa)	Mean No. DRUs per molecule (from M_P)	Unsaturated glucuronates per 1000 DRUs
NaHA Solid 2	0	5010	~ 13219	< 0.10
	5	2340	~ 6174	0.10
	10	1820	~ 4802	0.20
	25	480	~ 1266	0.73
NaHA Solid 3	0	5890	~ 15541	< 0.10
	5	3090	~ 8153	0.11
	10	1910	~ 5040	0.15
	25	530	~ 1398	0.56

{DRU = hyaluronan anhydro-disaccharide repeating unit, MW ~ 379 g/mole}

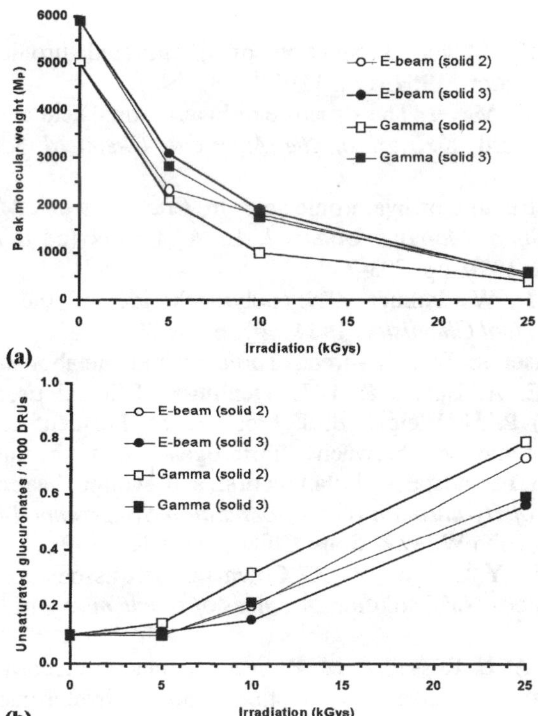

Figure 7. Effects of increasing irradiation levels on: (a) Peak molecular weight (M_P); and (b) unsaturated glucuronate levels per 1000 DRUs.

Gamma-irradiation generally results in a higher degree of depolymerisation of NaHA solids than electron beam-irradiation at the same irradiation level. As in the case of the data in Table 1, non-reducing unsaturated sugar levels are significantly lower than the theoretical reducing sugar levels based upon the M_P data (there being one reducing and one non-reducing end to each hyaluronan molecule). Again, this indicates that not all depolymerisation has occurred via the β-elimination mechanism detailed in Figure 2.

CONCLUSIONS

- Sodium hyaluronate (NaHA) is depolymerised by both gamma- and electron beam-irradiation, extent of depolymerisation being dependent upon irradiation intensity.
- Gamma-irradiation generally results in a higher degree of depolymerisation of NaHA than electron beam-irradiation at the same irradiation level.
- NaHA solutions (1% w/v) are depolymerised to a much greater extent than NaHA solids by gamma-irradiation at the same irradiation level.
- Measured reducing sugar levels correlate reasonably well with theoretical levels (calculated based upon determined peak molecular weights, M_P).
- Measured non-reducing unsaturated glucuronate levels were consistently lower than measured / theoretical reducing sugar levels, indicating that not all non-reducing ends created by depolymerisation contain Warren assay active unsaturated sugars.

REFERENCES

[1] A. Linker & K. Meyer, Production of unsaturated uronides by bacterial hyaluronidases, *Nature*, **1954**, 174, 1192-1193.

[2] B. Weissman & K. Meyer, The structure of hyalbiuronic acid and hyaluronic acid from umbilical cord, *Journal of the American Chemical Society*, **1954**, 76, 1753-1757.

[3] T. C. Laurent, Structure of hyaluronic acid, In: *Chemistry and Molecular Biology of the Extracellular Matrix, Volume 2*, E. A. Balazs (ed.), Academic Press, New York, USA, **1970**, pp. 703-732.

[4] K. Meyer & J. W. Palmer, The polysaccharide of the vitreous humor, *Journal of Biological Chemistry*, **1934**, 107, 629-634.

[5] (a) J. R. E. Fraser & T. C. Laurent, Turnover and metabolism of hyaluronan, pp. 41-53; (b) E. A. Balazs & J. L. Denlinger, Clinical uses of hyaluronan, pp. 265-280; (c) P. H. Weigel, S. J. Frost, R. D. LeBoeuf & C. T. McGary, The specific interaction between fibrin(ogen) and hyaluronan: possible consequences in haemostasis, inflammation and wound healing, pp. 247-264; In: *The Biology of Hyaluronan (Ciba Foundation Symposium 143)*, D. Evered & J. Whelan (eds.), John Wiley & Sons, Chichester, UK, **1989**.

[6] R. E. Turner, P. Y. Lin & M. K. Cowman, Self-association of hyaluronate segments in aqueous NaCl solution, *Archives of Biochemistry & Biophysics*, **1988**, 265, 484-495.

[7] M. F. McCarty, A. L. Russell & M. P. Seed, Sulphated glycosaminoglycans and glucosamine may synergize in promoting synovial hyaluronic acid synthesis, *Medical Hypotheses*, **2000**, 54, 798-802.

[8] R. A. Greenwald, Oxygen radicals, inflammation, and arthritis – pathophysiological considerations and implications for treatment, *Seminars in Arthritis & Rheumatism*, **1991**, 20, 219-240.

[9] P. H. Weigel, V. C. Hascall & M. Tammi, Hyaluronan synthases, *Journal of Biological Chemistry,* **1997**, 272, 13997-14000.

[10] B. J. Parsons, Chemical aspects of free radical reactions in connective tissue, In: *Free Radical Damage and its Control,* C. A. Rice-Evans & R. H. Burdon (eds.), Elsevier, Amsterdam, The Netherlands, **1994**, pp. 281-300.

[11] S. Al-Assaf, G. O. Phillips, G. O. Deeble, B. J. Parsons, H. Starnes & C. von Sonntag, The enhanced stability of the cross-linked hylan structure to hydroxyl (OH) radicals compared with the uncross-linked hyaluronan, *Radiation Physics & Chemistry,* **1995**, 46, 207-217.

[12] U. B. G. Laurent & R. K. Reed, Turnover of hyaluronan in the tissues, *Advanced Drug Delivery Reviews,* **1991**, 7, 237-256.

[13] H. Cho Tun, *Investigation of Carbohydrate Carbonates and Free Radical Degradation of Hyaluronic Acid,* PhD Thesis, University of Birmingham, Birmingham, UK, **1971**.

[14] J. F. Kennedy & H. Cho Tun, The degradation of hyaluronic acid by ferrous ions, *Carbohydrate Research,* **1972**, 22, 43-51.

[15] (a) C. J. Knill, Y. Latif, J. F. Kennedy & D. C. Ellwood, Effect of metal ions on the rheological flow profiles of hyaluronate solutions, pp. 175-180; (b) M. D. Rees, C. L. Hawkins & M. J. Davies, Polysaccharide fragmentation induced by hydroxyl radicals and hypochlorite, pp. 151-160; In: *Hyaluronan – Volume 1: Chemical, Biochemical and Biological Aspects,* J. F. Kennedy, G. O. Phillips, P. A. Williams & V. C. Hascall (eds.), Woodhead, Cambridge, UK, **2002**.

[16] M. P. C. da Silva, *Studies of Chemical and Radiolytic Changes to Sodium Hyaluronate,* PhD Thesis, University of Birmingham, Birmingham, UK, **1992**.

[17] G. O. Matsumura & W. Pigman, Catalytic role of copper and iron ions in the depolymerization of hyaluronic acid by ascorbic acid, *Archives of Biochemistry & Biophysics,* **1965**, 110, 526-533.

[18] G. O. Matsumura, A. Herp & W. Pigman, Depolymerization of hyaluronic acid by autoxidants and radiations, *Radiation Research,* **1966**, 28, 735-752.

[19] A. Caputo, Depolymerization of hyaluronic acid by x-rays, *Nature,* 1957, 179, 1133-1134.

[20] E. A. Balazs, T. C. Laurent, A. F. Howe & L. Varga, Irradiation of mucopolysaccharides with ultraviolet light and electrons, *Radiation Research,* **1959**, 11, 149-154.

[21] E. Hvidberg, Sv. A. Kvorning, A. Schmidt & J. Schou, Effect of ultraviolet irradiation on hyaluronic acid *in vitro, Acta Pharmacologica et Toxicologica,* **1959**, 15, 356-364.

[22] E. A. Balazs, J. V. Davies, G. O. Phillips & M. D. Young, Transient intermediates in the radiolysis of hyaluronic acid, *Radiation Research,* **1967**, 31, 243-255.

[23] L. Lapcik & J. Schurz, Photochemical degradation of hyaluronic-acid by singlet oxygen, *Colloid & Polymer Science,* **1991**, 269, 633-635.

[24] J. S. Moore, G. O. Phillips, J. V. Davies & K. S. Dodgson, Reactions of connective tissue and related polyanions with hydrated electrons and hydroxyl radicals, *Carbohydrate Research,* **1970**, 12, 253-260.

[25] M. Lal, Radiation-induced depolymerization of hyaluronic-acid (HA) in aqueous-solutions at pH 7.4, *Journal of Radioanalytical & Nuclear Chemistry,* **1985**, 92, 105-112.

[26] D. J. Deeble, G. O. Phillips, E. Bothe, H. P. Schuchmann & C. von Sonntag, The radiation-induced degradation of hyaluronic acid, *Radiation Physics & Chemistry,* **1991**, 37, 115-118.

[27] G. O. Phillips, Molecular transformations in connective tissue hyaluronic acid, In: *Viscoelasticity of Biomaterials (ACS Symposium Series 489)*, W. G. Glasser & H. Hatakeyama (eds.), ACS, Washington, USA, **1992**, pp. 168-183.

[28] P. Chabrecek, L. Soltes, Z. Kallay & I. Novak, Gel-permeation chromatographic characterization of sodium hyaluronate and its fractions prepared by ultrasonic degradation, *Chromatographia*, **1990**, 30, 201-204.

[29] J. Kiss, β-Eliminative degradation of carbohydrates containing uronic acid residues, *Advances in Carbohydrate Chemistry & Biochemistry*, **1974**, 29, 229-303.

[30] S. Al-Assaf, C. L. Hawkins, B. J. Parsons, M. J. Davies & G. O. Phillips, Identification of radicals from hyaluronan (hyaluronic acid) and cross-linked derivatives using electron paramagnetic resonance spectroscopy, *Carbohydrate Polymers*, **1999**, 38, 17-22.

[31] C. L. Hawkins & M. J. Davies, Direct detection and identification of radicals generated during the hydroxy radical-induced degradation of hyaluronic acid and related materials, *Free Radical Biology & Medicine*, **1996**, 21, 275-290.

[32] C. L. Hawkins & M. J. Davies, Degradation of hyaluronic acid, poly- and monosaccharides and model compounds by hypochlorite: evidence for radical intermediates and fragmentation, *Free Radical Biology & Medicine*, **1998**, 24, 1396-1410.

[33] M. J. Davies, Detection and identification of macromolecule-derived radicals by EPR spin-trapping, *Research on Chemical Intermediates*, **1993**, 19, 669-679.

[34] D. J. Wedlock & G. O. Phillips, Depolymerisation of sodium hyaluronate during freeze drying, *International Journal of Biological Macromolecules*, **1983**, 5, 186-188.

[35] B. C. Gilbert, D. M. King & C. B. Thomas, The oxidation of some polysaccharides by the hydroxyl radical: an esr investigation, *Carbohydrate Research*, **1984**, 125, 217-235.

[36] (a) Isomedix, Radiation sterilization dose audit, pp. 79-81; (b) R. Calhoun, G. M. Sullivan & C. B. Williams, Validation issues for electron beam systems, pp. 83-90; In: *Sterilization of Medical Devices*, A. F. Booth (ed.), Interpharm Press, Buffalo Grove, USA, **1999**.

[37] C. A. White & J. F. Kennedy, Manual and automated spectrophotometric techniques for the detection and assay of carbohydrates and related molecules, In: *Techniques in Carbohydrate Metabolism (Techniques in the Life Sciences – Biochemistry – Volume B3)*, H. L. Kornberg, J. C. Metcalfe, D. H. Northcote, C. I. Pogson & K. F. Tipton (eds.), Elsevier, Shannon, Ireland, **1981**, B312.

[38] M. F. Chaplin, Monosaccharides, In: *Carbohydrate Analysis – A Practical Approach, 2nd Edition,* M. F. Chaplin & J. F. Kennedy (eds.), Oxford University Press, Oxford, UK, **1994**, pp. 1-41.

[39] L. Warren, The thiobartituric acid assay of sialic acids, *Journal of Biological Chemistry,* **1959**, 234, 1971-1975.

[40] A. Weissbach & J. Hurwitz, The formation of 2-keto-3-deoxyheptonic acid in extracts of *Escherichia coli* B, *Journal of Biological Chemistry,* **1959**, 234, 705-709.

[41] V. S. Waravdekar & L. D. Saslaw, A method estimation of 2-deoxyribose, *Biochimica et Biophysica Acta,* **1957**, 24, 439.

[42] P. Bernfeld, Amylases, α and β, In: *Methods in Enzymology – Volume 1,* S. P. Colowick & N. D. Kaplan (eds.), Academic Press, New York, USA, **1955**, pp. 149-158.

PART 4

VIRAL ASPECTS OF TISSUES FOR TRANSPLANTATION

VIRUSES AND THEIR RELEVANCE FOR GAMMA IRRADIATION STERILISATION OF ALLOGENIC TISSUE TRANSPLANTS

Axel Pruss [1], Rüdiger von Versen [2] and Georg Pauli [3]

[1] *Institute for Transfusion Medicine (Tissue Bank), University Hospital Charité Schumannstr. 20/21, 10117 Berlin, Germany {E-mail: axel.pruss@charite.de}*

[2] *German Institute for Cell and Tissue Replacement, Köpenicker Str. 325, 12555 Berlin, Germany {E-mail: R_vonVersen@dizg.de}*

[3] *Robert Koch-Institute, Retrovirology, Nordufer 20, 13353 Berlin, Germany {E-mail: PauliG@rki.de}*

ABSTRACT

In the production of tissues intended for transplantation, the first steps are selection and screening of donors for markers of virus infections by serology (Anti-HIV, Anti-HCV, Anti-HBc, HBsAg, TPHA). We investigated the effect of gamma irradiation on the sterilisation of human bone diaphysis transplants to evaluate its impact to the virus safety of transplants. Relevant as well as model viruses were included in the study to determine the dose necessary to achieve a reduction factor for the infectivity titres of at least 4 $\log_{10}$ at a temperature of -30 ± 5°C. The following relevant viruses were used: human immunodeficiency virus type 2 (HIV-2), hepatitis A virus (HAV), and poliovirus (PV-1), and the following model viruses: pseudorabies virus (PRV) for human herpesviruses, bovine virus diarrhoe virus (BVDV) for HCV, and bovine parvovirus (BPV) for parvovirus B19. The $\log_{10}$ reduction was measured by cytopathogenic effects (CPE) after virus titration and given as $TCID_{50}$/mL.

A first approach in the irradiation experiments was to determine the D_{10} values (kGy) for the different viruses (virus inactivation kinetics: BPV 7.3; PV-1 7.1; HIV-2 7.1; HAV 5.3; PRV 5.3; BVDV < 3.0 kGy). Based on these results, virus inactivation was studied in experimentally contaminated human bone transplants (femoral diaphyses). For BPV, the most resistant one of the viruses studied, a dose of approximately 34 kGy was necessary to achieve a reduction of infectivity titres of 4 $\log_{10}$ $TCID_{50}$/mL. We therefore recommend generally a dose of 34.0 kGy for the sterilisation of frozen bone transplants.

For some tissues (e.g. large bone transplants, tendons, cartilage, menisci) a problem exists: the necessity to (i) achieve sterility of the transplants and to provide virus safety, and to (ii) preserve the biomechanical properties of the transplant. Different irradiation doses in combination with single donor HIV/HCV/HBV-NAT tests (detection limits: 20-600 genome equivalents/mL) should be discussed to achieve a compromise (proposal: 25 kGy) and the remaining rest risk.

KEYWORDS

Gamma irradiation; virus; inactivation; bone tissue transplants

INTRODUCTION

Despite increased efforts in tissue engineering to develop alternative forms of transplants, allogenic bone transplantation continues to be an indispensable tool in the treatment of large bone defects. Approx. 75,000 autogenous and 30,000 allogenic bone transplantations are performed in Germany each year, and there is a further need for approx. 20,000 transplants per year [1]. In the USA, the country with the most bone transplantations world-wide, the number of allogenic bone transplants amounts to somewhere between 650,000 and 800,000 a year [2].

Apart from the clinical-functional aspects of allogenic bone transplantation, the transplant recipient's risk of infections constitutes a focus of scientific and medical interest. The main concern in this is transmission of viruses and bacterial pathogens. Problematic viruses of recent years up until today are the Human Immunodeficiency Virus (HIV), the Hepatitis B virus (HBV) and the Hepatitis C virus (HCV). These three viruses are transmitted via blood and blood products, i.e. potentially also through tissue transplants. In the following we will give a brief overview of these three viruses and other clinically relevant viruses.

RELEVANT VIRUSES

Human Immunodeficiency Virus (HIV)

The coated Human Immunodeficiency Virus (HIV, type 1 with 3 subtypes: M(ajor), N(ew), O(utlier), subtype M with variants A-I; type 2) belongs to the group of Retro viruses (family of Lenti viruses). It was first described in the early eighties and isolated in 1983 by the working groups of Montagnier and Gallo. This approx. 100 nm large, coated, single-strand RNA virus causes the acquired immune deficiency syndrome (AIDS), which is characterised by opportunistic infections (affection of CD4 cells including failure of the immune system), which in the final stage causes death. The current rate of infection world-wide is estimated at approx. 30-40 million people, while the majority of infections is found in Africa (15 million people) [3].

The virus is transmitted in the majority of cases through homosexual and heterosexual intercourse, intravenous drug use, intrauterine/connatal infection, injury and other transmission from infectious blood (transfusions, transplants!), sperm, saliva and other bodily secretions. The current therapy focuses above all on the symptoms and consists of the treatment of the opportunistic infections (including CMV chorioretinitis, CMV pneumonia, HSV infections, zoster multiplex, oral leukoplakia) and the clinical pictures associated with AIDS (including encephalopathy, wasting syndrome, Karposi's sarcoma, B-cell lymphoma, cervix carcinoma). Due to the heterogeneous nature of the virus (types, subtypes, variants), no vaccination is yet available.

The risk of transmission through blood transfusion in Germany is currently estimated at < 1:5,000,000 [4,5]. The introduction of the HIV-NAT would decrease the rest risk to < 1:11,000,000 [6]. This risk is likely to be at a similar level for organ donors, taking into account the fact that organ/tissue donors are usually 'first-time donors' and that an increase in prevalence is to be expected in this group [7,8].

The transmission of HIV is one of the main topics of focus in allogenic bone transplantation. In the USA, the transmission of HIV via the sterilised removal of an allogenic bone transplant was first reported in 1988. The transplant had been inserted during a spinal operation in 1984 [9]. The transplant had been stored at -80°C for 24 days, and no sterilisation process was carried out. The donor had not been tested for HIV antibodies.

In 1985, a total of 58 tissue and organ transplants, including 28 bone transplants, were removed from a multiple organ donor who was infected with HIV, but who had not been diagnosed with the virus. 25 of the transplants were lyophilised and treated with 30% ethanol. 3 transplants were transferred without any disinfection process. The three recipients were infected with HIV-1 [10].

HIV transmission through bone transplants has also been reported in Germany. In 1984, 16 bone transplants were removed from a male donor under sterile conditions and preserved at -80°C without undergoing virus inactivation. It was not possible to conduct an HIV test, since at that time, no suitable test kits were available. From November 1984 to May 1985, 12 patients received transplants from this donor. Four of these recipients were infected with HIV [11].

Since then, no further cases of transmission have been recorded in the international literature. However, it is to be assumed that the risk of infection continues to be present, particularly in the African/Asian regions, despite comprehensive safety measures (laboratory diagnostics, anamnesis/clinic, inactivation procedures).

Human hepatitis B virus (HBV)

The human hepatitis B virus is a hepadna virus and has a (partially) double-strand DNA genome. This virus leads to the hepatitis B virus, which is also known as 'serum hepatitis' or 'transfusion hepatitis', due to the way in which it is transmitted. The first clinical descriptions of the 'jaundice epidemics', the cause of which is known today as being transmitted hepatitis B viruses, were recorded in 1883 (smallpox vaccination scheme) and in 1938 (measles vaccinations). The infectious virus particles are approx. 42 nm in size and have a spherical form with a coat. They are also known as 'Dane particles' after D.S. Dane, who discovered them in 1970. The HBsAg (hepatitis B virus surface antigen) is anchored as a viral protein in the coating membrane of the virus, which develops from the membrane of the endoplasmatic reticulum. This protein is also present as a free HBsAg particle (approx. 20 nm) in the serum. Providing evidence of HBsAg is one of the most fundamental diagnostic procedures in the detection of an HBV infection. The interior contains an icosaedrical capside with a diameter of approx. 24 nm. This consists of 180 units of the HBcAg (hepatitis B virus core antigen) and contains the DNA genome. Evidence of antibodies against this protein is also used when diagnosing the HBV.

Today, around 300 million people in the world are affected by chronic hepatitis B. In Germany, there are approximately 50,000 infections per year, while approx. 0.5% of the population is HBsAg-positive [12]. The rate is particularly high among intravenous drug addicts, homosexuals and prostitutes. Transmission usually occurs via contagious blood and blood products, inadequately sterilised medical appliances and sexual intercourse. Sperm, saliva, tear fluid and other secretions are also potentially infectious. The target organ of the viruses is the liver (target cells: hepatocytes). The clinical picture (abdominal pain, fever, pain in the joints, skin rash, jaundice) is triggered by the subsequent immune reaction. The process can take on a chronic form (active or persistent). One of the most dangerous after-effects is the development of a primary liver cell carcinoma. The current risk in Germany of being infected with the HBV via a transfusion is between 1:100,000 and 1:1,000,000 [4-6].

Although currently only one HBV transmission is known to have occurred via a bone transplant [13], this virus must without doubt be regarded as relevant, particularly taking into account the way in which transmission occurs, when assessing the risk from bone tissue donors. This approach is supported by a case that recently came to light of an

HBV transmission via an erythrocyte concentrate. The donor of the banked blood had been tested HBsAg-negative. The HBV-DNA study of the reserve sample using NAT (real time PCR) was positive (approx. 2,000 copies/mL). The test for HBsAg only became positive in a limited way 2 months after the infectious erythrocyte concentrate had been donated [14].

Human hepatitis C virus (HCV)

Following the discovery of the patho-physiological connection between hepatitis A and hepatitis B and the molecular biological and serological description of the viruses which cause these diseases, a large number of so-called NonA/NonB hepatitides remain for which it has not been possible to find the causing agent. Only in 1989, following chimpanzee infections with subsequent DNA isolation and characterisation was an antigen tested that reacted with sera from human convalescents and chronically infected patients with NonA/NonB hepatitis. The corresponding cDNA was sequentialised and the virus (since then referred to as the hepatitis C virus) was identified as a new genus of the flaviviridae. The hepatitis C virus is a single-strand RNA virus of approx. 60-70 nm in size. It consists of a capside and a coat with spikes. To date, it has not been possible to provide a clear morphological picture. The epidemiology is similar to the HBV. Risk groups are again intravenous drug addicts, dialysis patients, homosexuals and prison inmates. Transmission occurs via infectious blood and blood products, sperm, saliva and other exudates. In Germany, around 0.6% of the population are seropositive, while the figure in the USA is 1.5-4.4%. Each year, 20,000 – 50,000 new infections are recorded in Germany. The course of clinically visible hepatitis is usually easier than hepatitis B, but can also lead to cirrhosis and a primary liver cell carcinoma. As a central laboratory diagnostic test, an IgG-ELISA is used against structural and non-structural antigens (C100, C33, C22). Due to the long diagnostic window, the HCV-NAT is of great importance, although it does not offer absolute security. In therapeutic terms, alpha-interferon in particular is used, in some cases in combination with ribavirin [12]. Following the introduction of the HCV-NAT for blood donors in April 1999, the risk of transmission of HCV through blood products in Germany is above 1:13,000,000 [4,6].

In 1992, transmission of HCV via an allogenic bone transplant was recorded for the first time. The female donor of the bone transplant had been infected in 1985 via the transfusion of infectious fresh plasma with HCV. In 1990, a femoral head was removed from this patient during a hip operation and was used as a transplant following 8 weeks of refrigeration without virus inactivation. A short time later, the recipient of the femoral head transplant developed a positive HCV status [15].

A short time ago, a further HCV transmission case was published in the USA. Here, a total of 91 organ and tissue transplants were removed in 2000 from an anti-HCV-negative multiple organ donor. This was used to supply 40 patients with transplants. An HCV infection was detected in the organ recipients and in 5 tissue recipients (1 vena saphena recipient, 1 tendon recipient, and 3 bone/tendon recipients). The reserve sample of the donor, which was tested in July 2002, showed a positive result in the HCV genome test (AMPLICOR HCV test, version 2.0, Roche), but was again negative in the antibody test (ORTHO HCV version 3.0 ELISA) [16]. Among the 16 recipients who were given irradiated bone transplants (dosage: 15 kGy), no HCV infection was detected. Several studies indicate that this dosage, in view of the D_{10} value of flaviviruses (BVDV: <3 kGy) is sufficient for a reduction of 5 $\log_{10}$ degrees and that it is therefore highly likely that irradiation prevented infection among these recipients [17].

Additional relevant viruses

Viruses that are transmitted via blood, and therefore also via bone transplants, and that can cause severe illness, are dealt with as a final topic.

The human parvovirus B19 is a non-coated, small (22 nm), single-strand DNA virus that can lead to Sticker's disease in humans. When in utero infection occurs, the parvoviruses may cause a foetal hydrops (in 10 – 15% of all cases), or trigger an abortion. Persistent infections cause arthritis, athropathy and aplastic anaemia. The virus is transmitted via aerosol infection, and the blood group substance P has been described as the cellular receptor. As a result, the virus is found predominantly in erythroblasts, whose virus-related destruction leads to the temporary, continuous obstruction of the erythropoeisis with haemolysis, lasting about 10 days and leading subsequently to anaemia. The central pathogenic element is a lack of O_2. The virus is very resistant against environmental influences and inactivation measures. Diagnosis is usually made using serological and molecular biological tests. The therapy is symptom-based. When the Hb value is low (> 10 g/dL), an intra-uterine transfusion may be necessary.

Due to the high infection rate of the population (approx. 50%), the risk of transmission of parvoviruses is relatively high and is estimated to be 1:200,000 for blood transfusions [12]. Since parvoviruses persist in bone marrow in particular, the transfer of bone transplants containing bone marrow, from which no blood cells have been removed, presents a particularly high risk of infection. The transmission of the human parvovirus B19 through blood products has been recorded several times [18,19]. Reports on cases of infection through bone transplants do not yet exist, although the potential risk is not insignificant, due to the target cells of the viruses.

The non-coated hepatitis A virus (HAV), approx. 28 nm in size, contains a single-strand RNA and belongs to the picornaviridae group. It remains infectious outside the human body under normal environmental conditions for 4 weeks. The hepatitis A virus is stable in ether and is very heat-resistant. To inactivate the virus, a heat treatment is required at 100°C for 5 minutes. 30-minute treatment at 60°C is insufficient. The virus is transmitted fecally and orally, e.g. via food products and drinking water. In most cases without later consequences, it leads to curable hepatitis [12]. To date, no cases of transmission via bone transplants are known.

The human cytomegalievirus (CMV) is approx. 180 nm in size, with a double-strand DNA, and belongs to the herpes virus group. The virus is usually transmitted via aerosol infection, but transmission is also possible via blood transfusion and organ transplants. The leukocyte depletion of banked blood (3-4 log) leads to a significant decrease in CMV, but offers no absolute security [20]. For risk patients (foetuses, newborn babies, immune suppressed patients), therefore, the transfused blood products should also indicate an antibody-negative CMV status. In Germany, approx. 55% of the population is infected. The course of the disease is similar to mononucleosis, and can also include interstitial pneumonia. Intra-uterine infections are particularly severe, and can lead to development defects in the foetus and to the infection of immune-suppressed patients, for whom CMV pneumonia is of particular concern. The use of ganciclovir and cidofovir in therapy is usually successful. Due to the high rate of infection among the population and the usually bland course of the infection, the transmission of CMV by allogenic bone transplantation is a less important factor. However, attention should be paid to the CMV status (antibody determination) of the tissue donor, in particular when transferring large transplants containing blood cells to CMV-Ak-negative pregnant women and immune-suppressed patients (cave: reactivation).

The human T-cell leukaemia virus 1 (HTLV-1, retro virus, coated single-strand RNA virus) is only of relatively minor importance in Germany from an epidemiological point of view. However, a Swedish working group has reported transmission via allogenic bones preserved through freezing [21].

Virus selection for validation studies

The appropriate existing norms and guidelines should be used when selecting test viruses for validation studies, and they should represent a spectrum of the most clinically relevant viruses.

The Bundesamt für Arzneimittel und Medizinprodukte (BfArM), the Federal Ministry for medicines and medical products, and the Paul-Ehrlich-Institut (PEI) [22] require that the following be used when validating the inactivation of viruses:

- Human immune deficiency virus (e.g. HIV-2)
- A model virus for the hepatitis C virus (e.g. bovines virus diarrhoea virus, BVDV)
- A coated DNA virus (e.g. pseudorabies virus, approx. 150 nm, an analogue of the human herpes virus 3, syn. varicella-zoster virus)
- A non-coated virus (e.g. hepatitis A virus)

These viruses must be included in the studies and should be supplemented by the poliomyelitis virus type 1 (picornaviridae family, genus enterovirus, approx. 30 nm, single-strand, non-coated RNA virus), as a standard test virus in disinfection agent tests, and the non-coated, 20 nm bovine parvovirus (BPV, as a model virus for the human parvovirus B19). BPV has been the subject of numerous studies on resistance against physical and chemical influences [23,24]. For a long time, it has been used as a highly resistant test virus when testing the validity and effectiveness of chemical disinfection processes.

LABORATORY METHODS TO DETECT VIRUSES

According to the relevant guidelines and recommendations regarding allogenic bone transplants, the following parameters are to be set for the medical laboratory diagnosis of tissue donors as minimum requirements:

- Antibodies against (HIV-1/2)
- Antibodies against HCV
- Antibodies against the HBcore protein HbsAg
- Antibodies against treponema pallidum

For living donors, the surrogate marker for potential liver infection should also be specified in addition to the ALAT (alanin aminotransferase).

In view of the limited opportunities for a comprehensive anamnesis or clinical examinations, it is recommended that for multiple organ and cadaver donors, additional tests be carried out to detect the HIV, HBV and HCV genome using suitable nucleic acid amplification techniques (NAT, e.g. polymerase chain reaction, PCR).

On the one hand, the 'diagnostic window' is reduced (see Table 1), while on the other, the greatest possible security should be demanded due to the fundamentally 'elective' character of bone transplantation. In the view of the author, this should also include carrying out the NAT examination solely on the individual donor sample, i.e. no pools should be created. This significantly decreases the risk. The reduction of a 96 pool (the maximum pool size permitted for blood donations in Germany) to a test on individual donors leads to the following reduction of risk of virus transmission per 10 million erythrocyte concentrate transfusions: HIV: 0.47-0.62 to 0.010-0.045; HBV: 11-13 to 3.3-5.1; HCV: 1.7-2.0 to 0.5-0.8 [25].

Table 1. Seroconversions and diagnostic windows: Anti-HIV, anti-HCV and HBsAg [26,27].

	Anti-HIV	Anti-HCV	HBsAg
Seroconversions (in 4.8 million blood donors)	23	49	37
'window period' serology (days)	22 (6-38)	66 (38-94)	56 (25-109)
'window period' NAT (days)	8-10	6-9	30-66
Probability of a 'window period' donation/10^6 donations	4.3	16	25

Studies on the questions of infectious doses have shown that the smallest virus quantities are already enough to lead to infection. A 50% infection rate has been given for HBV and HCV with 10 (1-100) viral genome equivalents (geq) and for HIV with 1,000 (100-10,000) viral genome equivalents [25].

In order to classify the values of the NAT examinations, particularly in view of the virus validation studies, the current detection limits for the individual NAT for HIV, HCV and HBV must be defined. A comparison of the NAT techniques currently used for the three relevant viruses is given in Table 2.

Table 2. Sensitivities of NAT assays (geq/mL) for HIV, HBV and HCV genome detection [25,28,29].

Virus	Assay	50% detection limit	95% detection limit
HIV RNA genotype B	AmpliScreen 1.5*	7.1 (4.7-10.7)	126 (67-311)
	Gen-Probe TMA#	3.6 (2.6-5.0)	31 (20-52)
	NucliSens-AmpliScreen 1.5§	4.6 (3.2-6.7)	37 (23-69)
	Cobas AmpliScreen 1.5	6.6 (4.4-10.1)	92 (49-238)
HCV RNA genotype 1	AmpliScreen 2.0*	14 (11-19)	126 (83-225)
	Gen-Probe TMA#	7.9 (6.3-9.9)	85 (64-118)
	NucliSens-AmpliScreen 2.0§	4.3 (3.1-5.9)	21 (13-44)
	Quiagen-HCV-Amplicor 2.0	18 (10-32)	144 (74-402)
HBV**	AmpliScreen 2.0*	14	126
	Gen-Probe TMA#	7.9	85
	NucliSens-AmpliScreen 2.0§	4.3	21

* Roche, Molecular Systems, Pleasanton, CA, USA; # Gen-Probe, San Diego, CA, USA; § Roche-Kit combined with Extractor-Kit from bioMérieux; ** Current data indicate a comparability with the HCV-PCR technique. The in-house PCR of the Institute for Virology at the Charité hospital has a 'detection limit' for HBV of 268 geq/mL.

In-house NAT sometimes show higher detection limits (up to 600 genome equivalents (geq)/mL). The person requesting the tests must therefore be informed previously about the sensitivity of the NAT assay offered.

A first case was recently published in which an HCV transmission via a thrombocyte concentrate occurred, despite a negative HCV-PCR [30]. Furthermore, the detection limits also depend directly on the virus genotype to be indicated. It could be possible to use post-mortem blood for the NAT [8,31].

TARGET PARAMETERS FOR VIRUS INACTIVATION STUDIES

The appropriate CEN norm for virus inactivation attempts, the ´Sterilization of medical devices utilizing tissue - validation of the inactivation of viruses and other transmissible agents´ [CEN, 1994] demands a reduction of over 6 $\log_{10}$ degrees for all reduction factors (anamnesis, donor selection, infection serology, preparative steps, inactivation steps). The German requirements for ´Validation studies to indicate the virus safety of medications from human blood or plasma´ [22] stress that the process used must contain a step that reduces the content of the virus in question by $\geq 4.0 \log_{10}$.

The 'European Agency for the Evaluation of Medicinal Products' makes a similar statement in its guideline 'The Design, Contribution and Interpretation of Studies Validating the Inactivation and Removal of Viruses' [32]. General requirements regarding levels of the overall reduction factor are not included here, since they depend, among other things, on theoretical virus load in initial material (donor selection).

Taking into account the detection limits for the current NAT techniques for an individual donor sample (20-600 geq/mL, corresponding to 1.3-2.8 $\log_{10}$ geq/mL), it can be assumed that a virus inactivation procedure which leads to a reduction in infectiousness (worst case: 1 geq = 1 infectious virus = 1 $TCID_{50}$) by 4 $\log_{10}$ degrees ($TCID_{50}$ (tissue culture infectious dose 50)/mL) offers the greatest possible security in terms of the transmission of HIV, HBV and HCV via allogenic bone transplants, when the virus genome detection test previously carried out using validated procedures from the individual donor sample is negative. Simultaneously, a reduction, potentially in the tissue of other existing viral pathogens or non-viral pathogens, is also achieved.

GAMMA IRRADIATION OF BONE TRANSPLANTS

The germ-destroying effect of ultra short-wave, extremely high energy, electromagnetic gamma irradiation, that is created when radioactive elements (e.g. cobalt-60) disintegrate, consists above all in hitting the cell nuclei in terms of their genetic information, setting defects and in so doing, to prevent the later reproduction of the pathogen. Cell destroying secondary effects such as the formation of radicals are also an influential factor. The irradiation dose is applied in many stages by circling the Co-60 radiation source several times, and is only possible in authorised institutions.

There are currently no clear specifications regarding reduction factors of viruses in human tissue. The range of the doses reported extends to 89 kGy [33]. The guidelines that essentially apply to the reduction of bacteria, fungi and spores, and the recommendations on the 'Industrial sterilisation of medical products' [34] require an overall reduction by 6 $\log_{10}$ degrees and the achievement of a 'sterility assurance level' (SAL; the probability of the presence of a living microorganism following completion of the sterilisation measures) of 10^{-6}. The SAL is however not very suitable for the evaluation of human initial materials, since the required standardisations and test conditions for medical products for transplants of human origin usually cannot be guaranteed, or other standards apply.

The International Atomic Energy Association (IAEA) recommends a standard dose of 25 kGy [35], which is currently used in most tissue banks that have integrated a gamma irradiation process into production. In the USA, bone tissue is sometimes irradiated with 15 kGy. This measure is used essentially to reduce the microbiological bio burden and not to inactivate the virus.

The entry of the D_{10}-value (the required irradiation dose to reduce initial viral load by 90%, or 1 log_{10}-stage) must be demanded as a standard for virus inactivation studies for comparability purposes. Definitive statements on the efficiency of the process when sterilising contaminated bone tissue are not yet available. As a result, the validation study regarding the gamma irradiation of bone tissue transplants is presented here [17].

VIRUS INACTIVATION IN BONE TISSUE TRANSPLANTS [17]

Material and methods

Femoral diaphyses specimen, virus contamination

The manufacturing process for diaphysis transplants included the following steps: preparation of diaphyses from human femurs after removing all attached muscles and connective tissue from the bone surface; sawing of diaphyses into segments of 75 mm length (belt saw, Bizerba); washing of the bone marrow canal several times with physiological salt solution (0.9% NaCl, B-Braun, Melsungen/Germany) in order to remove blood from the tissue.

It is known that irradiation at low temperatures positively affects the biomedical properties of implants [36]. To use an experimental design as close as possible to that used in the production process, all inactivation experiments by irradiation were therefore performed in a Styrofoam box filled with dry ice. Temperature validation (frozen diaphysis on dry ice with medium outside the irradiation facility) showed that the virus suspension inside the diaphysis reached a temperature of -30 ± 5°C during irradiation.

For the inactivation experiments the lower open end of each diaphysis was tightly sealed with bone cement (Palacos R™, Heraeus Kulzer, Wehrheim/Germany). The exterior surface of the diaphyses was coated with a film of Ethic-R bone wax™ (Ethicon; thickness approx. 0.1 mm) in order to close any small canals or holes on the bone surface. That the bone was watertight was confirmed by filling the bone marrow canal with 5 mL *aqua ad iniectabilia* for 24 hours. After removing the water, the diaphyses were transferred into a plastic vessel containing 250 mL of water, fixed with styroflex in an upright position and frozen at -21°C. Then 5 mL of a suspension of cell-free virus were pipetted into the cavity of the diaphyses, the open end was sealed with bone wax and the diaphyses stored at -21°C. All steps were carried out under sterile conditions and safety precautions in a biological safety cabinet.

The virus-contaminated diaphyses were transported to the gamma irradiation facility of Gamma-Service Produktbestrahlung GmbH (Radeberg/Germany), irradiated (conditions see below) with doses ranging from 30.6–35.4 kGy deduced from the D_{10} values of the most resistant viruses (BPV, HIV-2, PV-1). One contaminated but not irradiated diaphysis per virus investigated was retained as control to monitor storing and transportation conditions. The titre of viruses in the control sample was used as reference for calculating the experimentally determined inactivation factors. After irradiation all samples were returned to the virological laboratory under the cooling conditions described above. After sample thawing the virus titres were determined.

Determination of gamma ray dose distribution, irradiation procedure

For technical reasons, it is difficult to measure the X-ray dose directly in specimens, i.e. in the experimentally contaminated diaphyses. Therefore a dose-distribution study within and on the surfaces of the box used in the irradiation experiments was performed in the gamma irradiation facility, using a cobalt source (^{60}Co). Validation took place in a Styrofoam box (length: 34 cm, width: 25 cm, height: 34 cm, weight: 1.46 kg, sample load: 6 plastic flasks with diaphyses or paraffin phantoms) at room temperature and a dose of 1 kGy/h. The following measure points were chosen to determine the dose distribution: two diaphyses and four paraffin phantoms corresponding in size and shape to diaphyses were used to study the dose distribution inside the box and the influence of mass variation. Dosimeters were placed (i) inside the diaphyses or phantoms (an alanine pellet dosimeter at each end); (ii) on the surface of samples (4 alanine foil dosimeters for each sample); (iii) on the inner surface of the box (4 alanine foil dosimeters); (iv) on the outer surface of the box (4 alanine foil dosimeters). All procedures followed EN 552 [37] without major changes.

Determination of virus inactivation kinetics and the decimal reduction value (D_{10} value)

Eight 15 mL Nunc tubes were filled with 5 mL of cell-free virus suspension each, transferred into a plastic vessel, fixed with styroflex, and placed into a Styrofoam box containing dry ice, which was transported to the irradiation facility in a transport container according to EN 829 [38]. One tube was removed from each group to be used as control and placed into a plastic vessel on dry ice. The remaining seven vials in the box were exposed to the ^{60}Co source. At defined intervals (1 kGy/h) one vial each was removed from the box and stored in a plastic vessel on dry ice (temperature in the virus suspension -30 ± 5°C) until virus titration, which took place in the virological laboratory after transportation and thawing as described above.

The results of the inactivation kinetics of the different viruses showed a linear relationship between the logarithm of the infectivity titres and the radiation dose applied. To evaluate the results of irradiation, the following equation was used:

$$N(D) = N_0 \times 10^{(-D/D_{10})^n}$$

N_0 is the virus titre (log) before and N(D) after irradiation with the dose D. D_{10} is the dose necessary to reduce the titre of the infectious agent by a factor of 1 $\log_{10}$. n is the correction factor introduced to avoid a systematic error.

Cell cultures/cell lines, virus titrations, virus titre calculation

The cell lines investigated were obtained from stocks by the Robert Koch-Institut Berlin and are registered there and documented [17]. The respective viruses were obtained from the supernatants of cultivated infected cells following procedures reported elsewhere [39]. The cell debris was discarded after centrifugation and the viruses obtained were frozen in aliquots at -70°C.

After preparing 10-fold dilutions of virus suspensions (supernatants/homogenates) with cell culture medium, 100 µL of each dilution were pipetted into each of four or eight wells of a 96-well micro titre plate. Each of these contained 100 µL medium with 1–5 × 10^4 cells suitable for the cultivation of the respective virus. The micro titre plate was covered, incubated at 37°C until the virus control showed a cytopathogenic effect (CPE).

Determination of the virus content was done by end-point titration (micro titre plate, four- or eightfold preparation). Regarding the procedure of titration or the methodology of the reduction factor calculation, see the specifications of the publication 'Requirements of validation studies as evidence of the virus safety of drugs from human blood or plasma' [22]. Viruses can cause different forms of the CPE (cell lysis, cell fusion, inclusion bodies, syncytia formation, transformation etc.). The cytopathogenic effects were observed over several days by means of inverse transmitted light microscopy from always the same investigator while a second person read and confirmed them before the concluding evaluation. When toxic effects occurred in the supernatant examined, the appropriate suspension dilution was assessed as 'inactivated' in the sense of a *detection limit*.

A quantitative measurement of the virus reduction using the decrease of the virus genome in the experimental preparation is problematic due to false positive results which are to be expected (cross-reactivity with genome from blood cells or bone tissue cells) as well as the missing information regarding infectiosity [40]. The titre reduction indicates the degree of virus inactivation. The titre was given as the virus dilution when 50% of the cell culture showed a cytopathogenic effect ($TCID_{50}$ = tissue culture infectious dose 50%). The titre was calculated according to Reed and Münch [41] and/or Spearman and Kärber [42].

RESULTS

Gamma ray dose distribution

Validation of the dose distribution in the Styrofoam box used for irradiation in the production process of diaphyses showed that the values in and on the phantoms as well as in and on the box varied within a range of approximately 3%. From the results obtained by the dose distribution experiments it was possible to calculate the effective dose in the internal cavity of the diaphyses to be approximately 98% of that on the surface of the box. This value was within the range of variation determined at the different measuring points. From these results it seemed feasible to determine the irradiation dose during the inactivation experiments on the surface of the box via dosimeter.

Suspension test (kinetics)

The virus inactivation kinetics were performed in frozen suspensions (-30 ± 5°C), using an experimental design as close as possible to that used in the production process. The titre of the virus suspension used as transport and incubation control served as reference. As expected, a linear relationship between the reduction factor and the irradiation dose was observed.

From the regression curve the D_{10} values for the different viruses were calculated (Table 3). BVDV showed the highest sensitivity (D_{10} value ≤ 3 kGy) to irradiation, whereas BPV was the most resistant virus (D_{10} value 7.3 kGy). From the D_{10} values the doses necessary to reduce virus titres by 4 $\log_{10}$ and 6 $\log_{10}$ were calculated (Table 3).

Carrier test (diaphyses)

Data of inactivation kinetics studies with frozen virus suspensions suggested that for most of the viruses investigated a reduction factor of 4 $\log_{10}$ can be achieved using an irradiation dose of approximately 30 kGy. The reduction factors obtained for virus suspensions were verified in a model system using virus-contaminated diaphyses. At least two independent experiments were performed for each of the viruses. In general, the reduction factors obtained corresponded to those calculated from the D_{10} values (Table 4). With the model system of contaminated diaphyses it was also shown that parvoviruses were the most resistant viruses regarding irradiation, with reduction factors of approx. 4 $\log_{10}$ when a dose of 34 kGy was used.

Table 3. D_{10} values and calculated doses to reduce virus titres by 4 or 6 $\log_{10}$.

Virus	D_{10}-value* (kGy)	4 $\log_{10}$ reduction (kGy)	6 $\log_{10}$ reduction (kGy)
BVDV	≤ 3.0	≤ 12.0	≤ 18.0
PRV	5.3	21.2	31.8
HAV	5.3	21.2	31.8
HIV-2	7.1	28.4	42.6
PV-1	7.1	28.4	42.6
BPV	7.3	29.2	43.8

* D_{10} values were calculated for frozen virus suspensions in plastic tubes

Table 4. Comparison of experimental data for the inactivation of different viruses in experimentally contaminated diaphyses with the calculated D_{10} values.

Virus	Dose (kGy)	Reduction factor ($\log_{10}$) determined experimentally	Reduction factor ($\log_{10}$) calculated from D_{10} value
BPV	31.3	3.1	4.4
BPV	31.3	3.7	4.4
BPV	33.7	3.7	4.7
BPV	33.9	4.1	4.7
BPV	35.4	4.8	4.9
PRV	30.6	≥ 4.4	5.9
PRV	35.4	≥ 5.5	6.8
PV-1	30.6	5.9	5.8
PV-1	35.4	≥ 8.1	6.8
BVDV	33.7	≥ 6.5	≥ 11.2
BVDV	35.4	≥ 5.6	≥ 11.8
HAV	33.7	≥ 7.2	7.2
HAV	35.4	≥ 7.7	7.5
HIV-2	33.7	≥ 4.1	6.3
HIV-2	35.4	≥ 4.0	6.7

DISCUSSION, CONCLUSIONS & RECOMMENDATIONS

One of the significant risks from allogenic bone transplantation is the transmission of viral and bacterial pathogens [43]. This is above all dependent on the strength of the blood flow in the tissue, and is therefore relevant particularly for spongy tissue and complete corticalis transplants. However, ligament and tendon transplants also present a high risk of transmission of non-viral pathogens in particular. Secondary contamination when removing post-mortal bacterial colonies in the tissue is the focal issue here [44]. In recent years, transmissions of the human immune deficiency virus, the HTLV-1, and the hepatitis C virus via allogenic bone transplants have been reported. Transmission of non-viral pathogens (mainly spore-forming bacillae and staphylococci) is more frequent. Regrettably, a case of transplant-related death occurred in 2001 following transplant of a femur condyle infected with *clostridium sordelli* [2]. Since allogenic bone transplants are usually used electively, and vital indications, as for organ transplants, are only given in exceptional cases, all existing measures must be employed in order to minimise or prevent the risk of infection via bone transplants.

The first measure to minimise the risk is the recording of the donor anamnesis and the detailed clinical examination. In a similar way to the exclusion criteria for blood donors, the focus here is on the detection of infection risks. To this purpose, the international Tissue Banking Standards of the AATB and EATB/EAMST [45] offer valuable guidance in the form of a comprehensive anamnesis questionnaire and special standards for the clinical examination. A further central criterion for minimising risk is the study of the tissue donor in the medical laboratory. This serves above all to detect viral infections and includes infection-serological parameters (anti-HIV, anti-HCV, anti-HBc, HBsAg, TPHA, ALAT). It is particularly necessary that testing be requested for the HIV, HBV and the HCV genome using the nucleic acid amplification technique. Considering the elective nature of bone transplantation and the donor/recipient key, that varies considerably from blood and organ donations (e.g. 1 whole blood donor -> 2 recipients [EK, FFP], 1 organ donor -> approx. 5 recipients, 1 tissue donor -> approx. 100 recipients), the NAT must be carried out in its most sensitive form, however. This means that only individual donor samples with adequate volumina may be tested. Considering the current detection limit of the NAT tests (95% CI: 20-600 geq/mL) and their ´diagnostic window´ (HIV/HCV approx. 10 days, HBV approx. 30 days), a significant degree of security has already been created with regard to the three most relevant viruses (HIV/HBV/HCV).

In order to close the existing gap in residual risk, while at the same time achieving a reduction in the tissue of the potential existence of other viral and non-viral pathogens, inactivation procedures should be included in the production process of allogenic bone tissue transplants. This leads on the one hand to a shortening of the quarantine storage time, and avoids requesting bone tissue a second time from living donors and the subsequent testing of an organ recipient (of the tissue donor), while at the same time ensuring the almost complete destruction of germs.

There are currently different inactivation procedures available according to the type of tissue. The chemical procedure, involving treatment with peracetic acid/ethanol (peracetic acid 2%, ethanol 96%, aqua ad injectabilia, 2:1:1, 4 hours, 200 mbar, agitation) is particularly suitable for small bone transplants with a layer thickness of $\leq$ 15 mm, and for soft tissue (ligaments, tendons, amnion) [46,47]. Larger transplants (epiphyses, femur heads, entire extremity bones) can also be treated using this procedure, but it must be ensured using suitable methods (e.g. bore holes) that the sterilisation agent penetrates the tissue to an adequate degree [48].

The use of the Marburg bone bank system is currently limited to thermo-disinfection (effective treatment: at least 82.5°C in the femur head centre for at least 15 minutes) of femur head transplants that have been aseptically removed during hip operations. In Germany, the method is used extensively and is applied by approx. 90 bone banks in clinics [1]. The method has been validated as virus inactivating [49], but is not able to inactivate spores and spore-forming bacilli to a sufficient degree [50]. For the sake of completeness, please note that other inactivation procedures are also in use. These usually involve combination methods (Tutogen® process: osmosis, H_2O_2, aceton, gamma rays) or procedures that include the use of ethylene oxide treatment [51].

Gamma irradiation (irradiation level: -30°C, ^{60}Co-source) is very well suited for penetrating the tissue for all transplant forms. The main problem, however, is the alteration of the biological properties, which becomes higher with the increasing influence of the irradiation. With an irradiation dose of 15 kGy, no biomechanical changes in the bone transplants have been observed. With a dose of up to 25 kGy, only slight changes have been observed. Irradiation doses of above 30 kGy lead to significant changes [52]. New procedures, with which a protection of the bone protein has been provided using ascorbic acid substances, allows the use of irradiation doses of up to 50 kGy [53,54]. These doses lead to complete germ destruction and the highest level of security. However, no experiments on bone transplants have been carried out.

The question of which potency the procedures have in terms of virus inactivation has so far hardly been tested. In view of the valid norms and guidelines, and the minimisation of risk through laboratory diagnostic measures (including NAT), the author regards a procedure as being 'suitable' when a reduction in virus levels of 4 $\log_{10}$ degrees ($TCID_{50}$/ml) is achieved. In the inactivation kinetics studies as well as in the model experiment using contaminated diaphyses, bovine parvovirus showed the highest resistance to irradiation (D_{10} of 7.3 kGy). It was unexpected that BVDV revealed the highest sensitivity to gamma irradiation (D_{10} of $\leq$ 3 kGy). All other viruses showed D_{10} values around 5-7 kGy. Combining the results of the virus-inactivation kinetics and of the diaphysis model system, a reduction factor of 4 $\log_{10}$ is obtained for parvovirus using a dose of approximately 34 kGy. Parvovirus was revealed to be the virus with the lowest sensitivity to gamma irradiation (small single-stranded DNA genome); all other viruses showed a lower D_{10} value indicating a higher sensitivity to gamma irradiation.

Estimations on the doses of gamma irradiation necessary to inactivate HIV and HCV in bone tissue are in agreement with the results obtained in our investigations. Conrad *et al.* [15] calculated that a dose of 17 kGy is sufficient to completely inactivate Hepatitis C virus in allografts. The D_{10} values resulting from the irradiation tests corresponded largely to those recorded in the literature to date. To inactivate HIV, doses of 2.5 kGy to over 25 kGy [55,56] were recorded. In tests for which the HIV-1 genome as an indication of inactivation using NAT was used, gamma irradiation with a dose of 30-40 kGy was required until the HIV-1 sequences could no longer be amplified [57]. Other working groups examined the D_{10} in relation to the temperature. The D_{10} value for HIV-1 was between 7.2 kGy at room temp. and 8.3 kGy at -80°C [58]. This would mean for a reduction in virus infectiousness by 4 $\log_{10}$ degrees, 28.8 kGy (RT) and 33.2 kGy (-80°C) would be necessary. These results correlate with tests carried out by the author, who recorded a D_{10}-value for HIV of 7.1 kGy (-30°C), i.e. 28.4 kGy for 4 $\log_{10}$ degrees.

High irradiation resistance of parvovirus (PV) was not surprising, since it is very small. In order to achieve greatest possible security, i.e. to also detect viruses the size of PV, and taking into account stronger absorption of irradiation by bone tissue, in general, a dose of 34 kGy for allogenic bone transplants can be recommended. Incidentally, for years, comparable doses have been used successfully in tissue banks in Poland [59].

In view of the reduction of biological properties for high irradiation doses (> 25 kGy), the following compromise solution for calculating the dose is suggested:

- Anamnesis/clinical examination in accordance with AATB/EATB criteria,
- Infection-serological examinations (Anti-HIV, anti-HCV, anti-HBc, HBsAg, TPHA),
- HIV, HBV and HCV genome detection using NAT (e.g. PCR) from the individual donor sample *,
- Irradiation with 25 kGy **,

* No pooling (cave: donor/recipient key); pre-analytical factors, 'detection limit' (target: below 1,000 geq/mL). The use of post-mortal samples should in all cases be agreed with the laboratory!

** This recommendation is only given for temperatures of approximately -30°C, because virus infectivity is significantly influenced by the temperature prevalent during irradiation. With increased irradiation temperature, a lower dose is necessary for the same log reduction; that is, the same dose leads to a higher inactivation factor [58].

In terms of the D_{10}-values of the most clinically relevant viruses (max. 8.8 kGy for HIV [33]), the dose of 25 kGy achieves a reduction in viral infectiousness of approx. 3 log_{10} ($TCID_{50}$/mL). This reduction potency should be sufficient to completely inactivate any still existing HIV/HBV/HCV viruses with negative NAT (detection limits of the NAT: approx. 20-600 geq/mL).

In terms of other viral pathogens (e.g. parvoviruses, hepatitis A virus), a residual risk remains when a dose of 25 kGy is applied, of which the patient must be informed as part of the pre-operative support information.

ANNEX 1: PRIONS

The transfer of pathenogenic prions through allogenic Dura mater transplants has been known for some time [60,61]. However, to date, no cases of an infection of this nature through allogénic bone transplants or blood transfusions has been reported [62]. Furthermore, the risk of transmission of vCJD/CJD through bone tissue is currently graded as very low by the WHO [63].

However, the following measures should be taken as standard by every tissue bank:

- exclusion criteria for donors (anamnesis):
  recipients of extracts from human pituitary glands,
  recipients of dura mater transplants,
  family history of CJD/similar TSE,
  geographical risks
- clinical history, (autopsy report), (serology)
- avoid processing of high risk tissues (e.g. brain, eye, spine)
- disinfection of surgical instruments with contact of high risk regions (1 M NaOH, 1 hour)

The current inactivation methods generally in use are only able to inactivate pathological prions to a limited extent. The procedures suitable to achieve this cannot be recommended due to their effect on the biological properties of bone and soft part transplants. Here is an overview of the potencies of the inactivation procedures:

No/low effect (≤ 3 log_{10} reduction in 1 hour):

- aldehydes (formaldehyde, glutaraldehyde)
- ß-propiolactone
- heat (800°C)
- UV light and ionising irradiation (small protein, no DNA/RNA)
- detergents (e.g. ethanol, non-ionic)
- acetone, ethylene oxide, peracetic acid

Inactivation (≥ 3 log_{10} reduction in 1 hour):

- autoclave: 134°C, 60 minutes
- sodium hydroxide 1 M; 60 minutes
- sodium dodecylsulphate (SDS)
- guanidinium-thiocyanate 4 M, trichloracetate, phenolic 1%

REFERENCES

[1] von Garrel, T., Survey of 100 bone banks in Germany, *Personal Report*, **2003**.

[2] CDC, Center for Disease Control and Prevention, Update: Allograft-Associated Bacterial Infections. *MMWR*, **2002**, 51, 207-210.

[3] Modrow, S. and Falke, D., *Molekulare Virologie*, Spektrum Akademischer Verlag, Heidelberg, **1998**.

[4] Graul, A., Heiden, M., Gräf, K. and Keller-Stanislawski, I., Hämovigilanz in Deutschland – Berichte an das Paul-Ehrlich-Institut über Verdachtsfälle von Transfusionsreaktionen im Beobachtungszeitraum Januar 1995 bis Dezember 2002, *Transfus. Med. Hemother.*, **2003**, 30, 232-238.

[5] Bein, G. and Sachs, U., Unerwünschte Wirkungen, In: *Bundesärztekammer – Wissen-schaftlicher Beirat (Hrsg.) Leitlinien zur Therapie mit Blutkomponenten und Plasma-derivaten*, Deutscher Ärzte-Verlag, Köln, **2003**, pp. 259-282.

[6] Caspari, G. and Gerlich, W. H., Durch Blut übertragbare Infektionskrankheiten, In: *Transfusionsmedizin, Grundlagen - Therapie - Methodik. 3. überarbeitete und erweiterte Auflage*, Mueller-Eckhardt, C. and Kiefel, V. (eds.), Springer, Berlin, **2003**, pp. 599-645.

[7] Offergeld, R., Stark, K. and Hamouda, O., Infections among blood donors, *Bundesgesundheitsbl Gesundheitsforsch Gesundheitsschutz*, **2003**, 46, 775-779.

[8] Strong, D. M., NAT for Cadaveric Donors, *27th Annual Meeting, American Association of Tissue Banks*, San Diego, CA, USA, **2003**.

[9] CDC, Center for Disease Control and Prevention, Transmission of HIV through bone transplantation: case report and public health recommendations, *MMWR*, **1988**, 37, 597-599.

[10] Simonds, R. J., Holmberg, S. D., Hurwitz, R. L., Coleman, T. R., Bottenfield, S., Conley, L. J., Kohlenberg, S. H., Castro, K. G., Dahan, B. A., Schable, C. A., Rayfield, M. A. and Rogers, M. F., Transmission of human immunodeficiency virus type 1 from a seronegative organ and tissue donor, *N. Engl. J. Med.*, **1992**, 326, 726-732.

[11] Schratt, H. E., Regel, G., Kiesewetter, B. and Tscherne, H., HIV infection through bone transplants preserved through refrigeration, *Unfallchirurg*, **1996**, 99, 679-684.

[12] Hahn, H., Falke, D., Kaufmann, S. H. E. and Ullmann, U., *Medical Microbiology and Infectiology, 4th Edition*, Springer-Verlag, Berlin, **2001**.

[13] Shutkin, N. M., Homologous-serum hepatitis following the use of refrigerated bone-bank bone, *J. Bone Joint Surg.*, **1954**, 36A, 160-162.

[14] Meisel, H., Endres, A. S., Walther, H. U., Wend, U. C. and Gerlich, W. H., Transmission of hepatitis B virus 2 months prior to hepatitis surface antigen positivity of donor blood. *Transfus. Med. Hemother.*, **2003**, 30, 228-231.

[15] Conrad, E. U., Gretch, D. R., Obermeyer, K. R., Moogk, M. S., Sayers, M., Wilson, J. J. and Strong, D. M., Transmission of the hepatitis-C virus by tissue transplantation, *J. Bone Joint Surg. Am.*, **1995**, 77, 214-224.

[16] CDC, Center for Disease Control and Prevention. Hepatitis C Virus Transmission from an Antibody-Negative Organ and Tissue Donor, *MMWR*, **2003**, 52, 273-276.

[17] Pruss, A., Kao, M., Gohs, U., Koscielny, J. K., von Versen, R. and Pauli, G., Effects of gamma irradiation on human cortical bone transplants contaminated with enveloped and non-enveloped viruses, *Biologicals*, **2002**, 30, 205-213.

[18] Koenigbauer, U. F., Eastlund, T. and Day, J. W., Clinical illness due to parvovirus B19 infection after infusion of solvent/detergent-treated pooled plasma, *Transfusion*, **2000**, 40, 1203-1206.

[19] Matsui, H., Sugimoto, M., Shizuko Tsuij, S., Shima, M., Giddings, J. and Yoshioka, A., Transient hypoplastic anaemia caused by primary human parvovirus B19 infection in a previously untreated patient with haemophilia transfused with a plasma-derived, monoclonal antibody-purified factor VIII concentrate, *J. Paed. Haem. Onc.*, **1999**, 2, 74-76.

[20] Nichols, W. G., Price, T. H., Gooley, T., Corey, L. and Boeckh, M., Transfusion-transmitted cytomegalievirus infection after receipt of leukoreduced blood products, *Blood*, **2003**, 101, 4195-4200.

[21] Sanzén, L. and Carlsson, A, Transmission of human T-cell lymphotrophic virus type 1 by a deep-frozen bone allograft, *Acta Orthop. Scand.*, **1997**, 68, 72-74.

[22] Bundesinstitutes für Arzneimittel und Medizinprodukte (BfArM) and Paul-Ehrlich-Institut (PEI). Requirements for validation studies regarding the proof of virus safety among medications and human blood or plasma, *Bundesanzeiger*, **1994**, 84, 4742-4744.

[23] Mahnel, H., Differences in resistance between viruses of different groups against several chemical and physical decontamination procedures, *Infection*, **1979**, 7, 240-246.

[24] Bräuniger S., Peters, J., Borchers, U. and Kao, M., Further studies on thermal resistance of bovine parvovirus against moist and dry heat, *Int. J. Hyg. Environ. Health*, **2000**, 203, 71-75.

[25] Weusten, J. J. A. M., van Drimmelen, H. A. J. and Lelie, P. N., Mathematic modeling of the risk for HBV, HCV, and HIV transmission by window-phase donations not detected by NAT, *Transfusion*, **2002**, 42, 537-548.

[26] Müller-Breitkreuz, K., Results of viral marker screening of unpaid blood donations and probability of window period donations in 1997, *Vox Sang*, **2000**, 78, 149-157.

[27] Busch, M. P., Infectious risks of blood transfusions: known viruses and future threats, *27th Annual Meeting, American Association of Tissue Banks*, San Diego, CA, USA, **2003**.

[28] Lelie, P. N., van Drimmelen, H. A. J., Cuypers, T. M., Best, S. J., Stramer, S. L., Hyland, C., Allain, J. P., Moncharmont, P., Defer, C., Nübling, M., Glauser, A., da Silva Cardosos, M., Viret, J. F., Lankinen, M. H., Grillner, L., Withmüller, U., Coste, J., Schottstedt, V., Masecar, B. and Dax, E. M., Sensitivity of HCV RNA and HIV RNA blood screening assays, *Transfusion*, **2002**, 42, 527-536.

[29] Tomono, T., Status of NAT screening for HCV, HIV and HBV: experience in Japan, *Advances in Transfusion Safety*, Brown, F. and Seitz, R. (eds.), *Dev. Biol.*, **2003**, 108, 29-39.

[30] Schüttler, C. G., Caspari, G., Jursch, C. A., Willems, W. R., Gerlich, W. H. and Schaefer, S., Hepatitis C virus transmission by a blood donation negative in nucleic acid amplification tests for viral RNA, *Lancet*, **2000**, 355, 41-42.

[31] Anderson, S. M., Gocke, D., Belzer, S. W., Gamble, A., Campanella, C. and Bankowski, M. J., Detection of hepatitis C virus RNA in cadaveric serum or plasma samples using nucleic acid amplification, *27th Annual Meeting, American Association of Tissue Banks*, San Diego, CA, USA, **2003**, S-1.

[32] EAEMP, European Agency for the Evaluation of Medicinal Products, Note for guidance on virus validation studies: the design, contribution and interpretation of studies validating the inactivation and removal of viruses, Committee for Proprietary Medicinal Products (CPMP), *CPMP/BWP/268/95*, London, UK, **1996**.

[33] Campbell, D. G. and Li, P., Sterilisation of HIV with irradiation: Relevance to infected bone allografts, *Aust. N. Z. J. Surg.*, **1999**, 69, 517-521.

[34] IAEA, International Atomic Energy Agency, Sterility Assurance Level. In: *Guidelines for Industrial Radiation Sterilization of Disposable Medical Products (Cobalt-60 Gamma Irradiation), IAEA, TECDOC-539*, Vienna, Austria, **1990**.

[35] IAEA, International Atomic Energy Agency, *Code of Practice for the Radiation Sterilization of Tissue Allografts: Requirements for Validation and Routine Control. Preliminary version*, Boston, MA, USA, **2002**.

[36] Ascherl, R., Mogalla, M., Geissdörfer, K., Schmeller, M. L., Langhammer, H., Lechner, F. and Blümel, G., Experimental studies and clinical aspects of the refrigeration of allogenic spongiosa, *Orthopäde*, **1986**, 15, 22-29.

[37] Deutsches Institut für Normung, *DIN EN 552 Sterilisation of medical devices – Validation and routine control of sterilisation by irradiation*, Beuth-Verlag, Berlin, **1994**.

[38] Deutsches Institut für Normung, *DIN EN 829 Transport packaging for medical and biological test materials*, Beuth-Verlag, Berlin, **1996**.

[39] Scheidler, A., Rokos, K., Reuter, T., Ebermann, R. and Pauli, G., Inactivation of viruses by beta-propiolactone in human cryo-poor plasma and IgG concentrates, *Biologicals*, **1998**, 26, 135-144.

[40] Willkommen, H., Experience with the practise of infection assays and model viruses for evaluating the safety of blood products for patients, In: *Nosocomial Virus Infection – Recognition and Treatment*, Rabenau, H. F., Thraenhart, O. and Doerr, H. W. (eds.). Pabst Science Publishers, Lengerich, Berlin, **2001**, pp. 166-173.

[41] Reed, L. J. and Münch, H., A simple method for estimating fifty percent endpoints, *Am. J. Hyg.*, **1938**, 27, 493-497.

[42] Spearman, A. and Kärber, G., Essential features of biological and medical statistics, In: *Biometry*, Cavalli-Sforza, L. (ed.), Gustav Fischer-Verlag, Stuttgart, **1974**, pp. 171–173.

[43] Tomford, W. W., Transmission of disease through transplantation of musculoskeletal allografts, *J. Bone Joint Surg. Am.*, **1995**, 77, 1742-1754.

[44] Eastlund, T., Tissue transplant-transmitted bacterial infections: AATB investigation, *AATB Quality Assurance Workshop IV*, New Orleans, LA, USA, **2002**.

[45] EATB/EAMST, *Common standards for musculo skeletal tissue banking. OBIG-Transplant*, Vienna, Austria, **1997**.

[46] Pruss, A., Kao, M.,Kiesewetter, H., von Versen, R. and Pauli, G., Virus safety of avital bone tissue transplants: evaluation of sterilization steps of spongiosa cuboids using a peracetic acid-methanol mixture, *Biologicals*, **1999**, 27, 195-201.

[47] Pruss, A., Baumann, B.,Seibold, M., Kao, M., Tintelnot, K., von Versen, R., Radtke, H., Dörner, T., Pauli, G. and Göbel, U. B., Validation of thc sterilization procedure of allogeinic avital bone transplants using peracetic acid-ethanol, *Biologicals*, **2001**, 29, 59-66.

[48] Pruss A, Kao, M.,. von Garrel, T., Frommelt, L., Gürtler, L., Benedix, F. and Pauli, G., Virus inactivation in bone tissue transplants (femoral heads) by moist heat with the 'Marburg bone bank system', *Biologicals*, **2003**, 31, 75-82.

[49] Pruss A., Seibold, M., Benedix, F., Frommelt, L., von Garrel, T., Gürtler, L., Dörffel, Y., Pauli, G. and Göbel, U. B., Validation of the 'Marburg bone bank system' for thermodisinfection of allogenic femoral head transplants using selected bacteria, fungi, and spores, *Biologicals*, **2003**, 31, 287-294.

[50] Pruss, A., Jacker, H. J., Borchert, H. H., Seibold, M. and von Versen, R., Peracetic acid-ethanol treatment of bone allografts – a reliable sterilization method, *27th Annual Meeting, American Association of Tissue Banks*, San Diego, CA, USA, **2003**, S-22.

[51] Lomas, R., Drummond, O. and Kearney, J. N., Processing of whole femoral head allografts: a method for improving clinical efficacy and safety, *Cell and Tissue Banking*, **2000**, 1, 193-200.

[52] Tomford, W. W., Clinical experience of radiation effects with structural and non-structural grafts, *IAEA International High Level Expert Meeting 'Sterilization of Tissues using Ionising Radiations'*, Wrexham, Wales UK, **2003**.

[53] Grieb T., Forng, R. Y., Brown, R., Owolabi, T., Maddox, E., Mc Bain, A., Drohan, W. N., Mann, D. M. and Burgess, W. H., Effective use of gamma irradiation for pathogen inactivation of monoclonal antibody preparations, *Biologicals*, **2002**, 30, 207-216.

[54] Amareld, R. W., Wersocki, M., Drohan, M.,Drohan, W. N., Burgess, W. H., Mann, D. M. and Forng, R. Y., Controlled gamma-irradiation mediated pathogen inactivation of human urokinase preparations with significant recovery of enzymatic activity, *Biologicals*, **2003**, 31, 261-264.

[55] Spire , B., Dormont, D., Barre-Sinoussi, F., Montagnier, L. and Cherman, J. C., Inactivation of lymphadenopathy-associated virus by heat, gamma rays, and ultraviolet light, *Lancet*, **1985**, 1, 188-189.

[56] Salai M., Vonsover, A., Pritch, M., von Versen, R. and Horoszowski, H., Human immunodeficiency virus (HIV) inactivation of banked bone by gamma irradiation, *Ann. Transplant.*, **1997**, 2, 55-56.

[57] Fideler, B. M., Vangness, T., Moore, T., Li, Z. and. Rasheed, S., Effects of gamma irradiation on the human immunodeficiency virus. A study in frozen human bone-patellar ligament-bone grafts obtained from infected cadavera, *J. Bone Joint Surg. Am.*, **1994**, 76, 1032-1035.

[58] Hernigou P., Gras, G., Marinello, G. and Dormont, D., Influence of irradiation on the risk of transmission of HIV in bone grafts obtained from appropriately screened donors and followed by radiation sterilization, *Cell and Tissue Banking*, **2000**, 1, 279-289.

[59] Dziedzic-Goclawska, A., Personal report, **2003**.

[60] CDC, Center for Disease Control and Prevention, Creutzfeldt-Jakob disease in a patient receiving a cadaveric dura mater graft, *MMWR*, **1987**, 36, 324-325.

[61] CDC, Center for Disease Control and Prevention, Creutzfeldt-Jakob disease in a second patient who received a cadaveric dura mater graft, *MMWR*, **1989**, 38, 37-38.

[62] *Bundesärztekammer - Wissenschaftlicher Beirat*, Guidelines on therapy using blood components and plasma derivatives. Deutscher Ärzte-Verlag, Köln, **2003**, pp. 259-282.

[63] WHO, World Health Organisation – Health Technology and Pharmaceutical Cluster, Guidelines on Transmissible Spongiform Encephalopathies in relation to Biological and Pharmaceutical Products, *WHO/BCT/QSD/03.01*, Geneva, Switzerland, **2003**.

VIRAL INFECTIONS TRANSMITTED THROUGH TISSUE TRANSPLANTATION

Ted Eastlund [1,2]

[1] *Division of Transfusion Medicine, Department of Laboratory Medicine and Pathology University of Minnesota Medical School, Minneapolis, Minnesota 55455, USA*

[2] *Fairview-University Medical Centre, MMC 198/D251 Mayo 420 Delaware Street SE, Minneapolis, Minnesota 55455, USA*

ABSTRACT

The incidence of tissue allograft-transmitted infection is unknown and can best be inferred from prospective studies – that have not yet been performed and reported. Viral infections have been transmitted via tissue allografts such as bone, skin, cornea, and heart valves. Bone allografts have transmitted hepatitis C, human immunodeficiency virus (HIV-1), and human T-cell leukaemia virus. Corneas have transmitted rabies, hepatitis B virus, cytomegalovirus (CMV), and herpes simplex virus. Heart valves have been implicated in transmitting hepatitis B. HIV-1 and CMV have been transmitted by skin allografts. Use of comprehensive donor eligibility criteria; excluding potential donors with behaviours risky for HIV-1 and hepatitis infection, and donor blood testing have greatly reduced the risk. Recent reports of HIV transmitted from a seronegative donors prompts the addition of viral nucleic acid testing of the donor. During tissue processing, many allografts are exposed to disinfectants and sterilisation steps such as gamma irradiation, which further reduce or remove the risk of transmitted disease. Some viruses are fairly resistant to gamma irradiation and the high doses needed may be harmful to the tissue allografts. Because the effectiveness of some tissue grafts depends on cellular viability, not all can be subjected to sterilisation steps, and, therefore, the risk of infectious disease transmission remains. For these, preventing the transmission of viral infection relies mostly on careful donor selection and viral testing, but processing with mild disinfectant can be useful. To further assure safety in the use of allografts, the physician and hospital should select tissue banks that follow national professional standards as their source for allografts.

INTRODUCTION

Tissue transplantation therapy, which has been utilised for over 50 years [1], is a rapidly developing field carrying with it great promise for ameliorating or curing many diseases. One of its drawbacks, however, is the potential for donor-to-recipient disease transmission. This risk is greatly reduced by excluding donors at risk of carrying infection and by testing the donor for transmissible infectious disease. Aseptic surgical technique in a quality environment, when removing the tissue from the donor, when processing and storing the tissue and during implantation is critically important to prevent bacterial and fungal contamination. Non-viable tissue grafts such as bone can undergo disinfection and sterilisation steps. During the past two decades the disease transmission risk associated with tissue transplantation has been greatly reduced by implementation of standards set by professional organisations, such as the American

Association of Tissue Banks [2], the European Association of Tissue Banks (EATB) [3], the Eye Bank Association of America [4], and governmental regulations. However, the incidence of transplant-transmitted infection is unknown and the studies needed to determine this have not been performed.

Cadaveric donations (Table 1) and clinical transplants (Table 2) of cornea, bone, skin, heart valve, and other tissue allografts in the USA greatly exceed that of organs [5]. Organ transplantation flourished in the early 1980s following the discovery and introduction of cyclosporin as an effective immune suppressant.

Table 1. Cadaveric organ tissue donation in the USA.

Type of Donor Tissue	Donations Per Year
Cornea donors *	46,729
Bone, skin, or other tissue donor †	18,021
Organ donor ‡	6,082

* Eye Bank Association of America (EBAA) [4]
† American Association of Tissue Banks (AATB) [2]
‡ United Network for Organ Sharing (UNOS) [5]

Table 2. Estimated number of allografts transplanted annually in the USA.

Tissue	Transplants
CADAVER TISSUE *	
Bone	675,370
Cornea †	50,868
Skin (sq. ft.)	11,222
Heart Valve	5,500
Vessels	433
Pericardium	5,327

* American Association of Tissue Banks (AATB) [2]; Office of the Inspector General [6]
† Eye Bank Association of America – 2001

This brought a large supply of cadaveric donors available that could also be used for tissue donation. Unlike the limitations of organ transplants, tissue transplantation generally is not limited by HLA histocompatibility barriers [7] or by ABO blood group incompatibility [8]. No longer a scarce resource, the widened availability of tissue allografts encouraged new clinical use and brought attention not only to their effectiveness and advantages over autografts but also to their drawbacks, side effects, and complications. One complication of tissue transplantation has been transmission of viral infections of donor origin to the recipient [9,10]. Viral infections can be transmitted if the donor has a viral infection and the viral levels are too low for detection. In asymptomatic donors who are recently infected, a transient viremic phase can exist prior to development of a positive donor-screening test for antibodies. Preventing donor-to-recipient infectious disease transmission relies heavily on selecting safe donors not only through testing, but also by medical and social behaviour screening to select donors more likely to be free of transmissible infections. In addition, sterilisation with gamma irradiation can be important.

Some tissue allografts, e.g. corneas, heart valves and skin, need to remain viable and cannot be exposed to disinfectants or sterilised without an unacceptable loss of viability. Other grafts are non-viable, largely comprised of acellular connective tissue, and can be disinfected or sterilised more freely resulting in a greater assurance of sterility (Table 3). This review focuses on viral infections transmitted through transplanted tissues and steps taken for its prevention.

Table 3. Allograft characteristics affecting ability to transmit disease.

Nonviable Allograft	Viable Allograft
Type	
Bone	Heart valve and vessels
Dura mater	Cornea
Pericardium	Skin
Tendon	Marrow
Costal cartilage	Blood stem cells
Fascia	Vascularised organs
Ear ossicles	Semen and oocyte
	Foetal tissue
Characteristics	
Non-viable	Contains viable cells
Acellular	May be antibiotic treated
Connective tissue	Cannot be sterilised
Can be processed, sterilised	

VIRAL INFECTIONS TRANSMITTED BY CORNEAL ALLOGRAFT

Human Immunodeficiency Virus:

As thin avascular tissue, cornea comprises a well-hydrated transparent layer of connective tissue and a single-cell layer of viable endothelial cells. Consequently, it is not very immunogenic nor is it often rejected by the recipient, unless it becomes vascularised. Similarly it is not very efficient in transmitting viral infectious disease from the donor. Diseases transmitted through corneal transplants are listed in Table 4. The cornea is not efficient in carrying or transmitting HIV. Based on assumptions about HIV antibody test sensitivity, Goode *et al.* estimated that 3 per 10,000 cornea allografts would be from HIV-infected donors despite HIV antibody testing [11]. There have been several documented cases in which cadaveric organ and tissue donors were infected with HIV but the cornea recipients did not become infected. Although HIV has been isolated from tears, retina, cornea, aqueous humour, iris, and conjunctiva [12-16], HIV from infected cadaveric donors has not been transmitted to cornea recipients [17,18]. This should not be surprising because the inoculum of HIV is small in the relatively avascular, hypocellular cornea compared to that in an organ transplant or a blood transfusion from an HIV infected donor.

Table 4. Viral diseases transmitted by tissue allografts.

Allograft	Virus
Bone	Hepatitis C Hepatitis, unspecified type HIV HTLV
Tendon	Hepatitis C HIV
Cornea	Hepatitis B Rabies Herpes simplex virus Cytomegalovirus (?)
Heart Valve	Hepatitis B
Skin	HIV (?) Cytomegalovirus (?) Hepatitis C(?)
Saphenous vein	Hepatitis C

Hepatitis B Virus

Failure to transmit viral infection from hepatitis B surface antigen (HBsAg)-positive donors has been reported in two recipients of corneas. This suggests that the cornea is inefficient as a mode of HBV transmission, although in these cases the administration of hepatitis B immune globulin and vaccine to the recipient may have prevented infection [19].

Khalil *et al.* assessed the presence of HBsAg and HBV DNA in corneal buttons taken from HBsAg-positive donors [20]. They found HBV in a small percentage of corneas. Others studied 31 donors infected with HBV or HCV and were unable to detect HBV DNA or HCV RNA in the corneas [21]. Despite this inefficiency, HBV transmission by corneal transplantation has been reported.

In earlier reviews by O'Day [22] and Raber and Friedman [19] there were brief reports of hepatitis B transmission to cornea recipients from HBsAg positive donors. Two cases of recipient HBV infection after transplants from two different HBsAg-positive donors were eventually reported [23]. Corneal donations took place from two donors; one in 1984 from an alcoholic man and one in 1985 from an injecting drug user. Tests for HBV were performed on the donor retrospectively after recipients developed HBV infections. Both donors were positive for HBsAg. Only one of the two recipients of corneas from each of the two donors developed symptomatic HBV infection.

The use of current professional standards and federal regulations would have prevented these cases since exclusion of donors with hepatitis risk behaviours and testing for HBsAg are now required.

Rabies

Rabies virus infection in humans is often found in the cornea. Because of this, a cornea impression test has been useful for early diagnosis [24]. Corneal allografts are also capable of transmitting rabies. Seven cases of fatal rabies transmission from cornea transplantation have been reported in the US, France, Thailand and India during 1979-1988 and in Iran in 1994 [25-31]. The first case involved a 39-year-old man in the US with ascending paralysis [26] and the second involved a donor in France who died from paraplegia, encephalitis and myocarditis [27].

In 1997 Javadi *et al.* and Gode and Bhide each reported rabies developing in two patients who received corneal transplants from the same donor [25,31]. Each of these cadaveric donors had an obvious acute neurological illness clinically consistent with rabies. National professional standards used by tissue banks today prohibit the use of these donors and would have prevented these cases of rabies transmission.

Other Viruses

Cytomegalovirus does not appear to be readily transmitted by cornea transplantation from seropositive cadaveric donors to seronegative recipients. Of 25 seronegative patients receiving a corneal graft from a seropositive donor, only two seroconverted [32].

Herpes simplex virus, type 1, has been found widespread in corneal stromal cells but only one case of transmission by a cornea allograft has been reported. The infection caused corneal deterioration in the recipient by the fifth day after transplant [33]. HSV DNA was found in two of five cornea allografts from other donors [34].

VIRAL INFECTIONS TRANSMITTED BY BONE ALLOGRAFT

Hepatitis

Hepatitis has been reported from use of unprocessed refrigerated and frozen bone allografts, but not from bone grafts that were cleaned of cells and fat with water jetting and ethanol soaks prior to being freeze dried or treated with sterilants such as gamma irradiation or ethylene oxide. In 1954, prior to the availability of viral hepatitis testing of donors, a Yale medical student received a refrigerated bone graft to treat a depressed fracture of the proximal tibia and developed hepatitis with jaundice 10 weeks later [35]. The bone graft was obtained from the amputated leg of a patient with occlusive vascular disease and gangrene. Otherwise, the donor was in good health, with normal liver function tests and without a history of jaundice or liver disease. The donor had received blood transfusions 3 years previously.

Three reports from nearly a decade ago documented that hepatitis C virus (HCV) can be transmitted from donor-to-recipient through the use of frozen, unprocessed bone allografts [36,37,38]. In the first case, donor testing for HCV antibodies was not available. HCV was transmitted by the use of a femoral head allograft after it was donated by a donor undergoing hip arthroplasty and stored frozen for 2 months [36]. In a second report, HCV was transmitted from an infected cadaveric tissue donor through frozen, unprocessed bone and tendon grafts, but not through freeze-dried bone allografts that were treated with gamma irradiation [38]. In this study, the cadaver bone donor tested negative for HCV antibodies using the first generation test available at the time of donation in 1990, but stored serum tested positive when a new, more sensitive test was introduced in 1992. Testing for HCV RNA by polymerase chain reaction (PCR) was also positive. In a third brief report involving five HCV-infected organ and tissue cadaveric donors, a minority of the recipients of frozen bone allografts became infected with HCV [37].

In a more recent case, an HCV-infected organ and tissue donor was tested and found negative for HCV antibodies [39,40]. Despite this negative screening test for HCV, several organ and tissue recipients became infected. When blood samples from the donor were tested later for HCV RNA, the results were positive and this confirmed the link between the donor and multiple infected recipients. The donor had been recently infected and was viremic but had not yet produced detectable antibodies. Bone allografts from the same donor that had been treated with gamma irradiation did not transmit HCV. With the implementation of HCV RNA as a donor-screening test in the future, cases such as this would be prevented.

There have been no reports of HBV transmission through bone transplantation, although it has been recognised as a complication of organ, cornea, and heart valve transplantation. It is quite probable that there have been transmissions but none have been recognised and published.

Human Immunodeficiency Virus

HIV-1 has been transmitted through blood, tissues, and organs [41,42]. Viable HIV-1 can be recovered from bone, marrow, and tendons of patients with acquired immunodeficiency syndrome [43-46]. In 1984, a fatal HIV-1 infection was transmitted to a woman undergoing spinal fusion for scoliosis through the use of a frozen femoral head allograft several weeks after it had been donated during hip arthroplasty from a donor who had a history of intravenous drug abuse and who had an enlarged lymph node that had been biopsied the previous year [47]. Both the donor and the bone allograft recipient subsequently died of AIDS. A test for HIV-1 antibody was not available at the time of donation. This donor would not have been eligible to be a donor today due to his history of intravenous drug abuse and lymphadenopathy. There have been other cases of HIV infection in recipients of bone allograft derived from HIV-infected donors who were not tested for HIV at the time of donation. Prior to HIV antibody test availability in Germany, 12 recipients had frozen bone allografts from an infected cadaveric donor during November 1984 through January 1985 [48]. Only four of these recipients became HIV positive. Seven remained HIV-negative. In Taiwan a man donated a femoral head during hip replacement surgery but was not tested for HIV. The bone allograft was used in a 34-year-old woman in 1996 during knee reconstructive surgery. She seroconverted with HIV antibodies when tested five months later [49].

Another reported case of HIV transmission through the use of frozen bone allograft involved a seronegative but infected cadaveric donor but the test was new and not very sensitive. Multiple organs, corneas, bones, and connective tissues were transplanted [42]. Three organ recipients and three recipients of frozen bone and tendon allografts became HIV infected. These allografts had not been sterilised with gamma irradiation or ethylene oxide gas prior to use. The donor tested negative for HIV antibody at the time of donation in October 1985, which was a few months after the first, relatively insensitive HIV antibody testing kits became available. Between 1985 and 1991, there were several modifications that greatly improved test sensitivity. Prior to 1989, HIV antibody was detectable a median of 63 days after initial infection [50,51]. A study of HIV infected blood donors between March 1987 and 1991, when whole viral lysate enzyme immunoassays were used to detect HIV antibodies, showed an average seronegative window period of 45 days [52]. A report in 1992 showed HIV antibody test kits in use at that time detected twice as many infected individuals as did the test kits available in 1985 [53]. Since 1992, HIV antibody tests became even more sensitive, detecting IgM, the earliest form of antibodies, an average of 8-20 days earlier [54,55] and resulting in a seronegative window period of approximately 22 days [56,57]. Since then blood donor testing for HIV RNA by nucleic acid testing (NAT) has been implemented and has further reduced the risk of a transfusion and when validated and implemented for cadaver tissue donors will reduce the risk in tissue transplantation [58].

The prevalence of HIV antibodies in bone donors is low and when the medical history screening and selection processes are applied vigorously, it should not be greatly different from that of voluntary blood donors. This may be true for living bone donors [59-61] but not necessarily for cadaveric donors. Of 9000 living bone donors who donated femoral heads at the time of hip arthroplasty surgery, none were found to have confirmed positive tests for HIV-1 antibodies at the time of donation [59]. Prevalence of infectious disease markers in surgical bone donors was not different from that of blood donors, except for a higher prevalence of false positive syphilis tests and antibodies to HBV core protein [60]. Retesting of 1608 living bone donors 180 days later yielded none with confirmed positive HIV or HCV tests [61]. Of 5513 cadaver bone donors tested throughout the United States in 1992, there were three confirmed positive for HIV antibodies [62], but these three were from a single tissue bank that later disclosed accepting donors with risk factors for HIV. A more recent survey by AATB [63] revealed a higher prevalence rate of infectious disease markers than have been reported for blood donors, with rates ranging from 2 to 40 times higher (Table 5). The addition of viral nucleic acid testing, since 1999, in screening blood donors for HIV and HCV has further reduced the risk from blood donors. First time blood donors have a higher infectious disease marker rate and may be more similar to organ and tissue donors [64].

Because prospective tissue donors with HIV risk behaviours and positive tests for HIV are excluded and most bone graft processing removes blood and marrow cells and applies disinfectants and sterilants, the risk of HIV transmission by bone transplantation is now very remote, if not nearly absent [65]. The risk of transmitting HIV through bone grafting has been calculated to be less than one in a million grafts [66,67], and is even less if the graft has been subjected to processing and sterilisation steps using gamma irradiation or ethylene oxide. However, the HIV transmission risk is higher in the less frequently used frozen unprocessed bone allograft. An accurate estimate of the risk cannot be made until a more accurate determination of the prevalence of HIV infection in the donor and recipient population is available and prospective studies have been done on recipients.

Table 5. Donor exclusions for hepatitis and HIV risk behaviour.

HIV and Hepatitis Risk Behaviour Exclusions
Persons with clinical or laboratory evidence of HIV infection
Men who have had sex with another man even one time in past 5 years
Non-medical injections of drugs in past 5 years
Persons with haemophilia or related clotting disorders who have received human-derived clotting factor concentrates
Persons who engaged in sex for money or drugs in past 5 years
Persons who have had sex with any of the above in past 12 months
Exposure to blood suspected to be HIV or hepatitis infected through percutaneous inoculation of open-wound or mucous membrane contact in past 12 months
Inmates of prisons for at least 7 days in past 12 months
Tattoo in past 12 months

Human T-Lymphotrophic Virus

Asymptomatic HTLV-I infection has been transmitted by transplantation of a fresh-frozen unprocessed femoral head bone allograft [68]. A 62-year-old man became infected by HTLV from a blood transfusion in 1987 during hip surgery. One month later he developed fever a rash and a transient right-sided radial nerve palsy. Frozen sera obtained during this illness but tested later demonstrated HTLV seroconversion. In 1991 he donated a femoral head without anti-HTLV testing during a second surgery for a hip prosthesis. The unprocessed frozen femoral head was used as a graft in a different patient one month later. This bone graft recipient developed HTLV-I antibodies but had no HTLV-I associated disease.

INFECTIONS FROM CARTILAGE AND OSTEOCHONDRAL ALLOGRAFTS

Costal cartilage allografts are routinely disinfected or sterilised prior to their use as allografts and provided in a freeze-dried or frozen form. There have been no reports of processed costal cartilage transmitting infection from the donor to the recipient. Donald and Cole surveyed 312 surgeons who used cartilage allografts preserved by 8 different methods for facial reconstructive surgery [69]. They found a postoperative bacterial infection rate of 19% that was similar to the 16% reported following use of autologous cartilage.

VIRAL INFECTIONS TRANSMITTED BY TENDON ALLOGRAFTS

Viral Diseases

The use of the patellar tendon allograft to replace the knee's injured anterior cruciate ligament has become commonplace [70]. HIV has been isolated from tendons in HIV infected persons [43,71] and has been transmitted from a seronegative cadaveric tissue donor through a donated patellar tendon used in knee surgery [42]. HCV was transmitted to recipients of frozen tendon allografts from an anti-HCV positive cadaveric donor [38]. It is possible that HIV and HCV were harboured in the unprocessed bone blocks at either end of the tendon allograft. These allografts had not been processed to remove blood and marrow cells. Despite these cases, the risk to recipients is presumably low as long as donor-screening steps are applied as required by national standards [2,72] and federal regulations [73-75]. In addition, tendons can be treated with gamma irradiation to further reduce the risk of disease transmission.

Selecting donors without risk factors and without HCV antibodies make the risk of spreading HCV by transplant an exceedingly rare event. However, an early HCV infection in a cadaveric organ and tissue donor not yet producing antibodies was reported recently [39,40]. A patellar tendon allograft recipient developed acute, symptomatic hepatitis C in May 2002, six weeks after transplantation. No other potential sources of infection were identified. The tissue donor was anti-HCV negative but stored serum showed HCV RNA when tested later. Thirty-nine other persons received tissues or organs from this same donor. Early results of a partially completed investigation showed that of 18 recipients tested, six showed HCV infection including a lung recipient who became HCV RNA positive on day 4 and died of liver failure 14 months later. Presumably, the cadaveric donor was in a viremic stage early in infection prior to antibody development. To date cadaveric testing for HCV RNA is not available for routine use. HCV RNA should be considered for cadaveric tissue donors as soon as test reliability has been evaluated particularly using cadaveric samples obtained up to 24 hours after death.

VIRAL INFECTIONS TRANSMITTED BY CARDIOVASCULAR ALLOGRAFTS

Viral Diseases

The capacity of human heart valve allografts to transmit HBV was demonstrated in a study of thirty-one patients who received heart valve allografts from HBsAg-positive donors. Twenty-two recipients were HBsAg-positive prior to transplant or were immune to HBV and not susceptible to HBV infection. Of the nine recipients susceptible to HBV infection, only one developed HBV viral markers. None developed clinically apparent hepatitis. However, four susceptible recipients received hepatitis B immune globulin and one received HBV vaccine following transplant, which may have prevented infection [76]. Currently all donors are tested for HBsAg and if positive are excluded. Despite testing donors for HBsAg and anti-HBc, HBV transmission can still occur because some donors can have circulating HBV at levels not detectable in routine tests. Thijssen et al. found one of 676 heart valve allograft donors to have HBsAg detectable with routine screening tests [77]. In addition, they found 10% to have anti-HBc. Fifty-two of 63 donors with anti-HBc also had antibodies to HBV surface protein (anti-HBs), indicating a resolved HBV infection and a recovered, immune non-infectious status. Three of those with anti-HBc but without anti-HBs were positive for HBV DNA using a more sensitive liquid-phase DNA hybridisation assay. This would suggest a possible value of anti-HBc donor testing to prevent transmission of HBV; however, one study of blood donors has shown a lack of predictive value in preventing post-transfusion hepatitis [78]. More recently, however, several reports have confirmed that some anti-HBc positive donors will be positive for DNA and will transmit HBV [79].

Recently a case of HIV infection transmitted by use of a saphenous vein allograft was reported. The cadaveric donor had no known HIV risk factors or signs of hepatitis and had negative tests for HIV antibodies. Subsequent studies demonstrated HCV RNA in the donor serum that was the same serotype, 1a, as that found in the donor.

VIRAL INFECTIONS TRANSMITTED BY SKIN ALLOGRAFTS

Viral Infection

Viral disease transmission by skin allografts has been reported. Epidermal cells can be infected with HIV-1 and the epidermis of HIV-infected individuals can transmit HIV to white cells in vitro [80,81]. In one study HIV RNA was found in only one of twelve infected patients [82]. Clarke reported, in a brief letter, a weakly positive test for antibody to HIV-1 in a burn patient after receiving skin from an HIV-positive donor [83]. The results of donor testing were not known before the skin was used. The authors did not report whether other recipient risk factors were present or the results of confirmation testing. HIV transmission from skin allograft has been recently reviewed [84,85].

Transmission of hepatitis from skin allograft has not been reported although HCV nucleic acid has been demonstrated in skin from infected donors [38]. There are recent reports implying transmissibility of human cytomegalovirus (CMV). Animal models clearly demonstrate that skin grafts are capable of transmitting CMV [86-89]. Earlier studies by Kealey *et al.* showed that burn patients acquire CMV during hospitalisation and that blood transfusions may be a contributing factor [90]. A subsequent study eliminated blood transfusion as a contributor by studying patients who received skin allografts but no CMV-positive blood. They showed that CMV-negative burn patients who receive skin allografts from CMV-positive donors can seroconvert to become CMV-positive [91].

CMV resides in peripheral blood leukocytes in asymptomatic CMV antibody-positive donors long after their initial infection. Asymptomatic CMV-positive donors can transmit CMV infection through transfusion and transplantation if the recipient is CMV negative. Most healthy adult prospective donors have CMV antibodies, and therefore excluding CMV positive donors to prevent CMV transmission would exacerbate the already existing shortage of skin allografts and would not be practical. Testing donors for CMV antibody is not required by national professional standards.

Immunosuppressed individuals such as organ recipients have a high mortality and morbidity rate from transplant-transmitted-CMV of donor origin. The burned patient also acquires CMV infection but does not generally experience serious complications as regularly as organ recipients, perhaps because burn-related immunosuppression may be less profound than that produced by drugs used to prevent organ rejection. As burn patients begin to receive potent immunosuppressants such as cyclosporine to block rejection of skin allografts, CMV may become a more serious complication of burn care and related blood transfusion and skin allografting. Further studies of skin allograft recipients are needed to determine whether transmission of CMV by skin allograft is associated with symptomatic disease as seen in organ transplantation recipients, or is asymptomatic as generally seen in transfusion-transmitted CMV infections in immunocompetent blood transfusion recipients. It is thus premature to assume that it is beneficial to base selection of skin donors on CMV antibody testing.

EMERGING INFECTIOUS RISKS

Infectious risks of tissue transplantation have often been identified after first being recognised as blood transfusion-transmitted infections. Many real or theoretical risks of tissue transplantation can be considered by looking at the emerging infections that threaten to affect transfusions [92,93]. Most recently West Nile Virus (WNV) infection has swept through the US with nearly 4,000 human cases identified and 254 deaths in 2002 [94]. In addition to being mosquito-borne, WNV has been transmitted through organ transplantation, blood transfusions, transplacental intrauterine spread and percutaneous route from laceration and needlestick [94]. A new corona virus infection recently caused a severe acute respiratory syndrome (SARS) and spread across the continents [95]. Although the susceptibility of these viruses to gamma irradiation or other sterilants is unknown, the routine use of sterilisation may provide some protection from transmission by tissue transplantation.

REDUCING THE RISK THROUGH DONOR SELECTION

To minimise the risk of transmitting viral infectious disease, several important approaches are taken by transplanting surgeons and tissue banks. An initial approach by the surgeon is to judiciously use tissue allografts only when needed and from accredited organisations, use sterilised allografts whenever possible and consider use of autografts and alternative non-human graft material. However, the most important steps are exercised by the tissue bank in excluding those prospective donors who have known or suspected viral infections or are suspected by their behaviours to be at risk for HIV and hepatitis (Table 5). Tissue transplantation is generally considered a non-urgent surgical procedure, permitting a careful tissue donor selection process. Tissue donor selection by tissue banks has evolved to include a direct interview with the donor's next-of-kin concerning the donor's medical history and risk behaviours for HIV and hepatitis along with blood infectious disease testing, a physical examination and the results of an autopsy examination, if performed (Table 6). These donor selection steps are essential activities that result in a low risk of transmitting viruses. In addition, many tissue allografts can undergo further processing and exposure to disinfectants or sterilants, all of which further reduce the hazard of disease transmission [65]. Although there have been no carefully controlled studies of allograft recipients to determine the incidence of disease transmission, there is good reason to believe that established donor screening procedures, infectious disease testing and processing and sterilisation effectiveness to reduce or eliminate viruses, results in a very low risk of transmitting disease.

Table 6. Cadaver tissue donor selection steps to prevent virus transmission.

Voluntary donation without monetary inducement
Health History Review:
• Review of medical records
• Interview of next of kin
• Exclusion of those with infection, no HIV, hepatitis risk behaviour
Blood Tests:
• Hepatitis B surface antigen
• Antibody to HIV-1, HIV-2, HCV
• Antibody to HTLV-I, HTLV-II, syphilis†
• Antibody to HBV Core Antigen (for living donors only) ††
• Nucleus acid test (future)
Physical Examination:
• Unexplained jaundice
• Evidence of injectable drug use
• External signs of infection, including HIV
Autopsy Examination (if performed):
• Exclude those with infection
Maternal HIV testing and risk factor exclusion if donor < 18 months old

{† AATB, EBAA, UNOS & American Red Cross requirements; †† AATB Requirement}

Donor Selection

One important contribution to recipient safety is to seek voluntary, non-remunerated donors. Monetary inducement to the next-of-kin of cadaveric tissue donors is prohibited by professional standards [2] but it is being considered as a means of reducing the severe organ supply shortage in the US [96].

Monetary payment to donors for blood donation increases the risk of disease transmission [97]. Monetary incentives to donate may cause donors or their next of kin to be untruthful about the donor's health history information and to donate when they should not. Published data clearly show a 5-10 times increased incidence of donor-to-recipient post-transfusion hepatitis B virus infection with the use of paid blood donors [97]. In addition, there is an 11-15 times increased prevalence of HCV antibodies and 3-14 times increased prevalence of HIV antibodies in paid blood donors compared to voluntary blood donors [97]. HCV RNA was detected more often in clotting factor concentrates derived from paid donor plasma than volunteer donor plasma [98].

To minimise the risk of transmitting infectious disease, tissue donor eligibility requirements have been set by national professional standards [2-4,72], US Public Health Service guidelines [99,100], and US federal regulations [73-75]. Donor selection is an important first step taken to ensure that the resulting allograft is safe and effective. The cadaver tissue donor selection process includes a donor's medical and social history obtained from the next-of-kin and medical care providers, blood tests, a physical examination performed by tissue bank personnel, and an autopsy, if performed (Table 6). Preliminary donor selection is based on the donor's medical history and circumstances surrounding death. Donors are excluded if elements of the past medical and social history or recent hospitalisation indicate a risk of infection, malignant disease, or inadequate quality of donated organ or tissue. The living donor, the legal consenting next-of-kin or life partner of a cadaveric donor, or both, must be directly interviewed to determine whether HIV or hepatitis risk behaviours are present (Table 5). Persons with HIV and hepatitis risk behaviours are excluded from donation. The tissue bank physician makes the final determination of suitability of a cadaveric tissue donor as required by national professional standards [2,72]

Physical Examination of Donor

The next screening step is a limited physical examination of the tissue donor by procurement staff at the time of cadaveric donation [2,72]. The body is examined for signs of injecting drug use and signs of HIV, hepatitis or other infection or trauma over bodily sites that can affect the quality of donated tissue.

Blood Testing of Donor

Donor blood testing for disease markers plays an important role in reducing the risk of disease transmission. By eliminating prospective donors with infectious disease risk factors prior to blood testing, the risk of a seronegative but infected donor is minimised. Testing for HBsAg, anti-HIV, and anti-HCV is required by federal regulations [73-75] and national professional standard setting organizations [2,4,72]. Other tests required by standard setting organizations are syphilis and anti-HTLV-I/II [2,4,72].

HIV antigen (p24 antigen) testing of the donor is not performed by most tissue or organ banks. Large-scale studies of low risk [101] and high risk [102] blood donor populations demonstrated a lack of utility for HIV antigen screening. These blood donor studies and similar studies on smaller numbers of cadaver bone [103] and cornea [104] donors did not detect HIV infected donors beyond those already detected by testing for HIV antibodies. However, some regions of the world may have a higher prevalence of HIV infection and in these populations HIV antigen testing may be useful, especially if HIV RNA testing is not performed. One tissue bank has reported finding bone donors with negative tests for HIV antibodies but positive tests for HIV antigen. Presumably these HIV antigen tests were not false positive. The bone bank has added the precaution of treating all bone allografts with 2.5 Mrad gamma irradiation [105].

Studies are underway to determine whether testing donors for HIV RNA and HCV RNA by nucleic acid testing is practical. One study of 1424 cadaver bone donors showed that the use of HIV DNA (not HIV RNA) and p24 antigen blood testing did not detect additional HIV infected cadaveric bone donors [103]. All 1424 donors negative for HIV-1 antibodies were also negative for HIV DNA. This is what is expected since the appearance of HIV DNA in a recently infected person's blood does not occur sooner that the onset of detectable antibodies. On the other hand, HIV RNA appears earlier and has a greater potential benefit. Although HIV RNA testing is more sensitive than antibody assays, it may be premature to apply it routinely to cadaver donor testing due to the low HIV prevalence in the donor population, its uncertain predictive value, its false positive rate, and its false negative rate due to haemoglobin contamination and other potentially interfering substances in cadaveric post-mortem blood samples.

Testing of living blood donors for HIV and HCV RNA has markedly improved the safety of the blood supply even though screening has been done using pools of 16 to 24 samples [58]. Initially, viral nucleic acid testing was not feasible in blood donor screening applications due to lack of automation, time and space restrictions and cost. Recently, however, test systems are being used to test over 13 million blood donations annually in the United States: the Roche Molecular Systems COBAS AMPLISCREEN tests for HCV and HIV and the Gen-Probe/Chiron Pooled Plasma HIV-1/HCV Amplified Assay. Testing is being done on pooled samples using pools of 24 or 16. Testing of pooled samples reduces the number of tests required on a daily basis, the time to perform testing and the cost. It also takes into account the rapid rise of viral RNA in recently infected individuals so that pooling has a minimal impact on the sensitivity of these assays. The increased sensitivity of these systems over previously available PCR tests has also made this possible. Both systems have now been licensed for blood donor screening and efforts are underway to qualify them for organ and tissue donor screening. The same approach is now being tested in trials for HBV and WNV.

Many bone banks test donors for anti-HBc, a test originally introduced for blood donors as a surrogate for detecting non-A, non-B hepatitis carriers. The utility of this test as a surrogate has been diminished since adding specific tests for HCV (HCV antibodies and HCV RNA), the major cause of non-A, non-B hepatitis [78]. Although not required by AATB Standards and in the absence of HBV DNA testing, the use of anti-HBc in donor testing likely had a safety benefit in reducing HBV infections. Several reports have documented the presence of HBV in the sera of HBsAg-negative, anti-HBc-positive blood donors [79,106,107]. These reports also suggest that the addition of HBV DNA testing will increase the sensitivity of HBV detection but may not entirely replace the need to test for anti-HBc or HBsAg. For example, recipient directed lookback procedures have revealed recipients of HBsAg negative, NAT negative, anti HBc-positive blood components to have been infected with HBV [79].

Hemodilution of Donor Blood Sample

Massive blood loss and intravascular volume replacement by transfusion of blood, colloid, and crystalloid solutions can cause hemodilution and result in unreliable donor test results for infectious diseases [108]. In 1987 a case was reported of HIV transmission to multiple recipients of organs derived from an infected donor. Testing of the donor was negative when blood was sampled immediately after receiving blood transfusions amounting to two to three total blood volumes and an additional large volume of crystalloid solution over an eleven-hour period [109]. When a blood sample was obtained 48 hours later was tested for anti-HIV, it was positive due to intravascular replenishment of immunoglobulin from extravascular sites.

In 1993 US federal regulations [73] were first published, with subsequent modification [74,75], requiring quarantine of tissue from adult donors who had blood loss and received greater than two litres of blood and colloids within 48 hours of blood sampling or greater than one litre of crystalloid within a one hour of sampling. The donated tissue was not to be used unless a pre-transfusion sample was available for testing or an algorithm used to evaluate blood and colloid volumes administered in the 48 hours prior to sampling to ensure sufficient plasma dilution to alter test results has not occurred.

AATB Standards also require tissue banks to follow written procedures setting hemodilution limits to prevent use of false negative results when testing post-transfusion blood samples for infectious disease [2]. Acceptability limits must be part of written procedures. Standards of the American Red Cross Tissue Services require that in the case of blood loss and transfusion within 48 hours of death, a pre-infusion sample must be used [72]. A post-infusion sample may be used in patients with major clinical blood loss provided that the tissue bank physician has evaluated whether blood and crystalloid infusions have compensated for blood loss, estimated hemodilution is 50 percent or less of the total blood volume and the tissue bank physician gives written approval. The estimated amount of hemodilution depends upon the type of fluid infused and the amount of time elapsed since infusion.

Quality of Cadaver Donor Blood Samples

The testing of cadaveric tissue donor serum for viral markers may be complicated by false positive tests when sampling is delayed after death [104] or when there is haemolysis [53,104]. False positive results for HBsAg and p24 antigen due to haemolysis may be found depending upon which manufacturer's test kit is used [53]. Sample quality and presence of haemoglobin can also cause false negative results when testing for HIV and HCV by NAT [110,111]. Frozen storage and multiple freeze-thaws do not have a major effect on detectability of antibodies to infectious agents in serum, but they may reduce the reliability of testing for microbial nucleic acids by NAT. Busch *et al.* showed that multiple freeze thaws can reduce the detectability of HCV RNA by NAT [112]. Published studies are too few to draw any firm conclusions, other than a possible deleterious effect of frozen storage and freeze thawing on testing serum for HCV RNA and peripheral blood white cells for HIV DNA by NAT. Despite this, NAT was used to detect and retrospectively diagnose HIV infection using marrow specimens from an organ and tissue donor and frozen sera from organ recipients 5 years after the donation and transplantation [42]. HCV testing by NAT was also useful in confirming HCV infection in one cadaveric donor [38] and was essential in another [39]. Although NAT for hepatitis and HIV will be very useful, the tests are under development for organ and tissue donor testing.

Autopsy

Autopsies of donors are not generally required but for those tissues that can be stored, a final donor suitability determination is not made by the tissue bank physician until the results of the autopsy, if performed, have been reviewed [2,72]. Autopsy findings may not be reliable in detecting viral infections but results have disqualified donors with undiagnosed malignancy, widespread granulomas, abscess, and pneumonia.

Living Bone Donor Selection

Femoral heads can be donated by persons undergoing hip arthroplasty. Donor medical history screening and testing requirements are similar to those for cadaver bone donors, except that the donor medical history interview can be made directly and a retest for HIV and HCV antibodies is required 180 days following donation [2,72,99,100]. The retest aims to identify recently infected donors who were seronegative for HIV and HCV antibodies at the time of donation. This 180 day retest is required for semen and bone donors but not for living donors of blood, marrow, amnion, umbilical veins, or foetal tissue. Retesting the low risk voluntary living bone donors has not detected any additional infections despite thousands of donations and retests [61], whereas it may have utility in the higher risk population of paid semen donors. Most tissue banks have ceased collecting surgical femoral heads, partly due to lower quality, but also due to the difficulties of acquiring the 180-day sample for retesting. Testing of samples for HCV and HIV RNA from living donors at the time of donation would enhance safety and could eliminate the need for a 180-day retest.

REDUCING THE RISK DURING CADAVERIC TISSUE PROCESSING

Tissue allografts can be contaminated by bacteria or fungus during processing from environment surfaces and air, from personnel, from contaminated reagents, surgical instruments, supplies and processing equipment. In cases of HIV and HCV transmission by tissue allografts, the origin of the virus has been from infection of the donor. Electrolyte solutions purchased commercially [113] or deionised water prepared by the tissue bank [114] can be contaminated by bacteria and contaminate the tissue allograft when used for processing. In rare cases, there has been HBV contamination acquired during liquid nitrogen storage [115]. This has led to the widespread practice of using the vapour phase rather than submersion in the liquid phase. Viral contamination during tissue processing has not been reported.

The bone graft disinfection and sterilisation step most often used by bone banks in the United States is gamma irradiation at doses of 15 to 30 kGy [116]. For several types of tissue, processing steps involve elimination of blood cells and exposure to disinfectants such as alcohol and peroxide, both of which have viral inactivation properties. The use of sterilants such as gamma irradiation, ethylene oxide and heat, are more reliable for eradicating viable viruses.

SUMMARY

Transplantation of tissues has resulted in transmitting viral diseases from donor to recipient. When the first truly effective immunosuppressant, cyclosporin, became available in 1981, organ transplantation flourished. The numbers of organs transplanted each year grew rapidly leading to implementation of effective programs to develop public support for organ donation. As organ donation grew, so did tissue donation. With the rapid growth of transplantation, early cases of transplant transmitted viral diseases arose from infections of the donor but could have been prevented if tests were available. The use of new and more sensitive donor viral detection tests has reduced the risk of recipient infections.

Early cases of transmitted viral infection involved donors who had HIV and hepatitis risk behaviours. Exclusion of these as prospective donors has played an important role in reducing the risk of transmission. More recently, several cases have been reported where the infecting organism is bacterial and fungal instead of viral and the microbes did not arise from the donor but was a contaminant acquired during the procurement, processing or storage of the tissue. Newer national professional standards and US FDA regulations are addressing this area.

To prevent, or at least minimise, the risk of transmission of infectious disease, several approaches are important. Surgeons should use allografts only when needed and obtain them from accredited tissue banks. They should consider use of autografts, alternative nonhuman graft material, or processed and sterilised tissue allografts whenever possible tissue banks reduce the risk for rejecting those donors suspected to be at risk for HIV, hepatitis, or other viral or bacterial infections, performing a donor physical examination, reviewing autopsy reports, and the surgical removal of donated tissue using aseptic technique. Lastly, sterilisation of certain tissues can be very effective, but it is not universally applicable, because some viruses are relatively resistant to sterilants and because the clinical effectiveness of many tissues (such as skin, cornea, valves, veins) can be altered by sterilisation procedures [117]. Application of all these steps will assure a very low risk of transmitting disease from the donor to the recipient.

REFERENCES

[1] Strong, D.M., The US Navy tissue bank: fifty years on the cutting edge, *Cell Tissue Banking*,. **2000**, 1, 9-16.

[2] McLean, V. A., *Standards For Tissue Banking*, American Association of Tissue Banks (AATB), **2001**.

[3] European Association of Tissue Banks (EATB), *Standards For Tissue Banking*, Berlin, Germany, **2000**.

[4] Eye Bank Association of America (EBAA), *Medical Standards*. Washington, DC, USA, **2002**.

[5] United Network for Organ Sharing (UNOS), Cadaveric Donors and Cadaveric Organ Transplantation In The US, *UNOS Update*, **2001**, 10, 29.

[6] Office of the Inspector General, *Oversight of Tissue Banking*, US Department of Health and Human Services, Boston, MA, USA, **2001**, pp. 1-17.

[7] Choo, Y. and Eastlund, T., Tissue transplantation and HLA typing, *Tissue Cell Report*, **1996**, 3, 2-3.

[8] Eastlund, T., The histo-blood group ABO system and tissue transplantation, *Transfusion*, **1998**, 38, 975-988.

[9] Gottesdiener, K. M., Transplanted infections: donor to host transmission with the allograft, *Ann. Intern. Med.*, **1989**, 110, 1001-1010.

[10] Eastlund, T., Infectious disease transmission through cell, tissue and organ transplantation: reducing the risk through donor selection, *Cell Transplant.*, **1995**, 4, 455-477.

[11] Goode, S. M., Hertzmark, E. and Steinert, R. F., Accuracy of the ELISA test for screening corneal transplant donors, *Am. J. Ophthalmol.*, **1988**, 106, 463-466.

[12] Cantrill, H. L., Henry, K., Jackson, B., Erice, A., Ussery, F.M. and Balfour, H. H., Recovery of human immunodeficiency virus from ocular tissues in patients with acquired immune deficiency syndrome, *Ophthalmology*, **1988**, 95, 1458-1462.

[13] Fujikawa, L. S., Salahuddin, S. Z., Ablashi, D., Palestine, A. G., Masur, H., Nussenblatt, R. B. and Gallo, R. C., Human T-cell leukemia/lymphotropic virus type III in the conjunctival epithelium of a patient with AIDS, *Am. J. Ophthalmol.*, **1985**, 100, 507-509.

[14] Fujikawa, L. S., Salahuddin, S. Z., Palestine, A. G., Masur, H., Nussenblatt, R. B. and Gallo, R. C., Isolation of human T -lymphotropic virus type III from the tears of a patient with the acquired immunodeficiency syndrome, *Lancet*, **1985**, 2, 529-530.

[15] Heck, E., Petty, C., Palestine, A., Luckenbach, M., Salahuddin, S. Z., Nussenblatt, R. and Harris, W., ELISA HIV testing and viral culture in the screening of corneal tissue for transplant from medical examiner case, *Cornea*, **1989**, 8, 77-80.

[16] Salahuddin, S. Z., Palestine, A. G., Heck, E., Ablashi, D., Luckenbach, M., McCulley, J. P. and Nussenblatt, R. B., Isolation of the human T-cell leukemia/lymphotropic virus type III from the cornea, *Am. J. Ophthalmol.*, **1986**, 101, 149-152.

[17] Pepose, J. S., Macrae, S., Quinn, T. C. and Ward, J. W., Serologic markers after transplantation of corneas from donors infected with human immunodeficiency virus, *Am. J. Ophthalmol.*, **1987**, 103, 798-801.

[18] Schwarz, A., Hoffman, F., L'Age-Stehr, J., Tegzess, A. M. and Offermann, G., Human immunodeficiency virus transmission by organ donation: outcome in cornea and kidney recipients, *Transplantation*, **1987**, 44, 21-24.

[19] Raber, I. M. and Friedman, H. M., Hepatitis B surface antigen in corneal donors, *Am. J. Ophthalmol.*, **1987**, 104, 255-258.

[20] Khalil, A., Ayoub, M., Abdel-Wahab, S. E-D. and El-Salakawy, A., Assessment of the infectivity of corneal buttons taken from Hepatitis B surface antigen seropositive donors, *Brit. J. Ophthal.*, **1995**, 79, 6-9.

[21] Sengler, V., Reinhard, T., Adams, O., Gerlich, W. and Sundmacher, R., Testing of corneoscleral discs and their culture media of seropositive donors for hepatitis B and C virus genomes, *Graefes Arch. Clin. Exper. Ophthal.*, **2001**, 239, 783-787.

[22] O'Day, D. M., Diseases potentially transmitted through corneal transplantation, *Ophthalmology*, **1989**, 96, 1133-1138.

[23] Hoft, R. H., Pflugfelder, S. C., Forster, R. L., Ullman, S., Polack, F. M. and Schiff, E. R., Clinical evidence for hepatitis B transmission resulting from corneal transplantation, *Cornea*, **1997**, 16, 132-137.

[24] Koch, F. J., Sagartz, J. W., Davidson, D. E. and Lawhaswasdi, K., Diagnosis of human rabies by the cornea test, *Am. J. Clin. Path.*, **1975**, 63, 509-515.

[25] Gode, G. R. and Bhide, N. K., Two rabies deaths after corneal grafts from one donor, *Lancet*, **1988**, 2, 791.

[26] Centre for Disease Control, Human to human transmission of rabies by a corneal transplant – Idaho, *MMWR*, **1979**, 28, 109-111.

[27] Centre for Disease Control, Human to human transmission of rabies by a corneal transplant – France, *MMWR*, **1980**, 29, 25-26.

[28] Centre for Disease Control, Human to human transmission of rabies by a corneal transplant – Thailand, *MMWR*, **1981**, 30, 473-474.

[29] Houff, S. A., Burton, R. C., Wilson, R. W., Henson, T. E., London, W. T., Baer, G. M., Anderson, L. J., Winkler, W. G., Madden, D. L. and Sever, J. L., Human to human transmission of rabies virus by corneal transplant, *N. Engl. J. Med.*, **1979**, 300, 603-604.

[30] Baer, G. M., Shaddock, J. H., Houff, S. A., Harrison, A. K. and Gardner, J. J., Human rabies transmitted by cornea transplant, *Arch. Neurol.*, **1982**, 39, 103-107.

[31] Javadi, M. A., Fayaz, A., Mirdehghans, A. and Ainollahi, B., Transmission of rabies by cornea graft, *Cornea*, **1996**, 15, 431-433.

[32] Holland, E. J., Bennett, S. R., Brannian, R., Osborne, J. C., Goeken, J. A. and Krachmer, J. H., The risk of cytomegalovirus transmission by penetrating keratoplasty, *Am. J. Ophthalmol.*, **1988**, 105, 357-360.

[33] Cleator, G. M., Klapper, P. E., Dennett, C., Sullivan, A. L., Bonshek, R. E., Marcyniuk, B. and Tullo, A. B., Corneal donor infection by herpes simplex virus: herpes simplex virus DNA in donor corneas, *Cornea*, **1994**, 13, 294-304.

[34] Tullo, A. B., Marcyniuk, B., Bonshek, R., Dennett, C., Cleator, G. M., Lewis, A. G. and Klapper, P. E., Herpes virus in a corneal donor, *Eye*, **1990**, 4, 766-767.

[35] Shurkin, N. M., Homologous serum hepatitis following the use of refrigerated bone bank bone, *J. Bone Joint Surg.*, **1954**, 36A, 160-162.

[36] Eggen, B. M. and Nordbo, S. A., Transmission of HCV by organ transplantation, *N. Engl. J. Med.*, **1992**, 326, 411.

[37] Pereira, B. J., Milford, E. L., Kirkman, R. L., Levey, A. S., Tomford, W. W., Leibowitz, H., Rhodes, M., Quan, S. and Wilbur, J. C., Low risk of liver disease after tissue transplantation from donors with HCV, *Lancet*, **1993**, 341, 903-904.

[38] Conrad, E. U., Gretch, D., Obermeyer, K., Moogk, M., Sayers, M., Wilson, J. and Strong, D. M., The transmission of hepatitis C virus by tissue transplantation, *J. Bone Joint Surg.*, **1995**, 77-A, 214-224.

[39] Tugwell, B. D., Patel, P. R., Williams, I. T., Thomas, A., Homan, H., Hedberg, K. and Cieslak, P. R., Hepatitis C virus (HCV) transmission to tissue and organ recipients from an antibody negative donor – United States, *42nd Ann. Interscience Conf. Antimicrob. Agents Chemother. (ICAAC)*, San Diego, CA, USA, **2002**, LB-17.

[40] Centre for Disease Control, Hepatitis C virus transmission from an antibody-negative organ and tissue donor – United States, 2001-2002, *MMWR*, **2003**, 5, 273-276.

[41] Petersen, L. R., Simons, R. J. and Koistinen, J., HIV transmission through blood, tissue and organs, *AIDS*, **1993**, 7, 99-107.

[42] Simonds, R. J., Holmberg, S. D., Hurwitz, R. L. Coleman, T. R., Bottenfield, S., Conley, L. J., Kohlenberg, S. H., Castro, K. G., Dahan, B. A., Schable, C. A., *et. al.*, Transmission of human immunodeficiency virus type 1 from a seronegative organ and tissue donor, *N. Engl. J. Med.*, **1992**, 326, 726-732.

[43] Buck, B. E., Resnick, L., Shah, S. M. and Malinin, T. I., Human immunodeficiency virus cultured from bone. Implications for transplantation, *Clin. Orthop. Rel. Res.*, **1990**, 251, 250-253.

[44] Merz, H., Rytik, G., Muller, W. E. G. and Roder, W., Bestimmung einer HIV infektion in menschlinchen Knochen, *Unfallchirurg*, **1991**, 94, 47-49.

[45] Nyberg, M., Suni, J. and Haltia, M., Isolation of human immunodeficiency virus (HIV) at autopsy one to six days postmortem, *Am. J. Clin. Pathol.*, **1990**, 94, 422-425.

[46] Marthy, S. and Richter, M., Human immunodeficiency virus activity in rib allografts, *J. Oral Maxillofacial Surg.*, **1998**, 56, 474-476.

[47] Centre for Disease Control, Transmission of HIV through bone transplantation: case report and public health recommendations, *MMWR*, **1988**, 37, 587-599.

[48] Schratt, H. E., Regel, G., Kiesewetter, B. and Tscherne, H., HIV infection caused cold preserved bone transplants, *Unfallchirung*, **1996**, 99, 679-684.

[49] Li, C. M., Ho, Y. R. and Liu, Y. C., Transmission of human immunodeficiency virus through bone transplantation: a case report, *J. Formosan Med. Assoc.*, **2001**, 100, 350-351.

[50] Bowen, P. A., Lobel, S. A., Caruanna, R. J., Leffell, M. S., House, M. A., Rissing, J. P. and Humphries, A. L., Transmission of human immunodeficiency virus (HIV) by transplantation: clinical aspects and time course analysis of viral antigenemia and antibody production, *Ann. Intern. Med.*, **1988**, 108, 46-48.

[51] Horsburgh, C. R., Ou, C. Y., Jason, J., Holmberg, S. D., Longini, I. M., Schable, C., Mayer, K. H., Lifson, A. R., Schochetman, G., Ward, J. W., *et al.*, Duration of human immunodeficiency virus infection before detection of antibody, *Lancet*, **1989**, 2, 637-640.

[52] Petersen, L. R., Satten, G. A., Dodd, R., Busch, M., Kleinman, S., Grindon, A. and Lenes, B., Duration of time from onset of human immunodeficiency virus type 1 infectiousness to development of detectable antibody, *Transfusion*, **1994**, 34, 283-289.

[53] Nowicki, M. J., Reevaluation of anti-HIV seroprevalence among blood donors with contemporary screening assays, *Transfusion*, **1992**, 32, S32.

[54] Stramer, S. L., Heller, J. S., Coombs, R. W., Parry, J. V., Ho, D. D. and Allain, J. P., Markers of HIV infection prior to IgG antibody seropositivity, *JAMA*, **1989**, 262, 64-69.

[55] Zaaijer, H. L., Exel-Oehlers, P., Kraaijeveld, T., Altena, E. and Lelie, P. N., Early detection of HIV-1 by third generation assay, *Lancet*, **1992**, 340, 770-772.

[56] Busch, M. P., HIV and blood transfusions: focus on seroconversion, *Vox Sang.*, **1994**, 62, S3, 13-18.

[57] Busch, M. P., Lee, L. L. L., Satten, G. A., Henrard, D. R., Farzadegan, H., Nelson, K. E., Read, S., Dodd, R. Y. and Petersen, L. R., Time course of detection of viral and serological markers preceding human immunodeficiency virus type 1 seroconversion: implications for screening blood and tissue donors, *Transfusion*, **1995**, 35, 91-97.

[58] Stramer, S. L., Cagliotti, S. and Strong, D. M., NAT of the United States and Canadian blood supply, *Transfusion*, **2000**, 50, 1165-1168.

[59] Hamilton, J., Eastlund, T., Steckler, D. Prather, J. and Dodd, R., Low Prevalence of human immunodeficiency virus seropositivity in surgical bone donors. A survey of 20 regional surgical bone banks, *14th Annual Meeting, American Association of Tissue Banks*, Denver, CO, USA, **1990**.

[60] Scofield, C., Eastlund, T., Steckler, D., Larson, M., Schuller, R. and Petersen, J., Prevalence of infectious disease markers in surgical bone donors, *17th Annual Meeting, American Association of Tissue Banks*, Boston, MA, USA, **1993**.

[61] Scofield, C., Eastlund, T., Larson, N., Steckler, D., Metcalfe, J. and Korent, H., Retesting of 1,608 living tissue donors for HIV and HCV. An evaluation of results, *17th Annual Meeting, American Association of Tissue Banks*, Boston, MA, USA, **1993**.

[62] Eastlund, T., Strong, D. M. and Mowe, J., The prevalence of infectious disease markers in cadaveric tissue donors: 1992 AATB survey results and a review, *18th Annual Meeting, American Association of Tissue Banks*, San Francisco, CA, USA, **1994**.

[63] McLean, V. A., *Annual Registration Survey of Accredited Tissue Banks*, American Association of Tissue Banks (AATB), **2000**.

[64] Schreiber, G. B., Glynn, S. A., Domesyn, M. A., Wright, D. J., Tu, Y., Dodd, R. Y. and Murphy, E. L., Lapsed donors: an untapped resource, *Transfusion*, **2003**, 43, 17-24.

[65] Asselmeier, M. A., Caspari, R. B. and Bottenfield, S. A., Review of allograft processing and sterilization techniques and their role in transmission of human immunodeficiency virus, *Am. J. Sports Med.*, **1993**, 21,170-17.

[66] Buck, B. E., Malinin, T. I. and Brown, M. D., Bone transplantation and human immunodeficiency virus; an estimate of risk of acquired immunodeficiency syndrome (AIDS), *Clin. Orthop.*, **1989**, 240, 129-136.

[67] Carlson, E. R., Marx, R. E. and Buck, B. E., The potential for HIV transmission through allogeneic bore. A review of risks and safety, *Oral Surg. Oral Med. Oral Path.*, **1995**, 80, 17-23.

[68] Sanzen, L. and Carlsson, A., Transmission of human T-cell lymphotrophic virus-type I by a deep-frozen bone allograft, *Acta Orthop. Scand.*, **1997**, 68, 72-74.

[69] Donald, P. J. and Cole, A., Cartilage implantation in head and neck surgery: report of a national survey, *Otolaryngol. Head Neck Surg.*, **1982** 90, 85-89.

[70] Noyes, F. R., Barber, S. D. and Mangine, R. E., Bone patellar ligament-bone and fascia lata allografts for reconstruction of the anterior cruciate ligament, *J. Bone Joint Surg.*, **1990**, 72A, 1125-1136.

[71] Buck, B. E. and Malinin, T. I., Human bone and tissue allografts, *Clin. Orthop.*, **1994**, 303, 8-17.

[72] Campagnari, D. and O'Malley, J. (eds.), *Standards of the American Red Cross Tissue Services, 6th Edition*, American Red Cross Tissue Services, Washington; DC, USA, **1994**.

[73] Food and Drug Administration (FDA), Human tissue intended for transplantation: US Department of Health and Human Services, Rockville, Maryland, 21 CFR 1270, *Federal Register*, **1993**, 58, 65514-65521.

[74] Food and Drug Administration (FDA), Human tissue intended for transplantation: US Department of Health and Human Services, Rockville, Maryland, 21 CFR 1270, *Federal Register*, **1997**, 62, 40429-40447.

[75] Food and Drug Administration (FDA), Suitability determination for donors of human cellular and tissue-based products, US Department of Health and Human Services, 21 CFR Part 1271, *Federal Register*, **1999**, 64, 52696-52723.

[76] Morris, A., Strickett, M. G. and Barratt-Boyes, B. G., Use of aortic valve allografts from hepatitis B surface antigen positive donors, *Ann. Thorac. Surg.*, **1990**, 49, 802-805.

[77] Thijssen, E. J., Kroes, A. C., Box, E., Persijn, G. G. and Rothbarth, P. H., The significance of complete serological testing for hepatitis B in heart valve banking, *Transplantation*, **1993**, 56, 82-84.

[78] Blajchman, M., Feinman, S. and Bull, S., Results of a prospective randomized multicenter trial to evaluate the non-A, non-B surrogate tests (ALT and anti-HBc) to prevent post transfusion hepatitis, *Blood*, **1993**, 81, S1, 204A.

[79] Roth, W. K., Weber, M., Petersen, D., Drosten, C., Buhr, S., Sereis, W., Weichert, W., Hedges, D. and Seifried, C., NAT for HBV and anti-HBc testing increase blood safety. *Transfusion*, **2002**, 42, 869-875.

[80] Berger, R., Gartner, S., Rappersberger, K., Foster, C. A., Wolff, K. and Stingl, G., Isolation of human immunodeficiency type 1 from human epidermis: viral replication and transmission studies, *J. Invest. Dermatol.*, **1992**, 99, 271-277.

[81] Gala, J. L., Vandenbroucke, A. T., Vandercam, B., Pirnay, J. P., Delferriere, N. and Burtonboy, G., HIV-1 detection by nested PCR and viral culture in fresh and cryopreserved postmortem skin: potential implications for skin handling and allografting, *J. Clin. Path.*, **1997**, 50, 481-484.

[82] Kanitakis, J., Escach, S., Trepo, C. and Thivolet, J., Detection of human immunodeficiency virus – DNA and RNA in the skin of HIV-infected patients using the polymerase chain reaction, *J. Invest. Dermatol.*, **1991**, 97, 91-96.

[83] Clarke, J. A., HIV transmission and skin grafts, Lancet, **1987**, 1, 983.

[84] Van Baare, J., Mackie, D.P. and Middelkoop, E., HIV transmission by transplant of allograft skin: a review of the literature, *Burns*, **1997**, 23, 460.

[85] Pirnay, J. P., Vandenvelde, C., Duinslaeger, L., Reper, P. and Vanderkelen, A., HIV transmission by transplantation of allograft skin: a review of the literature, *Burns*, **1997**, 23, 1-5.

[86] Abecassis, M., Kaufman, D. and Stuart, F., Direct evidence that murine cytomegalovirus (CMV) can be transferred via skin grafting using novel oligonucleotides in the polymerase chain reaction (PCR), 19th Annual Meeting, American Society of Transplant Surgeons, Houston, TX., USA, **1993**.

[87] Abecassis, M., Transmission of CMV by skin allografts – a review, *Tissue Cell Rep.*, **1994**, 2, 15-20.

[88] Shelby, J. and Shanley, J., Transfer of murine cytomegalovirus by syngereic skin grafts, *Transplantation*, **1987**, 44, 318-320.

[89] Shelby, J., Saffle, J. R. and Kern, E. R., Transmission of cytomegalovirus infection in mice by skin graft, *J. Trauma*, **1998**, 28, 203-206.

[90] Kealey, G. P., Bale, J. F., Strauss, R. G. and Massanari, R. M., Cytomegalovirus infection in burn patients, *J. Burn Care Rehabil.*, **1987**, 8, 543-545.

[91] Cederna, P., Bale, J. F., Strauss, R. and Kealey, G. P., Transmission of cytomegalovirus infection by cadaveric allograft in burn patients, *24th Annual Meeting, American Burn Association*, Salt Lake City, UT, USA, **1994**, A143.

[92] Chamberland, M. E., Emerging infectious agents: do they pose a risk to the safety of transfused blood and blood products?, *Clin. Infect. Dis.*, **2002**, 34, 797-805.

[93] Strong, D. M. and Katz, L., Blood-bank testing for infectious diseases: how safe is blood transfusion?, *Trends Mol. Med.*, **2002**, 8, 355-358.

[94] Centre for Disease Control, Provisional surveillance summary of the West Nile Virus epidemic – United States, January – November, 2002, *MMWR*, **2002**, 51, 1129-1133.

[95] Centre for Disease Control, Update. Outbreak of severe acute respiratory syndrome-worldwide, 2003, *MMWR*, **2003**, 52, 241-248.

[96] Peters, T. G., Life or death. The issue of payments in cadaveric organ donation, *JAMA*, **1991**, 269, 1302-1305.

[97] Eastlund, T., Monetary blood donation incentives and the risk of transfusion-transmitted infection, *Transfusion*, **1998**, 38, 874-882.

[98] Markris, M., Garson, J. A., Ring, C. J., Tuke, P. W., Tedden, R. N. and Preston, F. E., Hepatitis C virus RNA in clotting factor concentrates and the development of hepatitis in recipients, *Blood*, **1993**, 81, 1898-1902.

[99] Centre for Disease Control, Semen banking, organ and tissue transplantation, and HIV antibody testing, *MMWR*, **1988**, 37, 57-63.

[100] Centre for Disease Control and Prevention, Guidelines for preventing transmission of human immunodeficiency virus through transplantation of human tissue and organ, *MMWR*, **1994**, 43, 1-17.

[101] Alter, H. J., Epstein, J. S., Swenson, S. G., Van Raden, M. J., Ward, J. W., Kaslow, R. A., Menitove, J. E., Klein, H. G., Sandler, S. G. and Sayers, M. H., Prevalence of human immunodeficiency virus type 1 p24 antigen in US blood donors – an assessment of the efficacy of testing in donor screening, *N. Engl. J. Med.*, **1990**, 323, 1312-1317.

[102] Busch, M. P., Taylor, P. E., Lenes, B. A., Kleinman, S. H., Stuart, M., Stevens, C. E., Tomasulo, P. A., Allain, J. P., Hollingsworth, C. G. and Mosley, J. W., Screening of selected male blood donors for p24 antigen of human immunodeficiency virus type 1, N. *Engl. J. Med.*, **1990**, 323, 1308-1312.

[103] Harrell, J., McCreedy, B. and Johnston, A., PCR vs p24 antigen testing for detection of HIV-1 in cadaveric blood specimens, *17th Annual Meeting, American Association of Tissue Banks*, Boston, MA, USA, **1993**.

[104] Pepose, J. S., Buerger, D. G., Paul, D. A., Quinn, T. C., Darragh, T. M. and Donegan, E., New developments in serologic screening of corneal donors for HIV-1 and hepatitis B virus infections, *Ophthalmology*, **1992**, 99, 879-888.

[105] Malhotra, R. and Morgan, D. A., p24 antigen screening to reduce the risk of HIV transmission by seronegative bone allograft donors, *National Med. J. India*, **2001**, 13, 190-192.

[106] Yotsuyanagi, H., Yasuda, K., Meriya, K., Shintani, Y., Fujie, H., Tsutsumi, T. Norjiri, N., Juji, T., Hoshino, H., Shimoda, K., Hiro, K., Ilino, S. and Koike, K., Frequent presence of HBV in the sera of HBsAg-negative anti-HBC-positive blood donors, *Transfusion*, **2001**, 41, 1093-1099.

[107] Wang, J., Lee, C., Chen, P., Wang, T. and Chen, D., Transfusion-transmitted HBV infection in an endemic area: the necessity of more sensitive screening for HBV carriers, *Transfusion*, **2002**, 42, 1592-1597.

[108] Eastlund, T., Hemodilution due to blood loss and transfusion and reliability of cadaver infectious disease testing, *Cell Tissue Bank*, **2000**, 1, 121-127.

[109] Centre for Disease Control and Prevention, Human immunodeficiency virus infection transmitted from an organ donor screened for HIV antibody – North Carolina, *MMWR*, **1987**, 36, 306-308.

[110] Adams, M., Lee, T. H., Busch, M. P., Heitman, J., Marsh, G. J., Gjerset, G. F. and Mosely, J. W., Rapid freezing of whole blood or buffy coat sample for polymerase chain reaction and cell culture analysis: application to detection of human immunodeficiency virus in blood donor and recipient repositories, *Transfusion*, **1993**, 33, 504-508.

[111] Comeau, A. M., Harris, J., McIntosh, K., Weiblen, B. J., Hoff, R. and Grady, G. F., Polymerase chain reaction in detecting HIV infection among seropositive infants: relation to clinical status and age and to results of other assays, *J. Acquir. Immun. Defic. Syndr.*, **1992**, 5, 271-278.

[112] Busch, M. P., Wilber, J. C., Johnson, P., Tobler, L. and Evans, C. S., Impact of specimen handling and storage on detection of hepatitis C virus RNA, *Transfusion*, **1992**, 33, 420-425.

[113] Centre for Disease Control and Prevention, Ochrobactrium anthropi meningitis associated with cadaveric pericardial tissue processed with a contaminated solution – Utah, 1994, *MMWR*, **1995**, 45, 671-673.

[114] Farrington, M., Matthews, I., Foreman, J. and Caffrey, E., Bone graft contamination from a water de-ionizer during processing in a bone bank, *J. Hosp. Infect.*, **1996**, 32, 61-64.

[115] Hawkins, A. E., Zuckerman, M. A., Briggs, M., Gilson, R. J., Goldstone, A. H., Brink, N. S. and Tedder, R. S., Hepatitis B nucleotide sequence analysis: linking an outbreak of acute hepatitis b to contamination of a cryopreservation tank, *Virol. Methods*, **1996**, 60, 81-88.

[116] Strong, D., Eastlund, T. and Mowe, J., Tissue bank activity in the United States – 1992. Report of annual registration of AATB inspected tissue banks, *Tissue Cell Rep.*, **1995**, 3, 8-10.

[117] Pruss, A., Hansen, A., Kao, M., Gurtler, L., Paul, G., Benedix, F. and Von Versen, R., Comparison of the efficacy of virus inactivation methods in allogeneic avital bone tissue transplants, *Cell Tissue Bank*, **2001**, 2, 201-215.

PART 5

MICROBIOLOGICAL ASPECTS OF TISSUES FOR TRANSPLANTATION

PROCESSES AND PRACTICES LEADING TO BACTERIAL INACTIVATION IN TISSUES

Martell Winters * and Jerry Nelson

Nelson Laboratories, Inc., 6280 South Redwood Road
Salt Lake City, Utah 84123, USA
{E-mail: mwinters@nelsonlabs.com}

ABSTRACT

Tissue banks have a long history of producing quality tissue to enhance or save the lives of people. For many years they have implemented practices that are intended to provide safe tissue, which is free from bacterial and viral contamination. Among these practices are efficient donor screening, aseptic handling, and reducing microorganisms with antibiotics. Only recently have most tissue banks begun to fully validate and understand these practices. Although they are, and have proven to be, proficient at bacterial inactivation, many tissue banks are attempting to increase the level of sterility assurance.

Few tissue banks are currently using a terminal sterilisation step such as radiation to achieve this increased sterility assurance level (SAL). Most of those who currently employ radiation are using an arbitrary dose that is not validated to any SAL and are using biological indicators, which provide no scientific information regarding the effectiveness of the radiation dose. It is the intent of this presentation to discuss the current practices of bacterial inactivation, and point out potential areas of improvement.

KEYWORDS

Bacterial reduction; sterilisation; sterility assurance level; SAL; bioburden

INTRODUCTION

Tissue banks have a long history of producing quality tissue to enhance or save the lives of people. For many years they have implemented practices that are intended to provide safe tissue, which is free from bacterial and viral contamination [1]. Under increased scrutiny from FDA and other regulatory bodies, many tissue banks have begun to perform extensive validation work on their existing processes, and to augment current practices by adding a terminal sterilisation step.

It is the intent of this paper to discuss the current practices of bacterial inactivation, and point out potential areas of improvement. Some of the practices discussed in this paper are well known and are already effective in their current form. In these cases they are mentioned briefly due to their importance in microbial reduction, but not discussed in depth.

It should be noted that during this discussion, references to bacteria generally apply to fungi (moulds and yeasts) as well.

BACTERIAL REDUCTION PRACTICES: PROCUREMENT & PROCESSING

Donor screening

Tissue banks have well-established donor screening procedures and questionnaires. This is the first line of defence against contaminated tissue. Questions and observations during screening often are able to eliminate potentially contaminated donors based on history or cause of death. Tissue banks are continually improving donor screening as new microorganisms emerge that could affect the ability of a bank to safely use the tissue. One additional benefit of donor screening is that since it helps to eliminate septic donors, it increases the likelihood that most contamination on the tissue is on the surface rather than imbedded inside the tissue. This makes many of the current bacterial reduction methods more successful.

Aseptic handling during recovery and processing

Most recovery sites have strict procedures regarding the recovery of tissue from donors. These procedures include wearing sterile gloves, having well-trained personnel, using a dedicated room or area for recovery, and following strict cleaning and sanitising steps before and after recovery.

Recovery cultures

Usually during recovery, swabs are taken of every piece of tissue in an attempt to determine the presence of objectionable organisms. No standardised list exists, although there are some recommendations, which are included in tissue bank standards. Tissue is accepted or rejected based on the results of the recovery culture. Much discussion has occurred as of late regarding the limitations of swab testing. Data has been presented which demonstrates both low recovery efficiency of organisms and great potential variability using the swabbing technique. Nevertheless, swab recovery cultures have historically proven to be helpful in eliminating tissue contaminated with objectionable organisms.

Tissue banks also have a list of 'semi-objectionable' organisms which, when found, do not cause the tissue to be discarded, but do require that the tissue undergo additional disinfection steps during processing. These organisms generally are not known to cause serious disease. One problem with the organisms on this list is that the additional disinfection steps implemented have seldom been validated to be effective against these organisms. When 'semi-objectionable' organisms are found, additional swab testing is often performed post-processing to verify that detected organisms have been eliminated.

Considering the inefficiencies of swab testing and the great importance placed on their results, surprisingly few tissues have made it through the post-processing swab testing and continued on to cause infection in a recipient. This would indicate that regardless of the current problems in tissue banking, for example the lack of extensive validation, great potential variability of swab test results, and the potential ineffectiveness of removing organisms from the product, the current practices of relying heavily on swab test results do generally provide safe tissues. The antibiotics that are commonly prescribed to patients post-operation may also play a part in keeping the infection rate typical with other types of operations.

To eliminate the problems with swab testing, the method can be validated and the uncertainty of the method determined so that results of the testing can be adjusted or better understood, but this is rarely performed.

Physical removal of extraneous tissue

All tissue and bone used for transplantation undergoes at least some degree of shaping and/or trimming. The physical removal of tissue that was exposed during the recovery process is helpful in eliminating contamination on the tissue, which originated from the environment and handling. Conversely, additional handling that occurs during processing can add additional contamination. This is why tissue banks must place great importance on the cleanliness of the processing rooms and the aseptic technique of the technicians involved.

Cleaning with water and other solutions

All tissue banks incorporate some type of water or buffer cleaning step which usually involves pressure washing and/or ultrasonic bathing. This is another example of the physical removal of organisms and it can be very efficient. It is possible to validate this step, usually with inoculated pieces of tissue, to show how it contributes to the overall sterility assurance of the process.

Antibiotic soaking

This is another step that is performed by most tissue banks. The type and number of antibiotics used will vary from bank to bank, but the desired result is the same. Broad-spectrum antibiotics are used in an attempt to reduce the number of remaining organisms on the tissue. This can be a very successful bacterial reduction method when properly performed. Many of the organisms that are of most concern tend to be susceptible to these antibiotics.

One potential issue with antibiotic soaking is the residuals that likely remain on the tissue. These can cause problems with subsequent test results if the residuals are not either removed or neutralised. Antibiotic residuals can cause a false negative microbial test. False negatives have been demonstrated to be a problem by some tissue banks.

Antibiotic soaking can be difficult to validate due to resistance pattern variability of the organisms present on the tissue. Some organisms may be completely unaffected while others may be easily killed.

Freeze drying or storage in saline

Although freeze drying and storage in saline are not meant to be bacterial reduction steps, they both can provide some degree of kill or bacteriostasis. Once freeze dried, microorganisms can survive for years, but the process of drying kills many organisms. Saline can also be efficient as a bacterial preservative.

BACTERIAL REDUCTION PRACTICES: TERMINAL STERILISATION

Tissue functionality and current standards

Some tissue banks have implemented a terminal sterilisation step to add additional safety to the tissue they release. Many of these have used radiation as that sterilisation step and have obtained successful results. As sterility assurance concerns increase, so will the number of tissue banks looking at this option.

Issues related to tissue functionality have been and are being addressed as terminal radiation sterilisation becomes more common. Until recently, most decisions not to irradiate have been made based on personal preference, lack of validation options, or the results of little data. As of late, much information on the performance of irradiated tissue has been presented and there is apparently some degree of undesirable alteration of the tissue when it has been irradiated. The question to consider is, how much undesirable alteration of the tissue is acceptable in exchange for having a greater sterility assurance? Hopefully, future data will help to clarify this question. Per some tissue bank standards, the required radiation dose is not based on scientific data and may actually be significantly higher than actually needed. A dose of 15 kGy may provide too much or too little sterility assurance, depending on the tissue bioburden count and resistance. Another difference to be considered in selecting a sterilisation dose for tissue: irradiation is temperature (room temperature verses frozen in dry ice).

Biological indicators (BIs) of *Bacillus pumilus* are commonly used, and required, as a supposed indicator of sterility. There are two issues to consider with respect to using these BIs with radiation. First, the premise for using BIs is invalid for use with radiation compared with ethylene oxide and steam sterilisation. The premise for using BIs with other sterilisation modalities is that the organism on the BI is considered to be the most resistant organism (MRO) known for that sterilisation modality. For radiation, no MRO has been found or selected which is stable on a carrier. Even if one were selected, the required radiation dose to kill 6 to 12 logs of the organism would be much higher than what is currently being used. In fact it would likely be so high as to exclude most tissue types due to excessive damage. The second issue with BIs is that there is usually no link to demonstrate that it is an appropriate indicator for the tissue. This is because data is not available demonstrating that the BI is more difficult to sterilise than the tissue. Without this data, and continued demonstration that this is the case over time, it is scientifically inappropriate to use the BI as an indicator for the tissue.

Sterility assurance level (SAL)

The current terminology used for describing the sterility of a product is sterility assurance level (SAL). A SAL describes the probability of occurrence of a non-sterile product. Typical SALs for the pharmaceutical and medical device industries are 10^{-3} (one in 1,000 non-sterile) and 10^{-6} (one in 1,000,000 non-sterile) respectively. Most tissue banks are unable to provide a SAL for their tissue. This is either because they have not yet fully profiled or validated their existing processes, or because guidance on determining a SAL for tissue banks is not yet available. In some cases, even tissue banks that are employing a terminal radiation sterilisation step have not labelled their product as 'sterile' or documented a particular SAL.

Determining an exact SAL for tissue can be somewhat more complicated than for a medical device or pharmaceutical, however, it can be done. Two pathways exist for determining a radiation dose to be used to achieve a desired SAL and within each pathway there are various options. In the first pathway, the sterilisation dose is calculated based on the bioburden of the tissue, irrespective of what bioburden reduction processes are added prior to the irradiation step. This idea assumes that the previously performed bioburden reduction processes (e.g. cleaning and antibiotic rinses) have added no level of sterility assurance and thus, the radiation dose must provide the entire sterility assurance of the tissue. This pathway is the one chosen if a typical AAMI/ISO 11137 method is used to determine sterilisation dose (e.g. Method 1 or Method 2 [2]).

The second pathway involves a term used in other industries called 'polishing'. As the term implies, in this pathway the radiation dose merely adds the additional log reductions needed to achieve the desired SAL. For example, if it has been determined that the existing processes provide at least a 10^{-2} SAL, only four more log reductions are required to reach a 10^{-6} SAL. This idea allows the tissue bank to take advantage of the existing bioburden reduction processes and will result in an overall lower dose for sterilisation. The commonly used AAMI/ISO Methods 1 and 2 can easily be adjusted to achieve this polishing.

Several experts have been identified in the AAMI Radiation Sterilisation Working Group who are interested in providing assistance and/or guidance in using polishing in the tissue bank industry.

VIRAL REDUCTION

It may be obvious that most of this discussion is geared towards bacterial and not viral reduction. There are a couple of reasons for this. Viral contamination generally does not occur during recovery or processing. If it is present, it is usually present due to disease prior to death of the donor. Historical screening and testing performed on the donor are usually able to detect the presence of harmful viruses. However, window periods and other issues occasionally let a donor with viral contamination pass through the system undetected.

Any swabbing or bioburden testing will not detect viruses because the growth requirements are completely different from bacteria and fungi. Actually, each virus has different cell affinities and thus different growth requirements. This fact makes enumeration of viruses using a single screening procedure, like those for bacteria and fungi, impossible.

Another reason that this discussion is focused on bacteria and fungi rather than viruses is because most of the microorganism reduction steps that have been discussed are ineffective at killing or removing viruses. Physical removal of extraneous tissue is helpful, but it is arguable that if viral contamination is present on the removed tissue, it is also present on the remaining tissue. Freeze drying will reduce the viral contamination as it does with bacteria, depending on the type of virus. The other methods described previously are not known to have any affect on the viral population in or on tissue.

Terminal sterilisation is certainly a viable option to reduce any potential viral contamination that has passed through the normal detection methods. Irradiation is often discussed regarding viral kill in terms of 'It does not kill...', or 'It has proven ineffective at killing...', or 'After irradiation, high numbers of viruses were still present'. These types of comments are generally misleading.

Irradiation will and does kill viruses. The amount of radiation required to accomplish a log reduction of viruses (D-value) is often higher than for bacteria or fungi [3], but radiation will kill all viruses. The mode of action for bacterial and viral kill using radiation is generally the same: interaction with the nucleic acids that renders it unable to be read or duplicated. Since viruses contain a small fraction of nucleic acid compared to bacteria, the target for the radiation is smaller and therefore more difficult to hit.

One note of caution is that irradiation should not be used in place of tissue rejection when viral contamination is detected. Nonetheless, it should be recognised that when viral contamination passes through the system undetected, irradiation will assist in reducing the overall count of the viruses.

CONCLUSIONS

This discussion is not meant to be all-inclusive for the current and potential bacterial reduction steps involved in tissue banking. It is intended to point out where some of the major contributing steps lie. Advances in the cleanliness/sterilisation of tissue in the tissue bank industry are being made on a frequent basis. Many tissue banks are developing new means of sterilisation, as desires to provide additional sterility assurance increase. The overall awareness of the importance of increased sterility assurance has augmented sharply in the last few years and as a result, the overall safety of tissue being transplanted in the world has improved. Hopefully, the results of future investigations will assist to increase safety to the tissue bank industry.

REFERENCES

[1] AATB, *Standards for Tissue Banking*, AATB, VA, USA, **2001**, pp. 55, 56.

[2] ANSI/AAMI/ISO, *11137-1994 Sterilization of health care products - Requirements for validation and routine control – Radiation sterilization*, AAMI, VA, USA, **1994**, pp. 15-24.

[3] S. Block, *Disinfection, Sterilization, and Preservation, 5th Edition.*, Lea & Febiger, Philadelphia, PA, USA., **2001**, pp. 734.

NOVEL PATHOGEN INACTIVATION OF SOFT TISSUE ALLOGRAFTS USING OPTIMISED GAMMA IRRADIATION

Teri A. Grieb [1], Ren-Yo Forng [1], Jack Lin [1], Lloyd Wolfinbarger [2*], Jorge Sosa-Melgarej.[2], Chris Sharp [1], William N. Drohan [1] and Wilson H. Burgess [1*]

[1] *Clearant, Inc., 401 Professional Drive, Gaithersburg, MD 20879, USA {E-mail: wburgess@clearant.com}*

[2] *LifeNet, 1457 Miller Stone Road, Suite 10, Virginia Beach, VA 23455, USA {E-mail: lloyd_wolfinbarger@lifenet.org}*

ABSTRACT

Recent reports by the CDC have identified allograft-associated bacterial and viral infections in recipients of human allografts. There is a need for improved pathogen inactivation and terminal sterilisation of these grafts. We hypothesised that methods developed for gamma irradiation of plasma-derived therapeutic proteins could be applied successfully to a more complex biologic such as human allografts. The study included four groups of tibialis tendons: Non-irradiated controls; a group subjected to 18 kGy of conventional gamma irradiation; a group subjected to 50 kGy of controlled irradiation; and a group pretreated with radioprotectants prior to 50 kGy of controlled irradiation. Samples were tested for tensile strength and elasticity. In addition, biochemical analysis of collagen degradation was performed on each sample.

There was no significant difference in the tensile strength of tendons irradiated to 50 kGy using the optimised process with radioprotectants when compared to non-irradiated controls. Similarly, there was no change in Young's moduli between the groups. Biochemical analysis indicated there was no subtle collagen degradation following 50 kGy of irradiation in the presence of radioprotectants. Significant degradation is observed in tissue subjected to 50 kGy of irradiation without radioprotectants. Robust inactivation of bacteria (spore and vegetative), enveloped and non-enveloped virus was achieved with 50 kGy of optimised irradiation.

A new method of gamma irradiation can be used to provide a significant increase in the safety assurance of human allograft tissue. The inactivation of pathogens can be achieved without the loss of mechanical or biochemical integrity associated with existing irradiation protocols. Initial results of the first clinical experience with these grafts are described.

INTRODUCTION

Approximately 100,000 anterior cruciate ligament (ACL) tears occur annually, with the number expected to increase due to the rise in participation by the general population in sports [1]. Although autogenous bone-patellar tendon-bone grafts and hamstring tendons are currently the grafts of choice for ACL reconstruction, allografts have been advocated as successful alternatives to autografts. Allografts are gaining popularity, especially for multiple ligament injury and revision surgery [2]. Patellar tendons are the preferred allograft for ACL reconstruction but with a limited supply and advances in fixation techniques, Achilles [1], hamstring [1,3], and tibialis tendons [4,5] are viable substitutes. Allografts eliminate the disadvantages associated with autografts, including the perioperative pain and morbidity associated with harvesting the graft, muscle weakness in the area where the graft was obtained, and increased surgical time required to harvest and prepare the tissue prior to implantation. There are reports of a slower incorporation rate [6,7] and a higher instrument measured laxity [8] for allografts. However, the clinical relevance of these observations is not clear, as it is well documented that the clinical outcome for allografts is similar to that of autografts in ACL reconstruction [9-13]. A recurrent concern shared by both surgeons and patients with regard to the use of allografts is the potential risk of disease transmission from the cadaveric tissue.

Tissue banks take preventative measures to minimise the risk of disease transmission. They evaluate the donor's social and medical history, excluding all donors with high-risk behaviours in accordance to criteria established by the US Public Health Service [14]. Tissues are held in quarantine pending serological testing for infectious agents such as HIV-1, HIV-2, HTLV-I, HTLV-II, hepatitis B, hepatitis C, and syphilis [14]. Moreover, tissue banks rely on aseptic procurement and processing that involves removal of debris and organic matter, soaking in various disinfectant solutions, and monitoring for microbes at various stages of processing. Although it is recognised that the donor screening process provides an increased level of safety, screening is limited in that donors with viral infections in the window period might go undetected. Albeit rare, this limitation was realised when four recipients of grafts from an HIV-infected, seronegative donor became HIV positive [16] and more recently when five allograft recipients contracted hepatitis C from a window-period donor [17]. Another challenge faced by tissue processors is that screening donors for known pathogens does not offer protection against unknown or emerging pathogens, such as West Nile or SARS viruses.

Many orthopaedic surgeons have the impression that aseptic processing provides a sterile graft. In fact, aseptic processing alone cannot ensure sterility. The Centre for Disease Control and Prevention (CDC) reported recently that bacterial infections in tissue recipients could be traced to the allografts [18,19]. The number of allograft-associated infections nationwide is not known due, in part, to the challenge of distinguishing a graft-related infection from an infection due to the surgical procedure and to the current lack of a requirement from the Food and Drug Administration (FDA) to report infections from contaminated allografts. It is important to note that a majority of the reported allograft-associated infections have resulted from grafts (e.g. soft and osteochondral tissues) that due to their inherent properties are unable to withstand the rigorous mechanical and chemical processing commonly used for hard tissue grafts. Although methods vary among tissue banks, bone grafts are routinely subjected to high-pressure lavage to remove blood, marrow, and other cellular components; sonication; and treatments with chemicals, such as Betadine, detergents, hydrogen peroxide, ethanol, acetone, and isopropyl alcohol. Many of these processing methods compromise the structural integrity of soft tissue grafts. Therefore, minimal processing that often includes antibacterial soaks is employed to reduce contamination of soft tissues grafts.

Tissue processors, surgeons, and patients would benefit from a method that can terminally sterilise allografts without adversely affecting the functional integrity and biocompatibility of the tissue. Currently, there is not a general consensus as to a standard terminal sterilisation method for allografts. Gamma irradiation is effective in inactivating all types of microorganisms as well as both lipid-enveloped and non-enveloped viruses [20,21]. Low to moderate doses of gamma irradiation are used routinely to sterilise medical devices [22], animal sera used for tissue culture [23,24] and allograft tissues [25]. We report here the application of a terminal pathogen inactivation method for soft tissue allografts that involves the use of a relatively high dose (50 kGy) of gamma irradiation under well-defined conditions. This study demonstrates that 50 kGy of optimised gamma irradiation offers a viable terminal sterilisation method that does not alter the mechanical strength or collagen integrity of human tibialis tendons.

MATERIALS & METHODS

Tissue Preparation

Tibialis tendons obtained from donors with research consent were provided by LifeNet (Virginia Beach, VA, USA). Sixty anterior and posterior tendons from male and female donors ranging in age from 15 to 55 years old were assigned to four treatment groups. The tendons were left untreated (0 kGy, control group), were treated with 50 kGy of gamma irradiation under optimised conditions, were pre-treated with radioprotectants prior to 50 kGy of gamma irradiation under optimised conditions, or were treated with 18 kGy of conventional gamma irradiation in the absence of radioprotectants, a method currently used by some tissue banks. The tendons were either soaked in saline or a radioprotectant solution [2.2 M propylene glycol-USP, 3.1 M dimethyl sulfoxide (DMSO)-USP, 150 mM mannitol-USP, 100 mM trehalose (Spectrum Chemical, Brunswick, NJ, USA)] for 4 hours at 40°C with gentle agitation (LabLine Model 4628, Sheldon Manufacturing, Cornelius, OR, USA). The tendons were then soaked for an additional 24 hours at 4°C. The tendons were rinsed for 30 seconds in phosphate buffered saline and then packaged under vacuum in heat-sealed pouches. The tendons were stored at -80°C until gamma irradiated.

Gamma Irradiation

Tibialis tendons in the optimised 50 kGy groups with or without radioprotectants were irradiated under conditions controlled for a uniform dose distribution and a tight range in temperature and total dose. Pouches containing the tendons were placed into canisters (Infecon-3000, Com-Pac, Carbondale, IL, USA) that were then packaged with dry ice into coolers designed to maintain temperatures below -50°C during the irradiation process. Tendons in the 18 kGy conventional group were packaged on dry ice and irradiated using the standard practices of many tissue banks. The samples were shipped to a commercial Cobalt-60 gamma irradiation facility (Neutron Products, Dickerson, MD, USA) and irradiated to a minimum targeted dose of 18 kGy or 50 kGy. Far West radiochromic film dosimeters were used to measure the dose delivered to the tissue. The dosimetry was corrected with low-temperature response curves. The 0 kGy group was stored at -80°C for the duration of the irradiation.

Biomechanical Testing

Prior to mechanical testing, tendons were thawed at room temp. and rehydrated in saline prior to mechanical testing. The cross-sectional area was determined with the use of a specially designed measuring device. The device consisted of one piece that had a rectangular opening 4.5 mm wide and 30 mm deep. After inserting the tendon into the device, a second piece of the device was mounted onto the first piece, allowing for consistent measurements by conforming the tendons to a set thickness and width. Measurements were taken using an outside micrometer (Starrett, Athol, MA, USA).

Tensile tests were performed using a MTS Model 858 (Eden Prairie, MN, USA) servohydraulic mechanical test machine and specially designed cryogenic grips. One cryogrip was attached to the actuator and the other to the load cell. The grip was designed with reservoirs to hold a dry ice and ethanol mixture to freeze the tendons into the sinusoidal shape of the grip faces, allowing the tendons to be firmly held using only a minimal amount of pressure. The actuator was set so that there was a 50 mm gauge length. The samples were preloaded to 20 Newtons followed by loading in tension at a displacement rate of 5 mm/sec, corresponding to a strain rate of approximately 0.1/sec. The applied force was recorded as a function of time with data acquisition every 0.1 seconds. The displacement of the actuator and the applied load were converted to stress-strain curves for analyses. Tensile strength of each tendon was calculated by dividing peak force by average total tendon cross-sectional area. Young's modulus was calculated as the trend line slope of the linear, elastic region of the stress-strain curve.

Transmission Electron Microscopy

The tendons were cut into approximately 1 mm^3 blocks and fixed with 2.5% glutaraldehyde in 0.13 M phosphate buffer (pH 7.4) for 2 hours at room temp. The samples were post-fixed for 1.5 hours with 1% osmium tetroxide, dehydrated with a graded series of ethanol, and then embedded in araldite. Thin sections were cut with a Reichert OMU4 ultramicrotome and then stained with uranyl acetate and lead citrate. The resultant sections were examined in a JEOL 100CXII transmission electron microscope. Measurements of collagen crossbandings were determined with a Polaron magnifier on high power pictures.

Protease Sensitivity Assay

Chymotrypsin is a protease capable of cleaving and releasing fragments from denatured collagen but not intact collagen. The assay was performed essentially as described previously [25]. Briefly, on average 3 mg of tissue was removed from one end of each tendon prior to mechanical testing. The tissue samples were rinsed for 0.5 hours in phosphate buffered saline in order to reduce the radioprotectant concentration. To remove proteoglycans and soluble collagen, the samples were extracted with 4 M guanidine HCl containing Complete Protease Inhibitor Cocktail (Roche Diagnostics, Mannheim, Germany) at 40 times the tissue weight (v/w) for 65 hours at 4°C with agitation. The samples were washed three times with incubation buffer (0.1 M ammonium bicarbonate, pH 7.8). The tissue was digested in incubation buffer (100 times v/w tissue) that contained 100 mg/mL chymotrypsin for 24 hours at 37°C with gentle agitation. The digest was centrifuged and the supernatant and three washes with incubation buffer were combined. The tissue pellets and pooled supernatants were dried in a SpeedVac (Savant Instruments, Inc., Farmingdale, NY, USA). Once dry, the samples were subjected to amino acid analysis that was performed as described by Waters Associates (Milford, MA, USA). Samples were hydrolysed in 6 M HCl (Pierce, Rockford, IL, USA) *in vacuo* at 150°C for 1 hour. Amino acid compositions were determined based on reversed-phase separation of the phenylthiocarbamyl derivatives using a PICO-TAG amino acid analysis system (Waters Associates, Milford, MA, USA). Hydroxyproline was used to quantify the percent of chymotrypsin-sensitive, denatured collagen and chymotrypsin-resistant, intact collagen. Standard curves were generated using known concentrations of hydroxyproline. The percentage of denatured collagen was calculated by dividing the hydroxyproline in the supernatant by the sum of the hydroxyproline in the supernatant and pellet and then multiplying by 100.

Pathogen Inactivation

Tendons were either left untreated or treated with radioprotectants as described above. The tendons were frozen at -80°C and then pulverised in a freezer mill (Model 6850, Spex CertiPrep, Metuchen, NJ, USA). Virus or bacterial stocks (50 µL) were added to 0.1 g pulverised tendon and incubated with shaking at 4°C for 1 hour. Samples were then frozen at -80°C and irradiated as described above. Following irradiation, 950 µl of the appropriate media were added to the samples. The samples were vortexed for 30 seconds and then centrifuged at 3,000 rpm to pellet the tissue. Inactivation of porcine parvovirus (VR-742, American Type Culture Collection, ATCC, Manassas, VA, USA) and Sindbis virus (VR-68, ATCC) was quantified by standard $TCID_{50}$ assays as described in [26]. Inactivation of *Clostridium sordellii* (9714, ATCC) was quantified by incubating serial 10-fold dilutions in Reinforced Clostridial Medium (Becton Dickenson) followed by anaerobic culture. For one study, *C. sordellii* spores were irradiated in a lyophilised state to a dose of 25 or 50 kGy in order to determine their resistance to gamma irradiation.

Statistical Analysis

The effects of treatment were analysed by one-way ANOVA using Design-Expert 6 (Stat-Ease, Inc., Minneapolis, MN, USA) and Student's t- test. A p-value of 0.05 was used to determine statistical significance.

RESULTS

Biomechanical Properties

Biomechanical testing of soft tissue grafts presents a unique challenge in that grafts tend to fail prematurely at the grip-tendon interface due to stress concentrations and damage to surface fibres [27,28]. Although others [29] have included grip failures as long as the mean ultimate tensile strength was not statistically different from that of the midsubstance failures, we chose not to include the results of obvious grip failures which could lead to incorrect interpretation of the data. Therefore, the data presented is for midsubstance graft failures only.

Tibialis tendons were left untreated, were pretreated with a radioprotectant cocktail and then subjected to 50 kGy of gamma irradiation under optimised conditions, were irradiated to 50 kGy under optimised conditions in the absence of radioprotectants, or were irradiated to 18 kGy in the absence of radioprotectants under conventional conditions, a practice used by some tissue banks. The mean cross-sectional areas were 22.9 (± 5.7) mm^2, 20.6 (± 4.1) mm^2, 22.3 (± 2.6) mm^2, and 21.9 (± 3.7) mm^2 for the non-irradiated, 50 kGy with radioprotectants, 50 kGy without radioprotectants, and 18 kGy groups, respectively. The cross-sectional areas were not significantly different ($p = 0.574$) between the treatment groups. Following gamma irradiation, the tendons were tested in tension to failure (Figure 1). The non-irradiated, control group had a mean ultimate tensile stress of 84.5 (± 13.2) MPa. The tensile strengths of the 50 kGy with radioprotectants, 50 kGy without radioprotectants, and 18 kGy groups were 107%, 95%, and 114% of the control group with mean tensile stresses of 90.3 (± 16.4) MPa, 80.5 (± 17.3) MPa, and 96.5 (± 19.5) MPa, respectively. None of the groups differed significantly from the non-irradiated, control group ($p = 0.487$).

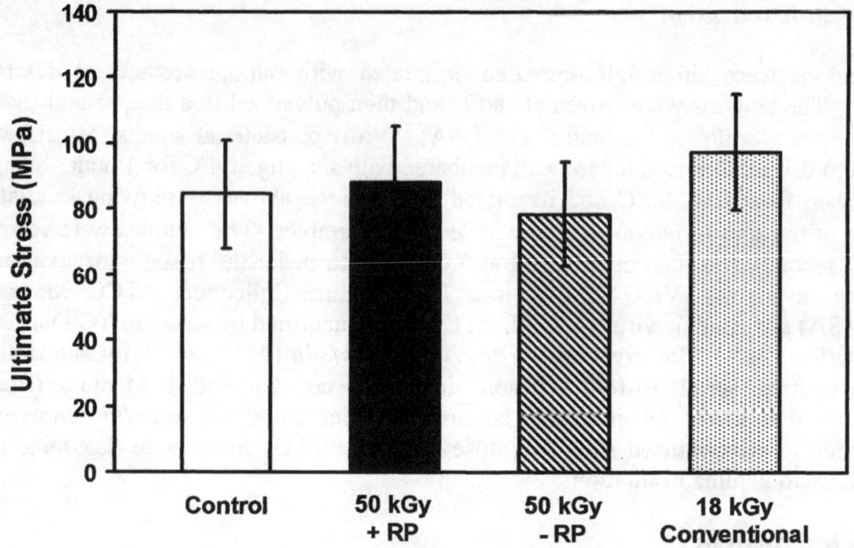

Figure 1. Mean tensile stress of non-irradiated and irradiated tibialis tendons. Tendons were either left untreated or subjected to gamma irradiation to 50 kGy using optimised conditions in the presence or absence of radioprotectants (RP) or to 18 kGy using conventional methods.

The Young's moduli of the tendons were also determined. The mean Young's moduli were 392.3 (± 94.5) MPa, 409.3 (± 67.4) MPa, 404.4 (± 41.0) MPa, and 438.5 (± 55.0) MPa for the 0 kGy, 50 kGy with radioprotectants, 50 kGy without radioprotectants, and 18 kGy groups, respectively (Figure 2). There is no statistically significant difference in the Young's Modulus between the groups ($p = 0.439$).

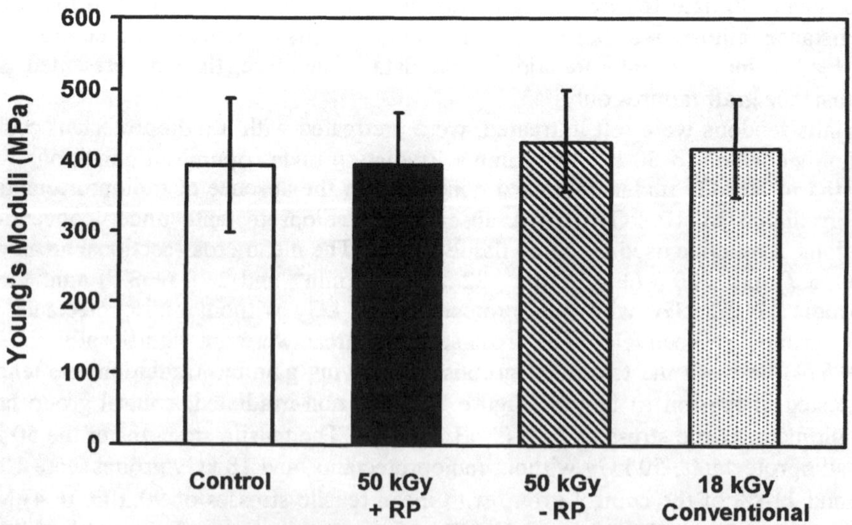

Figure 2. Mean Young's moduli determined for samples tested in Figure 1.

Transmission Electron Microscopy

Representative transmission electron micrographs for the non-irradiated control and irradiated groups are shown in Figure 3. The irradiated tendons had collagen fibre thicknesses within the parameters of the control group. The cross banding of the collagen ranged from 60.9 to 71 nm for the non-irradiated control. The tendons irradiated to 18 kGy, 50 kGy without radioprotectants, and 50 kGy with radioprotectants did not differ significantly from the control with cross bandings of 60.7 to 70.4 nm, 65.7 to 69.6 nm, and 62.7 to 70.5 nm, respectively ($p > 0.05$). The cross banding is due to the inherent collagen cross-links and is related to collagen strength. This data is consistent with the results of the tensile testing.

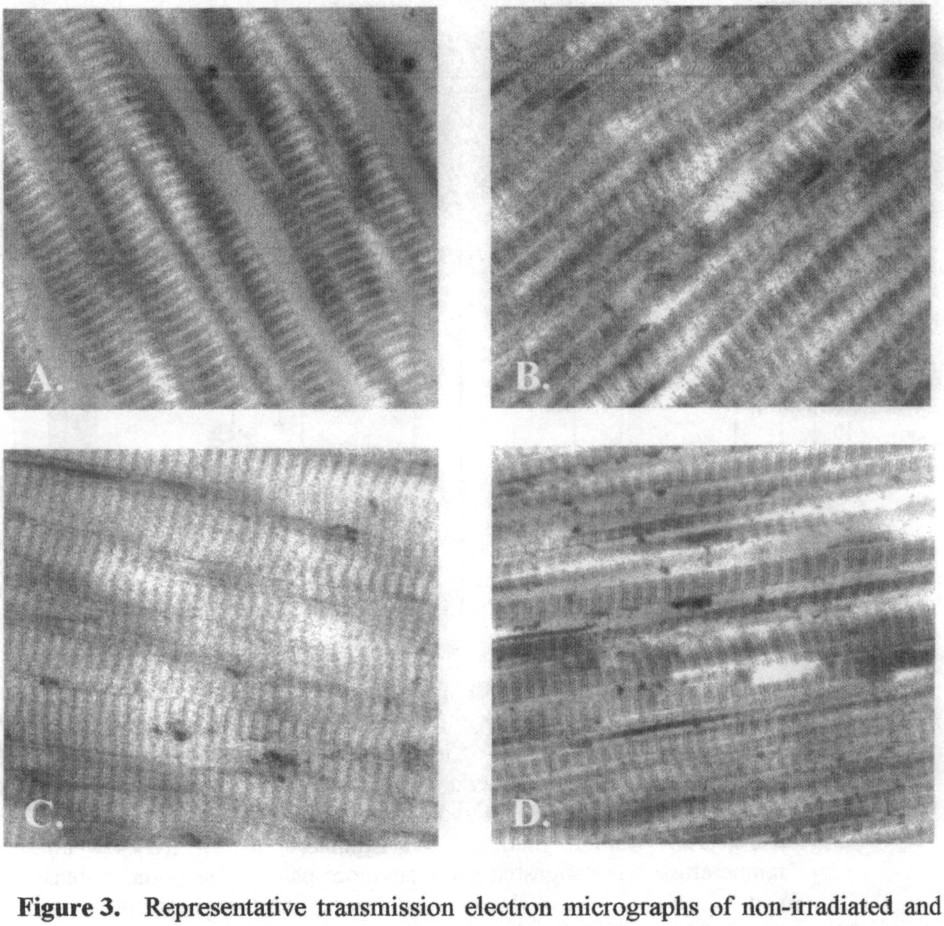

Figure 3. Representative transmission electron micrographs of non-irradiated and irradiated tendons. (A) Non-irradiated tendon, 49,227 x (magnification); (B) 18 kGy conventionally irradiated tendon, 49,400 x (mag.); (C) Tendon irradiated to 50 kGy under optimised condition without radioprotectants, 46,444 x (mag.); (D) Tendon irradiated to 50 kGy under optimised conditions with radioprotectants, 46,444 x. (mag.).

Protease Sensitivity

Prior to mechanical testing, a small piece of tissue was removed from an end of each tendon. The pieces of tendon were subjected to protease digestion using chymotrypsin in order to compare the extent of radiation-induced damage to the tendon collagen in the different treatment groups. Chymotrypsin is capable of cleaving and releasing fragments from denatured collagen but not intact collagen fibrils. Hydroxyproline, an amino acid unique to collagen, was used to quantify the chymotrypsin-sensitive collagen fragments released into the digestion buffer (i.e. supernatant) and the chymotrypsin-resistant collagen that was not damaged and remained associated with the matrix (i.e. pellet). The chymotrypsin digestion data are shown in Figure 4. A fifth treatment group, 50 kGy of irradiation at ambient temperature, was included in this study. This experimental group served as a positive control, demonstrating how severely damaged tissue performed in the assay.

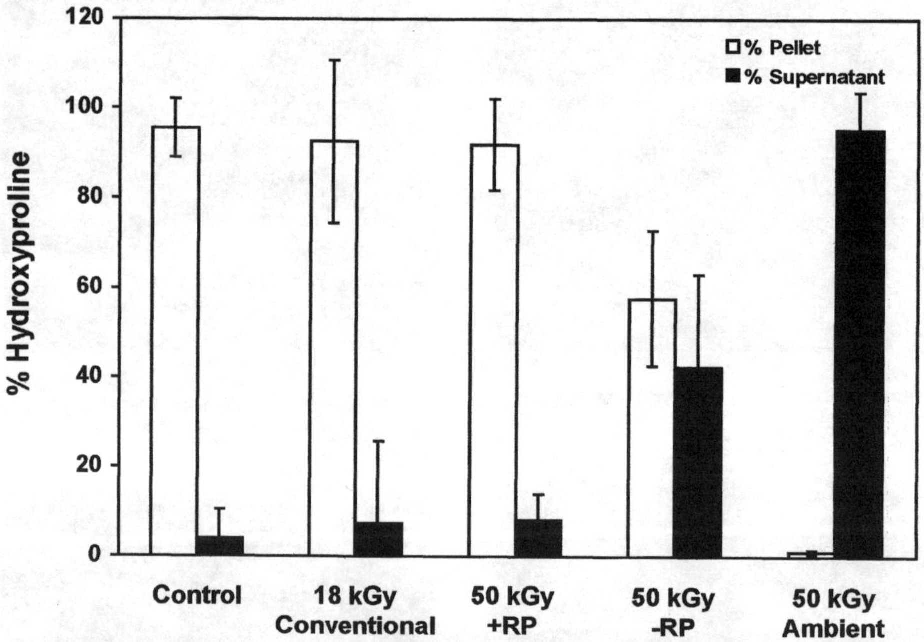

Figure 4. Damage to non-irradiated or irradiated tendon collagen. Tendons that were non-irradiated, irradiated to 50 kGy under optimised conditions in the presence or absence of radioprotectants (RP), irradiated to 18 kGy using conventional methods, or irradiated to 50 kGy at ambient temperature were digested with chymotrypsin. The percent denatured and intact collagen is shown as the percent of chymotrypsin-sensitive (supernatant) and chymotrypsin-resistant (pellet) collagen, respectively.

For the non-irradiated control, approximately 96% of the hydroxyproline was associated with the tissue pellet indicating that the collagen is intact; whereas, in the samples irradiated to 50 kGy at ambient temperature, 95% of the hydroxyproline was detected in the supernatant following the protease digestion. A significant increase in the collagen integrity was observed when 50 kGy of radiation was delivered under optimised conditions as compared to the ambient irradiation ($p > 0.001$). Approximately 60% of the hydroxyproline was detected in the pellet of the 50 kGy without radioprotectant group. Optimisation of the irradiation parameters alone, however, does not offer sufficient protection to the tendon collagen. When the tissue was pre-treated with the radioprotectant solution and then irradiated to 50 kGy under optimised conditions, complete protection of the collagen integrity was observed. Tendons treated with 50 kGy in the presence of radioprotectant had a similar protease profile to that of the 18 kGy samples even though nearly three times the dose of irradiation was delivered. This data provides insight into damage that can occur to the tendons that was not readily apparent when evaluating just the biomechanical structural properties of the grafts. This damage may be an indication of the longer-term stability of the grafts.

Pathogen Inactivation

The effect of gamma irradiation on the viability of the spore form of the anaerobe, *Clostridium sordellii* that led to the death of one tissue recipient is shown in Figure 5. Lyophilised spores were irradiated to a final dose of 0, 25 or 50 kGy. The spores were then resuspended and cultured as described above. Anaerobic growth could be detected in the non-irradiated preparation following up to 8 serial, 10-fold dilutions. Irradiation to 25 kGy resulted in 6 logs reduction in the concentration of live bacteria; however, growth could be detected in the stock and the first 10-fold dilution. No growth was detected in the sample that received 50 kGy of gamma irradiation, demonstrating that at least 8 logs of inactivation were achieved. Assuming log-linear inactivation kinetics, a dose of 50 kGy should be capable of providing up to 12 logs inactivation of *C. sordellii* based on the 6 logs reduction obtained with 25 kGy in this study.

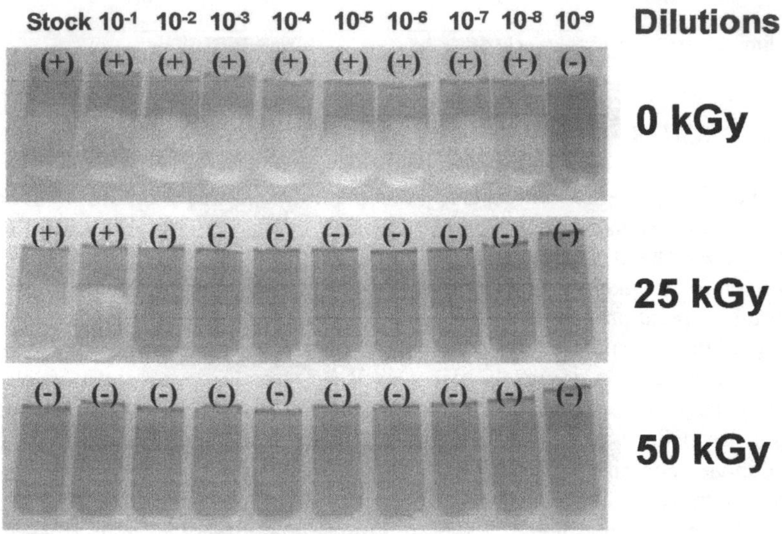

Figure 5. Anaerobic growth of *Clostridium sordellii*. Clostridium spores were either not irradiated (0 kGy) or irradiated to a final dose of 25 or 50 kGy. The spores were resuspended and cultured as a stock or series of ten-fold dilutions. Growth following 48 hours of culture is shown.

Inactivation data for porcine parvovirus, Sindbis, and *C. sordellii* individually mixed with untreated or radioprotectant treated pulverised tendons are shown in Table 1. The tendons were pulverised in order to maximise the surface area of tissue available for interaction with the pathogens.

For Sindbis virus, a model for hepatitis C, 50 kGy of gamma irradiation reduced infectivity to the limit of the assay (1.7 logs) for both the treated and untreated tendon samples. Because of the relatively low starting titre for this virus, ≥ 4.5 logs of inactivation can be claimed. For porcine parvovirus, the limit of the assay was not reached. Thus, 5 logs of inactivation are an accurate assessment of the capability of 50 kGy of irradiation to reduce parvovirus infectivity. Finally, ≥ 8 logs reduction in the colony forming units of the spore formulation of *C. sordellii* were achieved for both the treated and untreated tendon spiked samples. In addition, there was no growth of bacteria observed when the tissue samples were incubated for 8 days (data not shown). This result indicates that there were no residual, resistant bacteria in close association with the pulverised tendon.

Table 1. Pathogen inactivation in pulverised tendon untreated or treated with radioprotectants (RP).

Pathogens	0 kGy (Logs Recovered)		50 kGy (Logs Recovered)		Log Reduction	
	- RP	- RP	- RP	+ RP	- RP	+ RP
Sindbis virus	6.3	6.2	1.7 [a]	1.7 [a]	≥ 4.6	≥ 4.5
Porcine Parvovirus	8.7	8.6	3.3	3.7	5.4	4.9
C. sordellii	8.2	8.2	0	0	≥ 8.2	≥ 8.2

([a] limit of assay sensitivity reached)

DISCUSSION

The data presented in this study demonstrate that under optimised conditions, 50 kGy of gamma irradiation can be delivered to tendons without significantly altering either their mechanical properties or collagen integrity. Prior literature has established that conventional gamma irradiation has a dose-dependent, deleterious effect on the biomechanical properties of tissues [30-35]. Consequently, there is a negative perception, in general, by some surgeons and members of the tissue industry that high doses of gamma irradiation adversely affect graft integrity and efficacy. Much of the damage that occurs to the tissue results from free radicals and reactive oxygen species generated as a secondary effect of the ionising radiation. We demonstrate that when this secondary chemistry is controlled, the untoward effects to the tissue can be minimised. In this study, radiation-induced damage to the tissue is reduced by [14] pre-treating the tissue with radioprotectants that minimise the effects of any free radicals that are generated [30], performing the irradiation at dry ice temperature, a condition that should substantially limit the diffusion of free radicals, and [36] packaging the tissue in an optimised configuration that ensures a tight dose distribution and maintenance of low temperature for the duration of the irradiation.

Contrary to our results which showed no significant difference among the treatment groups, studies using either human or animal bone-patellar tendon-bone grafts have reported a decrease in mechanical properties following gamma irradiation to a total dose of at least 30 kGy at dry-ice temperatures [32,33,35,37]. Although reporting a 26% decrease in maximum force to failure for human bone-patellar tendon-bone grafts exposed to 40 kGy of gamma irradiation, Rasmussen *et al.* [37] concluded that the frozen irradiated grafts maintained biomechanical properties sufficient for ACL reconstruction given that they had strengths similar to that reported for the young human ACL. The maximum load to failure for human anterior cruciate ligaments is between 1725 N and 2160 N [1]. The tendons in this study had maximum loads to failure within this range regardless of the treatment.

The transmission electron microscopy data demonstrates that no change was observed in the tendon ultra-structure following gamma irradiation even at the 50 kGy dose. Our findings are in agreement with Voggenreiter *et al.* [38], who evaluated the surface structure of cortical bone grafts by scanning electron microscopy following different preservation and sterilisation methods. Irradiation to doses of 1, 5, 25, or 50 kGy did not have a deleterious effect on the surface structure of cortical bones and the fibrillary structure was similar to the fresh control.

Approximately 70-80% of the dry weight of tendons and ligaments is composed of Type I collagen [39]. Collagen is recognised as a critical component to the structural and functional integrity of all connective tissues. The triple-helical conformation of collagens distinguishes them from other proteins and makes them highly resistant to proteases. Therefore, the ability of a protease to cleave the tissue collagen serves as a good indicator of collagen damage. We show that collagen is almost completely denatured when tendons are irradiated to 50 kGy at ambient temperature in the absence of radioprotectants. Irradiation the tendons to 50 kGy at dry-ice temperature while utilising an optimised packaging configuration resulted in significant protection, nearly 60%, to the collagen. However, tendons treated with radioprotectants prior to the 50 kGy of optimised gamma irradiation exhibited a protease profile similar to that of non-irradiated tendons. This result is consistent with Hamer *et al* [40], who demonstrated that irradiating bone at dry-ice temperature reduced collagen damage as compared to the common practice of irradiating bone grafts at ambient temperature. The effects of gamma irradiation on purified collagen have been well studied [41-44]. Bailey *et al.* [41] demonstrated that the irradiation-induced damage to purified rat-tail tendon collagen was environment dependent and that low temperatures and a free radical scavenger could minimise the negative effects of gamma irradiation. Until now this approach had not been used specifically for sterilisation of tissue allografts with gamma irradiation.

The inconsistency in the mechanical strength and TEM data with the protease sensitivity data are likely due to differences in the sensitivity of the methods to distinguish damage. Gamma irradiation is known to induce both chain scission and cross-linking in proteins. Cheung *et al* [43] have speculated that since the collagen molecules are aligned, both native and potential radiation-induced cross-links could compensate for some breaks in the peptide chain, thereby making mechanical stress-strain analysis or TEM a less sensitive method of detecting the collagen damage. Furthermore, tissue strength differs from donor to donor and this variability might mask subtle changes in mechanical strength due to collagen damage.

The three pathogens chosen for study represent three different risks associated with allograft tissue. Sindbis is a lipid-enveloped virus that represents a model for pathogens that pose the greatest health risks, such as HIV, hepatitis C, and hepatitis B. Porcine parvovirus is non-enveloped and is a model for viruses such as hepatitis A and human parvovirus B-19. These pathogens do not present the health risks associated with the chronic infections of HIV and hepatitis C. They are, however, relatively resistant to other inactivation methods such as solvent detergent and heat pasteurisation. Consequently, the emerging pathogens that pose the greatest risk to the tissue recipient in the future could arise from non-enveloped viruses. *Clostridium sordellii* is the anaerobic bacteria linked to several recent infections and one death among tissue recipients. [45]. In general, spore forms of bacteria present the most resistance to inactivation by gamma irradiation and serve as the model for validation of medical device sterilisation. The 4.5 logs inactivation observed for Sindbis virus could have been higher if we had used a higher starting titre. In previous studies we observed inactivation of this magnitude with as little as 25 kGy of gamma irradiation [46]. Thus, based on log-linear inactivation kinetics, we would predict that at least 9 logs of inactivation could ultimately be demonstrated. In contrast, because the limit of the assay was not achieved for porcine parvovirus, this is an accurate assessment of the inactivation capability of 50 kGy for small (5,000 bp) non-enveloped viruses. This value compares favourably with any inactivation technology for this class of pathogen. The inactivation of *C. sordellii* obtained with 50 kGy of irradiation is important for a sterility claim for tissues, as the process must provide for 6 logs of extrapolated pathogen inactivation below a bioburden of Log 0. The studies presented here demonstrate that 6 logs of inactivation of *C. sordellii* are achieved with 25 kGy, which would require a starting bioburden of Log 0 in the product for a sterility claim. Thus, 50 kGy of gamma irradiation with radioprotectants is a new technique that can be applied to tissue to provide enhanced viral safety and sterility assurance without impairing tissue integrity.

Currently, sterilising soft tissue and certain composite grafts presents a unique challenge to the tissue industry. Aseptic processing alone does not eradicate contamination of allografts. Strict adherence to donor screening policies has increased the safety margin for allograft tissues. Buck *et al* [47] estimated that the risk of harvesting bone from a HIV-infected donor subsequent to thorough donor screening is 1 in 1,667,600; however, the risk increases to 1 in 161 when screening procedures are compromised. Although there might be significant variations in the rigor applied throughout the tissue industry, FDA inspectors have identified deficiencies in the screening and testing practices of a variety of tissue banks [48]. Most tissue banks do not rely solely on aseptic processing and screening. Many have implemented additional safety measures such as ethylene oxide gas, chemical treatments, or low-dose conventional gamma irradiation. None of these methods have gained industry-wide acceptance.

The effectiveness of EtO gas as a sterilising agent is limited due to its inability to penetrate larger, dense tissues [49], the generation of residual toxic by-products that are not easily removed from the grafts [50-53], and high rates of graft failures [50-52]. Consequently, this method has been abandoned by many tissue processors. Chemical inactivation methods also depend on complete penetration of the reagents throughout the entire graft. As a consequence, failure of the chemicals to reach even a small percentage of the graft results in residual sites of potential infectivity regardless of how effective the chemical is in the areas it does contact. Another disadvantage to relying solely on chemical inactivation is that when viral aggregates are present the chemicals may not be able to inactivate infectious particles at the core [54-57]. Unlike chemicals used to inactivate pathogens, the radioprotectants described in this study are used to minimise the adverse effects of gamma irradiation. Consequently, the failure of the radioprotectants to penetrate a certain percentage of the tissue would leave it susceptible to potential damage but would not alter the effectiveness of inactivating contaminating pathogens. In contrast to the previous two methods, gamma irradiation is capable of thoroughly penetrating grafts of all sizes and densities. Many tissue processors use doses of 10 kGy to 25 kGy. A dose of 25 kGy provides a sterility assurance level of 10^{-9} for most bacteria [36] but is not adequate for the inactivation of HIV, other more radioresistant viruses [20,31], and some microbial spores [54,55]. The dose of gamma irradiation required to inactivate HIV in tissue remains controversial, with recommended doses ranging from 30 kGy to 50 kGy [31,56-59]. Although there are discrepancies among the studies, the levels indicated are significantly higher than the doses routinely used by tissue processors today. The fact that window-period transmissions of infectious viruses have been documented emphasises the need for more effective inactivation treatments for human tissue.

The authors are unaware of other concerted efforts to minimise the dose-dependent effects of gamma irradiation in order to make it an effective method for terminally sterilising allograft tissues. We are able to deliver twice the accepted maximum dose of irradiation currently used by processors, thus providing substantially safer grafts that still maintain a biomechanical performance consistent with existing grafts. The data represents the initial mechanical performance of the grafts but does not indicate how the allografts will function mechanically when subjected to remodelling following implantation. For the present, we describe an effective terminal sterilisation procedure that increases the safety margin associated with allografts, particularly soft tissue and composite grafts that pose the greatest health risk due to existing processing constraints.

REFERENCES

[1] Graham, S. M. and Parker, R.D., Anterior cruciate ligament reconstruction using hamstring tendon grafts, *Clin. Orthop.*, **2002**, 402, 64-75.

[2] Jackson, D. W., Rosen, M. and Simon, T. M., Soft tissue allograft reconstruction: the knee, In: *Allografts and Orthopaedic Practice*, Czitrom, A. A. and Gross, A. E. (eds.), Williams and Wilkins, Baltimore, USA, **1992**, pp. 197-215.

[3] Boni, D. M. and Herriott, G. E., Hamstring tendon graft for anterior cruciate ligament reconstruction, *AORN J.*, **2002**, 76, 610-624.

[4] Caborn, D. N. and Selby, J. B., Allograft anterior tibialis tendon with bioarbsorbable interference screw fixation in anterior cruciate ligament reconstruction, *Arthroscopy*, **2002**, 18, 102-105.

[5] Haut Donahue, T. L., Howell, S. M., Hull, M. L. and Gregersen, C., A biomechanical evaluation of anterior and posterior tibialis tendons as suitable single-loop anterior cruciate ligament grafts, *Arthroscopy*, **2002**, 18, 589-597.

[6] Gazdag, A. R., Lane, J. M., Glaser, D. and Foster, R. A., Alternatives to autogenous bone graft: efficacy and indications, *J. Am. Acad. Orthop. Surg.*, **1995**, 3, 1-8.

[7] Tom, J. A. and Rodeo, S. A., Soft tissue allografts for knee reconstruction in sports medicine, *Clin. Orthop.*, **2002**, 402, 135-156.

[8] Westerheide, K. J., Flumhe, D. J., Francis, K. A., Irrgang, J. J., Fu, F. H., and Harner, C. D., Long tern follow up of allograft versus autograft bone patellar tendon bone ACL reconstruction, *AOSSM Meeting*, Orlando, FL, USA, **2002**.

[9] Harner, C. D., Olson, E., Irrgang, J.J., *et al.*, Allograft verses autograft anterior cruciate ligament reconstruction: 3 to 5-year outcome, *Clin. Orthop.*, **1996**, 324, 134-144.

[10] Miller, M. D. and Harner, C. D., The use of allograft. Techniques and results, *Clin. Sports Med.*, **1993**, 12, 757-770.

[11] Noyes, F. R., Barber-Westin, S. D.. Reconstruction of the anterior cruciate ligament with human allograft. Comparison of early and later results, *J. Bone Joint Surg. Am.*, **1996**, 78, 524-537.

[12] Saddemi, S. R., Frogameni, A. D., Fenton, P. J., *et al.*, Comparison of perioperative morbidity of anterior cruciate ligament autografts versus allografts, *Arthroscopy,* **1993**, 5, 519-524.

[13] Shelton, W. R., Papendick, L. and Dukes, A. D., Autograft versus allograft anterior cruciate ligament reconstruction, *Arthroscopy,* **1997**, 13, 446-449.

[14] American Association of Tissue Banks (AATB), *Standards for Tissue Banking. 10th Edition*, **2002**.

[15] Simonds, R. J., Holmberg, S. D., Hurwitz, R. L. *et al.*, Transmission of human immunodeficiency virus type 1 from a seronegative organ and tissue donor, *N. Engl. J. Med.*, **1992**, 326, 726-732.

[16] Cieslak, P. R., Hedberg, K., Thomas, A. R. *et al.*, Hepatitis C virus transmission from an antibody-negative organ and tissue donor -- United States, 2000-2002, *Morb. Mortal. Wkly Rep.*, **2003**, 52, 273-276.

[17] Archibald, L. K., Jernigan, D. B. and Kainer, M. A., Update: allograft-associated bacterial infections - United States 2002, *Morb. Mortal. Wkly Rep.*, **2002**, 51, 207-210.

[18] Lutz, B., Ratard, R., Dodson, D., et al., Septic arthritis following anterior cruciate ligament reconstruction using tendon allografts - Florida and Louisiana, 2000, *Morb. Mortal. Wkly Rep.*, **2001**, 50, 1081-1083.

[19] Sommer, R. E., The effects of ionizing radiation on fungi, *Manual on Radiation Sterilization of Medical and Biological Materials*, International Atomic Energy Agency (IAEA), Vienna, **1973**, pp. 73-79.

[20] Sullivan, R., Fassolitis, A. C., Larkin, E. P., *et al.*, Inactivation of thirty viruses by gamma radiation, *Appl. Microbiol.*, **1971**, 22, 61-65.

[21] Doue, B., Radiation doses and dose distribution during industrial sterilization by gamma rays and accelerated electron beams (Part I), *Med. Device Technol.*, **2001**, 32-35.

[22] Erickson, G. A., Landgraf, J. G., Wessman, S. J., *et al.*, Detection and elimination of adventitious agents in continuous cell lines, *Dev. Biol. Stand.*, **1989**, 70, 50-66.

[23] House, C., House, J. A. and Yedloutschnig, R. J., Inactivation of viral agents in bovine serum by gamma irradiation, *Can. J. Microbiol.*, **1990**, 36, 737-740.

[24] Wright, K. A. and Trump, J. G., Co-operative studies in the use of ionizing radiation for sterilization and preservation of biological tissue, In: *Sterilization and Preservation of Biological Tissues by Ionizing Radiation*, International Atomic Energy Agency (IAEA), Vienna, **1969**, pp. 107-18.

[25] Bank, R. A., Krikken, M., Beekman, B., *et al.*, A simplified measurement of degraded collagen in tissues: application in healthy, fibrillated and osteoarthritic cartilage, *Matrix Biol.*, **1997**, 16, 233-243.

[26] Charlick, D. A. and Caborn, D. N., Technical note: alternative soft-tissue graft preparation technique for cruciate ligament reconstruction, *Arthroscopy,* **2000**, 16, E20.

[27] Ker, R. F., Tensile fibres: strings and straps, In: *Biomechanics-Materials: A Practical Approach*, Vincent, J. F. (ed.), Oxford University Press, New York, **1991**, pp. 75-97.

[28] Matthews, G. L., Keegan, K. G. and Graham, H. L., Effects of tendon grip technique (frozen versus unfrozen) on in vitro surface strain measurements of the equine deep digital flexor tendon, *Am. J. Vet. Res.*, **1996**, 57, 111-115.

[29] Smith, C. W., Young, I. S. and Kearney, J. N., Mechanical properties of tendons: changes with sterilization and preservation, *J. Biomech. Eng.*, **1996**, 118, 56-61.

[30] Anderson, M. J., Keyak, J. H. and Skinner, H. B., Compressive mechanical properties of human cancellous bone after gamma irradiation, *J. Bone Joint Surg. Am.*, **1992**, 74, 747-752.

[31] Currey, J. D., Foreman, J., Laketic, I., *et al.*, Effects of ionizing radiation on the mechanical properties of human bone, *J. Orthop. Res.*, **1997**, 15, 111-117.

[32] Fideler, B. M., Vangsness, C. T., Lu, B., *et al.*, Gamma irradiation: effects on biomechanical properties of human bone-patellar tendon-bone allografts, *Am. J. Sports Med.*, **1995**, 23, 643-646.

[33] Gibbons, M. J., Butler, D. L., Grood, E. S., *et al.*, Effects of gamma irradiation on the initial mechanical and material properties of goat bone-patellar tendon-bone allografts, *J. Orthop. Res.*, **1991**, 9, 209-218.

[34] Hamer, A. J., Strachan, J. R., Black, M. M. *et al.*, Biochemical properties of cortical allograft bone using a new method of bone strength measurement. A comparison of fresh, fresh-frozen and irradiated bone, *J. Bone Joint Surg. Br.*, **1996**, 78, 363-368.

[35] Salehpour, A., Butler D., Proch, F., *et al.*, Dose-dependent response of gamma irradiation on mechanical properties and related biochemical composition of goat bone-patellar tendon-bone allografts, *J. Orthop. Res.*, **1995**, 13, 898-906.

[36] Angermann, P. and Jepsen, O. B., Procurement, banking and decontamination of bone and collagenous tissue allografts: guidelines for infection control, *J. Hosp. Infect.*, **1991**, 17, 159-169.

[37] Rasmussen, T. J., Feder, S. M., Butler, D. L. and Noyes, F. R., The effects of 4 Mrad of gamma irradiation on the initial mechanical properties of bone-patellar tendon-bone grafts, *Arthroscopy*, **1994**, 10, 188-197.

[38] Voggenreiter, G., Ascherl, R., Blumel, G. and Schmit-Neuerburg, K. P., Effects of preservation and sterilization on cortical bone grafts. A scanning electron microscopic study, *Arch. Orthop. Trauma Surg.*, **1994**, 113, 294-296.

[39] Woo, S. L. and Young, E. P.. Structure and function of tendons and ligaments, In: *Basic Orthopaedic Biomechanics*, Mow, V. C. and Hayes, W. C. (eds), Raven Press, New York, **1991**, pp.199-243.

[40] Hamer, A. J., Stockley, I. and Elson, R. A., Changes in allograft bone irradiated at different temperatures, *J. Bone Joint Surg. Am.*, **1999**, 81B, 342-344.

[41] Bailey, A. J. and Rhodes, D. N., Irradiation-induced crosslinking of collagen, *Radiation Research*, **1964**, 22, 606-621.

[42] Bowes, J. H. and Moss, J. A., The effect of gamma radiation on collagen, *Radiation Research*, **1962**, 16, 211-223.

[43] Cheung, D. T., Perelman, N., Tong, D. and Nimni, M. E., The effect of gamma-irradiation on collagen molecules, isolated alpha-chains, and crosslinked native fibers, *J. Biomed. Mater. Res.*, **1990**, 24, 581-589.

[44] Liu, B., Harrell, R., Davis, R. H., *et al.*, The effect of gamma irradiation on injectable human amnion collagen, *J. Biomed. Mater. Res.*, **1989**, 23, 833-844.

[45] Center for Disease Control and Prevention, Update: unexplained deaths following knee surgery - Minnesota, 2001, *Morb. Mortal. Wkly Rep.*, **2001**, 52, 273-276.

[46] Miekka, S. I., Forng, R. Y., Rohwer, R. G., *et al.*, Inactivation of viral and prion pathogens by gamma-irradiation under conditions that maintain the integrity of human albumin, *Vox. Sang.*, **2003**, 84, 36-44.

[47] Buck, B. E., Resnick, L., Shah, S. M. and Malinin, T. I., Human immunodeficiency virus cultured from bone. Implications for transplantation, *Clin. Orthop.*, **1990**, 251, 249-253.

[48] Office of the Inspector General, Department of Health and Human Services, *Oversight of tissue banking: Executive summary, OEI-01-00-0041*, **2001**.

[49] Cloward, R. B., Gas-sterilized cadaver bone grafts for spinal fusion operations. A simplified bone bank, *Spine,* **1980**, 5, 4-10.

[50] Jackson, D. W., Windler, G. E. and Simon, T. M., Intraarticular reaction associated with the use of freeze-dried, ethylene oxide-sterilized bone-patella tendon-bone allografts in the reconstruction of the anterior cruciate ligament, *Am. J. Sports Med.*, **1990**, 18, 1-10.

[51] Lawrence, W. H., Itoh, K., Turner, J. E. and Autian, J., Toxicity of ethylene chlorohydrin. II. Subacute toxicity and special tests, *J. Pharm. Sci.*, **1971**, 60, 1163-1168.

[52] Roberts, T. S., Drez, D., McCarthy, W. and Paine, R., Anterior cruciate ligament reconstruction using freeze-dried, ethylene oxide-sterilized, bone-patellar tendon-bone allografts. Two year results in thirty-six patients, *Am. J. Sports Med.*, **1991**, 19, 35-41.

[53] Parry, M. F. and Wallach, R., Ethylene glycol poisoning, *Am. J. Med.*, **1974**, 57, 143-150.

[54] Tsuji, K., Kane, M., Rahn, P. and Steindler, K., Cobalt-60 irradiation for sterilization of veterinary mastitis products containing antibiotics and steroids, *Radiat. Phys. Chem.*, **1981**, 18, 583-593.

[55] Wade, N., Division of biologics standards: scientific management questioned, *Science,* **1972**, 175, 966-970.

[56] Conway, B., Tomford, W. and Hirsch, M. S., Effects of gamma irradiation on HIV-1 in a bone allograft model, *Trans. Orthop. Res. Soc.*, **1990**, 15, 225.

[57] Wade, N., Division of biologics standards: in the matter of J. Anthony Morris, *Science,* **1972**, 175, 861-866.

[58] Conway, B., Tomford, W., Mankin, H. J., et al., Radiosensitivity of HIV-1 - potential application to sterilization of bone allografts, *AIDS,* **1991**, 5, 608-609.

[59] Smith, R. A., Ingels, J., Lochemes, J. J., et al., Gamma irradiation of HIV-1, *J. Orthop. Res.*, **2001**, 19, 815-819.

ESTABLISHING AN APPROPRIATE TERMINAL STERILISATION DOSE BASED UPON POST-PROCESSING BIOBURDEN LEVELS ON ALLOGRAFT TISSUE

Chad J. Ronholdt, Simon Bogdansky, and Tom F. Baker *

AlloSource, 6278 S. Troy Circle, Centennial, CO 80111, USA

ABSTRACT

It is essential to control bioburden levels on final allograft products in order to ensure that the aseptic process used to produce the final allograft products is under control and within the capabilities of the terminal sterilisation process. The purpose of this study was to determine if the current gamma radiation dose used by AlloSource is sufficient to eradicate post-processing bioburden levels found on final allograft products. A total of 128 allografts were selected for testing based upon a 6-month review of AlloSource allograft distribution activities. All allografts were analysed for aerobic, anaerobic, spore-forming bacteria and fungi using a destructive bioburden test method. Bioburden testing included the immersion of allografts in an extraction fluid and agitating for a defined period of time. The extraction fluid was processed via membrane filtration and the membrane was transferred to a solid agar plate, incubated and evaluated for microbial growth. 13 of the 128 allografts were positive for microbial growth after processing and prior to irradiation. Of the 13 positives, there were four distinct colony morphologies present: *Staphylococcus epidermidis, Enterococcus faecalis, Enterococcus faecium* and *Enterococcus durans*. The highest level of microorganisms found on any one allograft was 99 CFU (Achilles tendon and bone block). Based upon the AAMI/ISO Standard 11137 Method I calculation, the current gamma sterilisation dose (16–25 kGy) is capable of reducing our maximum bioburden (99 CFU) to a minimum SAL of 10^{-8} (1 in 100 million chance of a non-sterile product). This SAL estimation, although valid is based upon assumptions that are commonly made for medical devices and in some cases may not be appropriate for the terminal sterilisation of allografts, therefore AAMI/ISO Standard 11137 Method II B is being investigated as a more accurate estimation of the appropriate terminal sterilisation dose.

KEYWORDS

Sterilisation dose; allograft tissue; sterility assurance level (SAL); bioburden

INTRODUCTION

The safety of allografts has come under increased scrutiny due to a recent investigation by the Center for Disease Control and Prevention that has linked 54 cases of bacterial infections to allograft tissues [1-3]. Coupled with these infections, also comes increased regulatory pressure to ensure that the processes by which tissue retrieval, transport, processing and distribution are validated and in a state of control. Sterilisation of allograft tissues is a widely controversial topic within the tissue banking industry. Historically, two terminal sterilisation methods have been used to sterilise allograft tissues; ethylene oxide or gamma radiation [4,15].

Several scientific studies have been published that support the use of terminal sterilisation for allograft tissue and conversely, there are an equivalent number of publications that suggest that terminal sterilisation damages the functionality of the allografts and reduces biomechanical strength of the tissues [5-13].

The American Association of Tissue Banks (AATB) is a not-for-profit, peer group organisation founded in 1976 to facilitate the provision of transplantable cells and tissues of uniform high quality in quantities sufficient to meet national needs [14]. The guidelines published by the AATB, recommend using a radiation dose of at least 15 kGy (1.5 Mrad) for terminal sterilisation [15]. A percentage of allografts produced by AlloSource are sent out for terminal sterilisation via Cobalt 60 (gamma) radiation complying with the AATB guidelines (dose range of 16-25 kGy). The average bioburden that a typical production run dose setting of 18 kGy can successfully eradicate using a Sterility Assurance Level (SAL) of 10^{-3} is ~ 12,000 Colony Forming Units (CFU) using Method 1 of AAMI/ISO Standard 11137. Using the same dose but with a higher SAL (10^{-6}), only ~ 12 CFU can be eradicated [16,17]. Therefore it is essential that the bioburden level on post-processing allograft tissue be calculated to ensure that the bioburden from processing does not exceed the capabilities of the terminal sterilisation process.

MATERIALS & METHODS

One hundred and twenty eight traditional allograft products (cortical, cortical/cancellous and soft tissue) of varying packaging types were consented for research and selected for bioburden testing. The percentage of test articles was proportional to the quantity of final allografts released for distribution over a period of 6 months (Table 1). This historical comparison was performed to ensure samples were truly representative of current allograft products being distributed for transplantation.

Table 1. Test article sampling breakdown.

Class/Type	6-month Distribution		Test Article Breakdown	
	# of Allografts	%	# of Test Articles	%
C/C FD	4239	80.9%	105	82.0%
C/C FZ	157	3.0%	5	3.9%
S/NP FD	43	0.8%	6	4.7%
S/NP FZ	674	12.9%	11	8.6%
S/NP SL	128	2.4%	1	0.8%
Total	**5241**	**100.0%**	**128**	**100%**

(C/C = Cortical or Cancellous; FZ = Frozen; S/NP = Soft or non purged; SL= Saline; FD = Freeze Dried)

The greatest number of allografts (n = 105) was generated from the cortical/cancellous freeze-dried tissue category. Eleven allografts were selected from the soft frozen tissue category, six tissues were selected from the soft, freeze-dried tissue category, five tissues were selected from cortical/cancellous frozen category and finally, one allograft was tested from the soft tissue stored in saline category. The test articles were aseptically transferred to individual sterile jars and immersed with either 150 or 300 mL of Peptone Tween (Fluid D) solution (VWR, West Chester, PA, USA) depending upon the size of the test article.

The jars were sealed and manually agitated using a 90°-hand shake motion for 100 cycles. Four different microbial categories were evaluated for growth, aerobic, anaerobic, spore forming and fungal. For the aerobic, anaerobic and fungal test articles, 50 mL aliquots were removed from each test article, filtered through a membrane filter then aseptically transferred to a Soybean Casein Digest agar plate (Remel, Lenexa, KS, USA). For the spore-forming test articles, 10 mL aliquots were removed, heat shocked at 85°C for 10 minutes (kills viable microorganisms and sporulates spore-forming microorganisms), then filtered through a membrane filter and aseptically transferred to a Soybean Casein Digest agar plate. The aerobic and spore-forming bacterial plates were incubated aerobically at 30-35°C for 3 days then enumerated. Anaerobic plates were incubated anaerobically at 30-35°C for 3-4 days then enumerated. Fungal plates were incubated aerobically at 20-25°C for 4 days then enumerated. All positive test articles were subbed for isolation, Gram stained and identified using common microbiological techniques specific for each isolate.

RESULTS

Bioburden

The average bioburden for all 128 samples was 7 CFU/test article. The bioburden range on the test articles ranged from below the level of detection to a maximum of 99 CFU/test article (Achilles tendon and bone block). There was no microbial growth detected on any of the other microbial categories (fungi, spore-forming or anaerobic bacteria). The nature of the extraction test is such that '0 CFU' cannot be determined and the reported value is shown as less than the detection limit. In this study, the detection limit can vary depending on the volume of fluid used to perform the extraction (e.g. a femoral shaft would require more fluid than a tricortical wedge). Thus, the detection limit may be < 3 or < 30 depending upon the volume of fluid used (i.e. dilution factor). When calculating average bioburden values for aerobic organisms, the lower detection limit is used when no organisms are detected. The average CFU count is calculated based on the conservative assumption that 3 CFU were actually found on the grafts when they were reported as < 3 CFU. The average values for fungi, spores, and anaerobes are shown as 'N/A' because there were no organisms observed in any of the samples. Caution must be taken when using these values in subsequent applications. They must be considered theoretical worst-case values, not definitive values. Refer to Table 2 for a summary of the bioburden on the test articles.

Table 2. Summary of bioburden levels on post-processed allograft tissues.

Sample†	Aerobic			Fungi		Spore		Anaerobic	
	Min CFU	Max CFU	Avg. CFU	Min CFU	Max CFU	Min CFU	Max CFU	Min CFU	Max CFU
C/C FD	< 3	12	3	< 3	< 3	< 15	< 15	< 3	< 3
C/C FZ	< 3	68	20	< 3	< 3	< 15	< 30	< 3	< 3
S/NP FZ	21	99	39	< 3	< 30	< 15	< 30	< 3	< 30
S/NP FD	< 3	< 3	< 3	< 3	< 3	< 15	< 15	< 3	< 3
S/NP SL	< 3	< 3	< 3	< 3	< 3	< 15	< 15	< 3	< 3
TOTALS	**< 3**	**99**	**7**	No microorganisms observed					

† = C/C = Cortical or Cancellous; S/NP = Soft or non-purged;
FZ = Frozen; SL= Saline; FD = Freeze Dried

Microbial Identification

There were a total of 442 colonies observed on all 128 test articles processed for bioburden analysis. Of the 442 colonies, there were five colony morphologies typical of the entire microbial population. The representative colonies were streaked for isolation; gram stained and identified using common microbiological methods. MIDI (MIDI Labs, Newark, DE, USA) analysis confirmed the isolates identities as *Staphylococcus epidermidis*, *Enterococcus faecalis*, and *Enterococcus durans*. The remaining 2 isolates were non-viable, and could not be sub-cultured for positive identification.

DISCUSSION

The results from this study indicate that the average bioburden levels on post-processed allografts from all product categories, is 7 CFU with a maximum of 99 CFU and a minimum of no observable colonies. These values can be used to assess the adequacy of the dose (16–25 kGy) used to terminally sterilise allografts using gamma radiation. Each container of AlloSource final allograft products is sent for terminal sterilisation with two, *Bacillus pumilus* (ATCC 27142) biological indicator strips containing a minimum of 1,000 Spores per strip. The strips are sterilised along with the allografts then tested for microbial inactivation. Both strips must be negative for growth in order for the terminal sterilisation process to be considered valid.

Knowing the type of bioburden on your final product is as important as knowing how much there is. Several common microorganisms have published D-values, which indicate their resistance to radiation sterilisation [4]. A D-value is defined as the radiation dose required to kill 90% (1 Log) of a homogeneous microbial population where it is assumed that the death of microbes follow first order kinetics [16,17]. *B. pumilus* is typically used as the standard challenge microorganism for radiation sterilisation since it is stable on a biological indicator (BI) and one of the more resistant microorganisms to radiation. The D-value of B. pumilus can range from 2.6–3.3 kGy, whereas *Escherichia coli* has a D-value of 0.09 [4]. The importance of knowing the D-values is needed when calculating the robustness of your terminal sterilisation process. If your bioburden is loaded with *B. pumilus* spores vs. *E. coli* vegetative bacteria the dose needed to eradicate an equivalent amount of both microorganisms (e.g. 100 CFU) will be considerably higher for B. pumilus than it would for E. coli.

Given:

- The average bioburden count identified in this study is 7 CFU/test article
- The maximum bioburden count identified in this study is 99 CFU/test article
- Published D_{10} values for bioburden testing can be as high as 3.3 kGy, however the D_{10} value for BI's used in this study was 1.5 kGy
- The bioburden of the biological indicators is at least 1×10^3 Spores/strip
- The minimum recommended dose for the terminal sterilisation of allografts is 15 kGy

Formula:

- Dose = D_{10} * Log (average bioburden)

Where:

- **Bioburden is the population of viable microorganisms on a product**
- **D_{10} is the radiation dose required to kill 90% (1 Log) of a homogeneous microbial population**
- **Dose is the calculated level of radiation absorbed by the product**

Calculation 1:

The minimum dose for a 90% reduction (1 Log) of the average bioburden level is:

Dose = D_{10} * Log (average bioburden) = 1.5 kGy * Log 7 = 1.3 kGy

Calculation 2:

The minimum sterility assurance level (SAL) achievable using a maximum bioburden of 99 CFU and assuming a D_{10} value of 1.5 kGy is:

Dose = D_{10} * [Log (max bioburden) – Log SAL]

- Log SAL = (Dose/D_{10}) – Log (max bioburden)

Log SAL = 15 kGy/1.5 kGy – Log 99

SAL = 10^{-8}

Calculation 3:

The minimum dose for an SAL of 10^{-6} using the maximum bioburden level of 99 CFU and assuming a D_{10} value of 1.5 kGy is:

Dose = D_{10} * [Log (max bioburden) – Log SAL]

Dose = 1.5 kGy * (Log 99 – Log 10^{-6})

Dose = 11.98 kGy

The previous derivation of the minimum SAL achievable using our current bioburden and terminal sterilisation dose is dependent upon three fundamental assumptions that were made. The first assumption is that bioburden levels on your product are relatively low. The second assumption is that the bioburden is normally distributed. The final assumption is that you know exactly what the bioburden is so D-values can be calculated. The first assumption is the only valid assumption that we can make regarding the terminal sterilisation of human allograft tissue. This study indicated that our post-processing bioburden levels are low with a maximum value of 99 CFU and an average of 7 CFU. The second and third assumptions were not valid assumptions for purposes of this study. In the second assumption it assumes that you have a normally distributed microbial population on your product. Since allografts are made from multiple donors, from multiple retrieval sites the bioburden levels on each of the incoming donor tissues can fluctuate accordingly. The third assumption assumes that

you know exactly how much bioburden there is. In several cases the actual bioburden on the allograft test articles could not be determined and the dilution factor had to be reported as the lowest level of detection. Using these values to calculate the average bioburden value (7 CFU) will artificially skew your final number since you are substituting the dilution factor for actual colony counts that may not exist (i.e. 80 CFU for a dilution value of < 80).

The current dose calculation is a valid method if you can meet the aforementioned assumptions. This does not negate the fact that our current terminal sterilisation process is more than adequate to terminally sterilise final allograft products, rather it simply is not an accurate estimate of the true sterilisation dose required. In the effort to more accurately determine the appropriate dose a second method has been explored as outlined in AAMI/ISO 11137 Method II B (data not shown). In this method the dose is calculated based upon the number of positive sterility cultures actually observed not the amount of estimated bioburden present on your final product. This method is thought to be much more appropriate for the tissue banking industry.

CONCLUSIONS

This study has demonstrated that the average bioburden found on 128 human allograft test articles was 7 CFU. Following the calculations as defined in AAMI/ISO 11137 our current sterilisation process is theoretically capable of reducing our maximum bioburden (99 CFU) to a minimum SAL of 10^{-8} (a 1 in 100 million chance of a non-sterile product) assuming that the D-value of 1.5 kGy is representative of the average resistance of the bioburden routinely found on the products. It is this, among other assumptions that are outlined in Method I that really do not apply to the terminal sterilisation of allografts and as such, Method IIB is being investigated to better estimate the terminal sterilisation dose for allograft products.

ACKNOWLEDGEMENTS

The authors wish to thank Nelson Laboratories (Salt Lake City, UT, USA) for their help in generating the aforementioned bioburden counts as well as the microbial identifications.

The authors also wish to thank Martell Winters of Nelson Laboratories for his technical review of this manuscript.

REFERENCES

1. Archibald, L. K., Jernigan, D. B. and Kainer, M. A., Centre for Disease Control and Prevention, Update: Allograft-associated bacterial infections - United States, 2002, *MMWR*, **2002**, 51, 207-210.
2. Lutz, B., Ratard, R., Dodson, D., *et al.*, Septic arthritis following anterior cruciate ligament reconstruction using tendon allografts - Florida and Louisiana, 2000, *MMWR*, **2001**, 50,1081-1083.
3. Centre for Disease Control and Prevention, Update: Unexplained deaths following knee surgery - Minnesota, 2001, *MMWR*, **2001**, 50, 1080.
4. Yusof, N., Quality system for the radiation sterilisation of tissue allografts, In: *Advances in Tissue Banking, Volume 3*, Phillips, G. O., Strong, D. M., von Versen, R. and Nather, A. (eds.), World Scientific, River Edge, NJ, USA, **1999**, pp. 264-265.

5. Cloward, R. B., Gas-sterilized cadaver bone grafts for spinal fusion operations. A simplified bone bank, *Spine,* **1980**, 5, 4-10.
6. Jackson, D. W., Windeler, G.E. and Simon, T. M., Intraarticular reaction associated with the use of freeze-dried, ethylene oxide-sterilized bone patella tendon-bone allografts in the reconstruction of the anterior cruciate ligament, *Am. J. Sports Med.,* **1990**, 18, 1-10.
7. Lawrence, W. H., Ithoh, K., Turner, J. E. and Autian, J., Toxicity of ethylene chlorohydrin II. Subacute toxicity and special tests, *J. Pharm. Sci.,* **1971,** 60, 1163-1168.
8. Roberts, T. S., Drez, D., McCarthy, W. and Paine, R., Anterior cruciate ligament reconstruction using freeze-dried, ethylene oxide-sterilized, bone-patellar tendon-bone allografts. Two year results in thirty-six patients, *Am. J. Sports Med.,* **1991**, 19, 35-41.
9. Compressive mechanical properties of human cancellous bone after gamma irradiation, *J. Bone Joint Surg. Am.,* **1992**, 74, 747-752.
10. Currey, J. D., Foreman, J., Lakctic, I, *et al.,* Effects of ionizing radiation on the mechanical properties of human bone, *J. Orthop. Res.,* **1997**, 15, 111-117.
11. Fideler, B. M., Vangsness, C. T., Lu, B., Orlando, C. and Moore, T., Gamma irradiation: Effects on biomechanical properties of human bone-patellar tendon-bone allografts, *Am. J. Sports Med.,* **1995**, 23, 643-646.
12. Hamer, A. J., Strachan, J. R., Black, M. M., *et al.,* Biochemical properties of cortical allograft bone using a new method of bone strength measurement. A comparison of fresh, fresh-frozen and irradiated bone, *J. Bone Joint Surg. Br.,* **1996**, 78, 363-368.
13. Pelker, R. R., Friedlaender, G. E. and Markham, T. C., Biomechanical properties of bone allografts, *Clin. Orthop.,* **1983**, 174, 54-57.
14. American Association of Tissue Banks (AATB), website: http://www.aatb.org.
15. American Association of Tissue Banks, Standards for Tissue Banking. Kasprisin, D. and Woll J.E. (eds.), April 1, **2002**, pp. 55-57.
16. Sterilization of Medical Devices - Microbiological Methods - Part I: Estimation of population of microorganisms on products, *AAMI STANDARDS, ISO 11737,* **1995**.
17. Table B.1 - Radiation Dose (kGy) required to achieve a given SAL for different bioburdens having standard distribution of resistances, *AAMI STANDARDS, ISO 11737,* **1995**.

DETERMINATION OF MICROBIAL BIOBURDEN LEVELS ON PRE-PROCESSED ALLOGRAFT TISSUES

Chad J. Ronholdt and Simon Bogdansky

AlloSource, 6278 S. Troy Circle, Centennial, CO 80111, USA

ABSTRACT

Bioburden testing has been used in the pharmaceutical and medical device industries for years to evaluate the levels of microbial contamination on raw materials, components and final products. Recently tissue banks have turned to bioburden testing for microbial analysis of incoming and final allograft product as in the case of this study. The purpose of this study was to investigate what types as well as what concentrations of microorganisms are present on incoming allograft tissue. A total of 63 allograft tissues (pre-debridement), representing different tissue product families (e.g. Achilles tendon, femur, fibula, hemi-pelvis, iliac crest, tibia w/ patellar ligament and humerus) were selected for destructive bioburden analysis. The tissue was placed into sterile stomacher bags with a defined volume of extraction fluid and vigorously agitated for 1-minute. Exhaustive sessions were performed using the same tissue to ensure that all microbial contaminants were removed from the tissue. At the conclusion of each session, multiple 1 mL aliquots of the extraction fluid were removed, pour-plated and evaluated for microbial growth over a 7-day period. Results from the exhaustive bioburden testing sessions indicated a wide variety of microorganisms were present on incoming allograft tissues. A total of 37 different microorganisms were identified on the pre-processed allograft tissue. The most prominent Gram-positive microorganisms recovered were *Staphylococcus capitis* and *Staphylococcus epidermidis* (30 and 28% respectively). The most prevalent Gram-negative microorganisms recovered were *Klebsiella pneumonia* and *Morganella morganii* (36 and 22% respectively). The most prevalent fungal microorganisms recovered were *Candida parapsllosis* and *Candida albicans* (40 and 30% respectively). Average Colony Forming Units (CFU) per sample ranged from < 20 CFU to > 1.6 million CFU per tissue. The results from this study indicate that incoming allograft tissues can be heavily contaminated with a wide variety of microorganisms. Therefore every effort should be made to reduce the potential for cross-contamination at the time of retrieval, transport and processing.

KEYWORDS

Bioburden; microbial contamination; allograft tissue; cross-contamination; extraction

INTRODUCTION

The environment, incoming materials and/or human intervention are the most common sources of bioburden although the number of sources can be almost infinite. Knowing the level of bioburden on your incoming raw materials, components, solutions and equipment is a common activity within the medical device and pharmaceutical industries. Data generated by bioburden testing facilitates establishment of appropriate terminal sterilisation dose and is quite useful when designing an effective cleaning and inactivation process for products manufactured in an aseptic environment as allografts.

Tissue banking is unique in the sense that although allografts are processed in an aseptic environment the raw material (human tissue) is not sterile; there is an inherent level of bioburden associated with each tissue. Sources of this bioburden encompass the entire tissue banking process from donor retrieval, location of retrieval, tools used for retrieval to actual donor processing, processing environment and final packaging. Therefore it is essential that all processes involved in the production of allograft products be tightly controlled to minimise contribution of bioburden to the tissue [1].

Bioburden testing can be performed using several commonly accepted methods depending upon the article being tested [2]. For example most medical devices are tested using a full immersion method where each medical device is submerged in a defined volume of extraction fluid, agitated, incubated and then evaluated for microbial growth [2]. Liquid products are typically tested using membrane filtration and large, irregular products may be tested using surface agar contact plates, swabbing or a direct agar overlay method [2]. Each method is validated to recover a defined percent of microbial colony forming units (CFU). In most cases, this percent recovery is less than 100% since there are limitations of each of the aforementioned bioburden testing methods. Optimisation of these methods can come in the form of increasing agitation time or robustness, using a different extraction fluid among others. Typically, a series of exhaustive rinses are used to determine a conversion factor, which is then applied to process samples to adjust for the limitations of each recovery method [3]. This adjustment helps estimate the total population of microbial bioburden on your test article.

The purpose of this study was to evaluate pre-processing human cadaveric musculoskeletal tissue for bioburden levels as well as identify the most commonly isolated contaminants. Several pre-processing tissues were selected for testing using a fluid immersion bioburden test. The test articles were evaluated for aerobic, anaerobic and fungal bioburden counts per tissue. All positive cultures were identified to the genus and species level. What follows is a detailed account of the method and results generated from this bioburden study.

MATERIALS & METHODS

Thirteen unique traditional pre-processing tissue categories (cortical, cortical/cancellous and soft tissue) were consented for research and selected for bioburden testing (Table 1). These thirteen tissue categories represent 100% of all donor tissue AlloSource receives for traditional processing. In most cases six pre-processing tissues were taken for testing from each tissue category with the exception of ribs, costal cartilage, ulna and radius where 1, 2, 2, and 4 tissues were taken respectively, due to availability constraints.

Table 1. Sample tissues.

Achilles tendon	Humerus
Femur	Iliac crest / ileum
Fibula	Tibia, w/patellar tendon
Hemi-Pelvis	Ulna
Vertebral Bodies	Ribs
Hemi Pelvis	Costal Cartilage
Radius	

Prior to the start of the study, three tissue samples were tested to confirm that the intended bioburden method was an appropriate method to assess the allografts for contaminant microorganisms. Upon successful completion of the verification study, a total of 63 pre-processing tissues were sent to the testing laboratory (AppTec Laboratories, Atlanta, GA, USA).

Upon receipt, the tissue test articles were aseptically transferred to individual sterile stomacher bags (Seward, Northampton, UK) where defined volumes of Fluid D (Remel, Lenexa, KS, USA) between 100 and 600 mL were required depending upon the test article size. The bags were sealed and placed into a stomacher (Seward, Northampton, UK) and agitated for 1-minute. Triplicate 5 mL aliquots were removed (1 for aerobic, 1 for anaerobic and 1 for fungal microorganisms) for each of the 6 test articles and plated using the classic pour-plate method. Aerobic cultures were plated using Tryptic Soy Agar (TSA) plates (Remel, Lenexa, KS, USA) and incubated aerobically at 30-35°C for 2–4 days and 20-25°C for 3-5 days. Anaerobic cultures were also plated using TSA plates and were incubated anaerobically at 30-35°C for 3-5 days. Fungal growth was evaluated using Rose Bengal Agar (Remel, Lenexa, KS, USA) and incubated for 7 days at 20-25°C. All positive test articles were subbed for isolation and identified using common microbiological techniques specific for each isolate.

RESULTS

Bioburden Analysis

Prior to the start of the actual study a small number of allograft products were analysed to derive a correction factor to apply to the counts. This correction factor adjusts the number of Colony Forming Units (CFU) to an estimate of what the 'true' population of microorganisms are on a given test article. A correction factor is required since the extraction method does not recover 100% of the microorganisms present on the test article and this correction factor compensates for that loss. The correction factor that was calculated for this study was 3.5. The average CFU/test article was calculated using the actual colony count for each test article then multiplied by the respective dilution factor (the dilution factor was dependent upon the volume of Fluid D used to perform the test for each test article). Refer to the following tables (Table 2) for a summary of the bioburden counts (average and adjusted) from all 13-tissue categories.

Microbial Identification

Microbial identification was limited to the aerobic and fungal colonies with different morphologies since there were no observable colonies on the anaerobic plates. The identifications were broken down into distinct microbial categories (Gram positive, Gram negative and Fungal microorganisms) for simplicity. Refer to Tables 3-5.

DISCUSSION

Prior to the start of this bioburden study it was unclear as to the number and variation of microbial contamination on incoming donor tissue. The impression was that pre-processing culturing methods were giving an accurate assessment of the bioburden on the tissues, however after a comparison of contamination rates using a common culturing method (swab) and the bioburden results (data not shown) it was evident that the swab culturing method was not picking up all contaminants. Furthermore, the swab was unable to yield an accurate estimation of microbial contaminant concentration.

Table 2. Bioburden results.

Tissue Type	Average Colony Forming Units (CFU) per Sample			Adjusted Final CFU/Sample (Average CFU X 3.5)		
	Aerobic	Fungi	Anaerobic†	Aerobic	Fungi	Anaerobic
Humerus	151,220	105,270	< 80	529,270	368,445	N/A
Hemi	465,700	262,900	< 120	1,629,950	920,150	N/A
Tibia	67,933	49,672	< 120	237,766	173,852	N/A
Fibula	70,235	31,535	< 60	245,823	110,373	N/A
Ulna	60,390	56,000	< 60	211,365	196,000	N/A
Radius	42,015	40,753	< 40	147,053	142,636	N/A
Femur	32,925	19,355	< 120	115,238	67,743	N/A
Vertebral Bodies	24,553	9,668	< 120	85,936	33,838	N/A
Costal Cartilage	18,025	0	< 20	63,088	0	N/A
Achilles Tendon	9,020	2,047	< 60	31,570	7,165	N/A
Iliac Crest	8,555	7,860	< 60	29,943	27,510	N/A
Ribs	1,400	60	< 60	4,900	210	N/A

† = There were no observable colonies on the anaerobic plates, however due to the inherent properties of the assay the dilution factor must be recorded as the lowest level of detection.

All aerobic and fungal samples had colony counts in excess of 210 CFU, with the exception of the fungal counts for the costal cartilage samples, which were negative for growth. In all cases, there were no observable colonies present on the anaerobic samples, however due to the inherent properties of the method used the dilution factor must be recorded as the lowest level of detection (e.g. < 80). The data estimated that the maximum bioburden values were found on a hemi-pelvis in excess of 1.6 million aerobic CFU and > 900,000 fungal colonies.

Thirty-seven different colony morphologies were identified on all allograft samples tested. A majority of the contaminants identified were common environmental and human skin and enteric flora isolates suggesting that the area in which the tissues are retrieved and the methods by which they are retrieved can considerably contribute to elevated bioburden levels. The most commonly isolated microbial contaminants were *Staphylococcus capitis* and *Staphylococcus epidermidis* with percent recoveries of 30.0% and 27.5% respectively in the gram-positive category. Of more concern, are gram negative and fungal isolates since these microorganisms have higher virulence, greater pathogenicity and are commonly associated with human disease and infection. *Klebsiella pneumoniae*, *Morganella morganii*, *Candida parapsllosis* and *Candida albicans* had percent recoveries of 35.9%, 21.9%, 40% and 30%, respectively. Other isolates can be attributed to common environmental contamination (*Bacillus subtilis, Aspergillus varians*), gastrointestinal tract flora (*Enterobacter cloacae*, *Escherichia coli*) and typical human skin flora (*Corynebacterium renale*, *Micrococcus luteus*).

Table 3. Gram-positive isolates.

Sample Number	Contaminant ID	No. of Colonies per Sample	Percent of Total (%)
1	*Staphylococcus capitis*	36	30.0
2	*Staphylococcus epidermidis*	33	27.5
3	*Corynebacterium renale group*	5	4.2
4	*Corynebacterium species*	5	4.2
5	*Bacillus sphaericus*	5	4.2
6	*Alloiococcus otitidis*	4	3.3
7	*Micrococcus sedentarius*	4	3.3
8	*Streptococcus equi ssp equi*	3	2.5
9	*Lactococcus lactis ssp. Hordniae*	3	2.5
10	*Staphylococcus saprophyticus*	3	2.5
11	*Staphylococcus schleiferi*	3	2.5
12	*Streptococcus bovis*	2	1.7
13	*Streptococcus vestibularis*	2	1.7
14	*Staphylococcus haemolyticus*	3	2.5
15	*Micrococcus luteus*	2	1.7
16	*Bacillus subtilis*	2	1.7
17	*Staphylococcus warneri*	1	0.8
18	*Staphylococcus saccharolyticus*	1	0.8
19	*Rothia dentocariosa*	1	0.8
20	*Dienococcus proteolyticus*	1	0.8
21	*Bacillus megaterium*	1	0.8
	Total =	**120**	**1.000**

Knowing the level and type of contamination on incoming donor tissue has become a critical element in operations at our tissue bank since it can ultimately effect the final product. Based on this data we have made the assumption that any incoming donor tissue could contain a minimum of 1.6 million CFU of microorganisms. Also included in our assumption is that fungal colonies, yeast and enteric microorganisms may also be present at elevated levels. Using this assumption we have designed several projects and validations around the premise that assumes a worse case scenario for both type and number of microorganisms.

We have also taken significant steps to control the level of bioburden coming into our tissue bank as a direct result of this study. Knowing that the retrieval process is the first potential source of contamination we have instituted validated sanitisation procedures, validated the appropriate sterilisation equipment and formalised the retrieval process in the form of standard operating procedures. These steps have significantly reduced the bioburden levels at some retrieval sites (data not shown). Additional steps at our processing facility include validation of sanitisation methods for our cleanroom suites and equipment, optimisation of our cleaning and inactivation process among other projects. All of which, help reduce the possibility of introducing microbial contaminants into the allograft aseptic processes.

Table 4. Gram-negative isolates.

Sample Number	Contaminant ID	No. of Colonies per Sample	Percent of Total (%)
1	*Klebsiella pneumoniae ssp. pneumoniae*	23	35.9
2	*Morganella morganii*	14	21.9
3	*Enterobacter cloacae*	7	10.9
4	*Enterobacter sakazakil*	5	7.8
5	*Stentotrophomonas maltophilia*	3	4.7
6	*Pseudomonas aeruginosa*	3	4.7
7	*Acinetobacter iwoffi*	2	3.1
8	*Escherichia coli*	2	3.1
9	*Pseudomonas flourescens*	2	3.1
10	*Flavimonas oryzlhabitans*	1	1.6
11	*Pseudomonas species*	1	1.6
12	*Klebsiella pneumoniae ssp. pneumoniae*	1	1.6
	Total =	**64**	**1.000**

Table 5. Fungal isolates.

Sample Number	Contaminant ID	No. of Colonies per Sample	Percent of Total (%)
1	*Candida parapsllosis*	4	40.0
2	*Candida albicans GC subgroup A*	3	30.0
3	*Rhodotorula rubra*	2	20.0
4	*Aspergillus varians*	1	10.0
	Total =	**10**	**1.000**

Refer to Table 6 and Figure 1 for a summary of the distribution of all of the microorganisms found present on incoming pre-processed tissue regardless of microbial category.

Table 6. Overall percent CFU distribution.

Sample Number	Contaminant ID	No. of Colonies per Sample	Percent of Total (%)
1	*Staphylococcus capitis*	36	**22.0**
2	*Staphylococcus epidermidis*	32	**19.5**
3	*Klebsiela pneumoniae ssp. pneumoniae*	23	**14.0**
4	*Morganella morganii*	14	**8.5**
5	*Enterobacter cloacae*	7	**4.3**
6	*Enterobacter sakazakil*	5	**3.0**
7	*Corynebacterium renale group*	5	**3.0**
8	*Candida parapsllosis*	4	**2.4**
9	*Alloiococcus otitidis*	4	**2.4**
10	*Micrococcus sedentarius*	4	**2.4**
11	*Candida albicans GC subgroup A*	3	**1.8**
12	*Lactococcus lactis ssp. hordniae*	3	**1.8**
13	*Staphylococcus saprophyticus*	3	**1.8**
14	*Staphylococcus schleiferi*	3	**1.8**
15	*Escherichia coli*	2	**1.2**
16	*Streptococcus bovis*	2	**1.2**
17	*Streptococcus vestibularis*	2	**1.2**
18	*Staphylococcus haemolyticus*	2	**1.2**
19	*Micrococcus luteus*	2	**1.2**
20	*Bacillus subtilis*	2	**1.2**
21	*Staphylococcus warneri*	1	**0.6**
22	*Staphylococcus saccharolyticus*	1	**0.6**
23	*Rothia dentocariosa*	1	**0.6**
24	*Bacillus megaterium*	1	**0.6**
25	*Chryseobacterium meningosepticum*	1	**0.6**
26	*Aspergillus varians*	1	**0.6**
TOTALS =		**164**	**1.000**

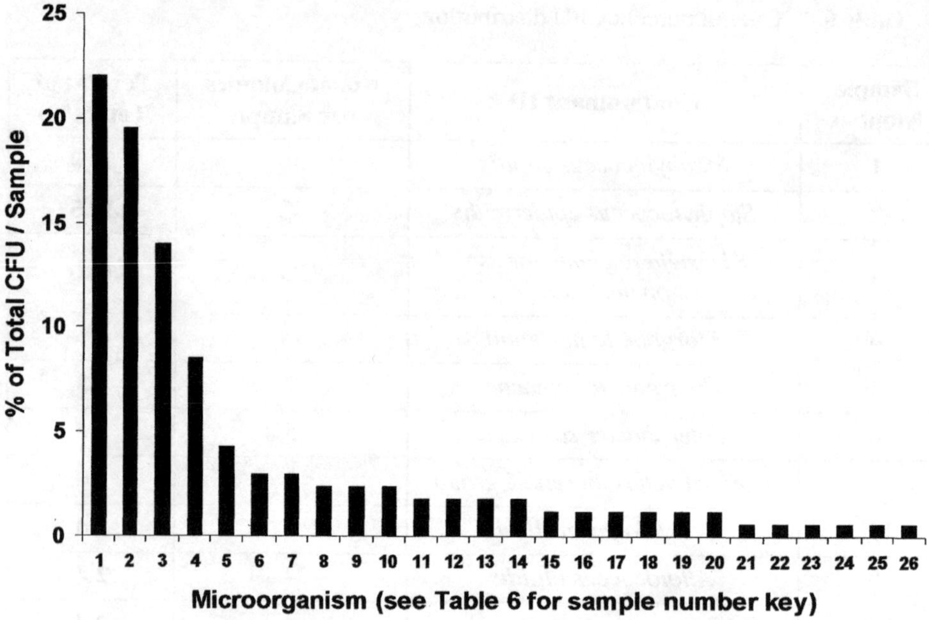

Figure 1. Percent distribution of microorganisms on cadaveric pre-processed human tissue (n = 164).

CONCLUSIONS

The data generated from this study indicated that incoming human cadaveric tissue might have bacterial, fungal and yeast colonies at levels in excess of one million colony forming units per tissue. Of the most commonly isolated were normal human skin flora (*Staphylococcus* and *Corynebacterium*) and gastrointestinal tract microorganisms (*Klebsiella* and *Enterobacter*) suggesting that retrieval techniques may significantly contribute to the bioburden on pre-processed tissue. Based upon this data, AlloSource has focused on reducing the sources of bioburden in several areas of the tissue banking process to lower the threat of microbial contamination on final allograft products.

ACKNOWLEDGEMENTS

The authors wish to thank AppTec Laboratories (Marletta, GA, USA) for help in generating the aforementioned bioburden counts as well as the microbial identifications.

REFERENCES

1. Panezai, A. K., Sterility testing under scrutiny, *Manufacturing Chemist (England)*, **1991**, 62, 19-21.
2. Carpenter, D., Coleman, M., *et al.*, Methods of bioburden evaluation, *Dev. Ind. Microbiol.*, **1981**, 22, 323-328.
3. Horn, W., Lugli, M, *et al.*, Method for estimating bioburden of non-sterile cotton Gauze, *J. Ind. Microbiol. Biotechnol.*, **1997**, 18, 15-17.
4. *AAMI STANDARDS, ISO 11737*, Sterilization of medical devices – microbiological methods – part I: estimation of population of microorganisms on products, **1995**.

BIOBURDEN ESTIMATION IN RELATION TO TISSUE PRODUCT QUALITY AND RADIATION DOSE VALIDATION

Norimah Yusof [1*], Abdul Rani Shamsudin [2], Hasim Mohamad [3], Asnah Hassan [1], Ang Cheng Yong [2] and Muhamad Nor Firdaus A Rahman [2]

[1] *Malaysian Institute for Nuclear Technology Research (MINT), Bangi, 43000 Kajang, Selangor, Malaysia {E-mail: norimah@mint.gov.my}*

[2] *National Tissue Bank, Hospital Universiti Sains Malaysia (HUSM), Kubang Kerian, Kelantan, Malaysia*

[3] *Dept. of Surgery, Hospital Kota Bharu, Kota Bharu, Kelantan, Malaysia*

ABSTRACT

Estimation of low bioburden or population of less than 100 viable microorganisms on amnion and bone allograft product is best measured by using a filtration method. From our routine quality control checks over several years, we found that the bioburden was reduced further after the tissues were properly handled and processed by trained staff with many years of experience. Bioburden of air-dried amnion was reduced to less than 100 cfu per item while bone showed bioburden around 10 cfu per item. Sterilisation dose validation exercises, carried out according to the three methods described in IAEA Code of Practices, requires skilled personnel with experience in conducting sterility test and dose mapping.

KEYWORDS

Bioburden; dose validation; sterilisation; tissues

INTRODUCTION

Bioburden is the count of viable microorganisms or contamination level on and in a product prior to sterilisation technology. In tissue banking a tissue processed from a screened donor is subjected to handling and processing according to standard procedures. The processing conducted by a skilled operator under a good hygienic environment will result in low bioburden. On the contrary failing to comply to the standard processing procedure, the processed tissue may have contaminants coming from various sources during processing and handling namely environment, tools, solution used or even operator himself, resulting in high bioburden. Therefore bioburden estimation can qualify whether the processing has been carried out properly, processed by either a trained operator and/ or whether the environment is well maintained. Three methods commonly used to estimate bioburden as described in ISO 11737-1 [1] are the Pour Plate method, Spread Plate method and Filtration method. The latter is more sensitive and particularly useful for tissue products of low bioburden.

Bioburden estimation is being used as a quality control (QC) tool to determine the hygienic level of medical products, which will then determine the effective irradiation doses for sterilisation. For instance, sterilisation at 25 kGy will only achieve SAL 10^{-6}

when a product has bioburden not exceeding 10^3 colony-forming units (cfu) per product unit [2]. High bioburden may require doses higher than 25 kGy, however in the case of tissue, its biological and physical properties should not be affected by the selected doses. Even though radiation can render the tissues sterile, endotoxins produced from membranes of gram –ve bacteria of high bioburden products will not be removed or inactivated. Toxins or any forms of protein require very high radiation doses (> 100 kGy) for inactivation. As for tissue products, donor screening and good processing procedures are emphasised to ensure the production of 'clean' tissue grafts with low bioburden, hence doses lower than 25 kGy are sufficient for sterilisation. Detrimental effects of radiation especially in soft tissues such as cartilage, tendon, skin and fascia lata can be minimised. Whatever sterilisation dose is used, a tissue banker must first conduct a dose validation exercise. Validation of sterilisation dose for medical products must be carried out according to methods as specified in ISO document No. 11137 [2]. Tissues, unlike medical products, are produced in a small number per processing batch and may have a different microbial population depending on the source of the tissue. Realising this, the International Atomic Energy Agency (IAEA) has come up with a Code of Practice offering several validation methods [3]. Bioburden determination is a prerequisite in all methods, followed by sterility testing for products after exposing at substerilisation or verification doses.

This paper highlights our work on bioburden of amnion and bone tissue and how the information on bioburden assisted us in monitoring as well as improving the processing. Our experience in using the IAEA Code of Practice (2002) [3] in dose validation exercise for radiation sterilisation of amnion is also presented. The code does not cover viruses as it is recommended that virus transmission must be dealt with through donor screening and virus inactivation processes. Viruses in general are more radiation resistant compared to bacteria and fungi.

MATERIALS & METHODS

Amnion and bone samples

Amniotic membranes were processed either as freeze-dried or air-dried at The National Tissue Bank, HUSM and MINT Tissue Bank using methods as established earlier by Mohamad and Yusof (1991) [4]. Packages were picked randomly from a processing/production batch for routine bioburden estimation. For the dose validation exercise, 10 amnion samples were required for bioburden testing, either from the same mother for sampling of small size pieces of 2.5 cm x 2.5 cm, 3 cm x 3 cm and 5 cm x 5 cm; or from several mothers for sampling of large size pieces of 10 cm x 10 cm. Another 10 samples were put aside for substerility testing after samples were exposed to a verification dose. Freeze-dried bone allograft was processed as described by Yusof *et al.* (1994) [5] except for bovine bone whereby it was processed as a demineralised graft. A small piece of bone was cut off either at an early stage of the processing or just before sterilisation and tested for bioburden estimation. However, dose validation exercise could not be carried out due to limited bone samples.

Sample item portion (SIP)

For amnion, the SIP is 1 when entire product package or unit or item was tested. For bone, about 20% by weight was taken. The adequacy of this selected SIP was demonstrated by taking 20 samples from one non-irradiated amnion and subjecting to a sterility test which must yield a minimum of 17 positive sterility tests (i.e. 85%) [2].

Filtration method

The outer pack of a sample was disinfected with 70% alcohol. The sample was transferred to a flask containing 100-200 mL 0.01% sterile saline polysorbate solution, mixed using a vortex mixer and kept aside for 15-20 minutes. The aliquot was filtered through a 0.2 μm membrane filter and rinsed with saline polysorbate solution. The membrane was placed on Tryptone Soya Agar (TSA) plate and incubated at 32-35°C for 5-7 days. Colony forming unit (cfu) for each sample was calculated. Correction factor was applied for the final cfu.

Pour plate method

Volume in the range of 0.1 to 1.0 mL of aliquot was mixed with molten agar at < 45°C, which was then allowed to solidify on a petri dish. The plate was incubated at 32-35°C for 5-7 days and the colonies were counted.

Spread plate method

Aliquot of < 1 mL was spread onto the surface of agar using a spreader. The plate was incubated at 32-35°C for 5-7 days and the colonies were counted.

Sterility test after verification dose

Method specified in ISO 11737-2 [6] was followed to conduct the sterility test on samples after being subjected to verification doses. Samples were handled aseptically and preferably conducted by trained staff. The outer pack was disinfected with 70% alcohol. Each sample was placed in Tryptone Soya Broth (TSB), incubated at 30 ± 2°C for 14 days. Samples were examined daily for any growth.

Radiation dose validation exercise

Three methods as described in the IAEA Code of Practice [3] were used:

Method (a)

For establishing specific sterilisation dose for Standard Distribution of Resistance and other microbial distribution for sample sizes between 10 and 100, Method 1 of ISO 11137:1995 [2] was adopted. Ten samples were taken for average bioburden estimation. Verification dose was obtained from Table 2 at the estimated initial bioburden to achieve sterility assurance level (SAL) of 10^{-1}. Verification dose should be delivered within ± 10% variation. Sterility test on ten samples should not yield more than one positive growth. When the exercise was valid, Table 2 was referred for sterilisation dose at SAL 10^{-6} and the closest bioburden number that was equal to or greater than the estimated bioburden.

Method (b)

For substantiation of 25 kGy sterilisation dose, the Method in ISO/TR 13409:1996 [7] was used. Ten samples were taken for average bioburden estimation. The verification dose SAL 10^{-1} is calculated using the following formula:

$$\text{Dose} = I + [S \times \log(\text{average SIP bioburden})]$$

Where I and S values were obtained from Table 3 in the Code of Practice at sample size 10.

Verification dose should be delivered within ± 10% variation. Sterility test on ten samples should not yield more than one positive growth.

Method (c)

For substantiation of a 25 kGy sterilisation dose, an alternative method AAMI TIR 27 has been developed. Ten samples were taken for average bioburden estimation. The verification dose was calculated using the following formula:

(i) For bioburden levels of 1 to 80 cfu per product

Step 1 $D_{lin} = 25\text{kGy}/(6 + \log \text{bioburden})$
Step 2 Verification dose = D_{lin} (log bioburden – log SAL_{VD})

(ii) For bioburden levels of 81 to 1000 cfu per product

Step 1 $TD_{10} = (\text{Dose}_{-6}\ \text{kGy} + \text{Dose}_{-2}\ \text{kGy})$
Step 2 Verification dose = 25kGy – [TD_{10} (log SAL_{VD} + 6)]

Where:

D_{lin} - is the D_{10} dose for a hypothetical survival curve which is linear between the coordinates (log N_0, 0 kGy) and (log 10^{-6}, 25 kGy),
N_0 - is bioburden or count prior to sterilisation,
SAL_{VD} - is the sterility assurance level at which the verification dose experiment is to be performed,
TD_{10} - is the hypothetical D_{10}

Dose_{-6} and Dose_{-2} are doses corresponding to SAL values of 10^{-6} and 10^{-2} respectively, obtained from Table B1 of ISO 11137 (1995) [2]. Verification dose should be delivered within ± 10% variation using GammaCell GC 4000A. Sterility test on ten samples should not yield more than one positive growth.

RESULTS & DISCUSSION

Validation of bioburden estimation

At the early stage of developing procedures of our tissue banks, we tested the three methods of bioburden estimation as described in ISO11373-1:1995 [1]. Table 1 shows the bioburden estimations of 10 pieces of freeze-dried amnion (cfu/product unit or item) using the filtration (21.3 cfu/item) compared to pour plate (82.4 cfu/item) method. Pour plate tended to give higher counts after multiplying with the volume correction factor, which resulted in a large variation in bioburden estimation. Filtration method provides true count when the whole aliquot was filtered.

We could not detect any growth of *Staphylococcus sp* (using Brian Heart Infusion Agar) and *Salmonella sp* (using Brilliant Green Agar). We observed little growth of fungi on MacConkey Agar as well as anaerobic microbes on TSA.

While in Table 2, bioburden of freeze-dried amnion could not be estimated by the Spread Plate method, even though the amnion samples used were purposely of large size (> 180 m^3) with considerable high counts as shown by filtration method (140.8 cfu/item).

Table 1. Bioburden estimations of freeze-dried amnion using filtration and pour Plate methods.

Sample no.	Bioburden (cfu/product unit or item)	
	Filtration	Pour plate
1.	36.3	64.3
2.	20	20
3.	25	32
4.	14	12
5.	10	93
6.	15	80
7.	27	160
8.	20	100
9.	24	180
Average Bioburden	21.3 ± 7.4	82.4 ± 55.4

Table 2. Bioburden estimation of freeze-dried amnion using filtration and spread plate methods.

Sample no.	Bioburden (cfu/product unit or item)	
	Filtration	Spread plate
1.	87	No growth
2.	153	No growth
3.	408	No growth
4.	44	No growth
5.	12	No growth
Average Bioburden	140.8 ± 141.7	-

The volume of aliquot taken for spread plate method was the limitation, as the very little volume used could not capture any growth/ count in tissue of low microbial population. From these experiments and supported by several other repeated studies, we decided to adopt the filtration method in our procedure of microbiological quality control on tissue products.

Validation of tissue processing

Microbiological quality control was conducted to validate the processing procedures of our tissues. In Table 3, bioburden of the washing solutions from every step of amnion processing was estimated. It shows that the microorganism count was reduced gradually with a series of washings. Six-step washing comprised of Wash 1 (to remove dirt and blood), Wash 2 with distilled water, shake with 0.05% Hypochlorite solution, wash with Saline 1, Saline 2 and Saline 3 was sufficient, as bioburden of amnion at the end of the processing stage was low, around 5 cfu/item. We managed to reduce the washing steps from nine to only six steps, therefore we could save processing time and labour. Amnion with gauze (AM23) gave high count and we decided to change to better quality gauze and gamma sterilised it before use.

Table 3. Microbial levels of washing solutions from several stages of the processing of amnion.

Processing Stage	Microbial levels (cfu/total washing solution)						
	A5 (FD)	A6 (AD)	A7 (AD)	A9 (AD)	A10 (AD)	A23 (AD)	A25 (FD)
Wash 1	>1000	>1000	>1000	>1000	>1000	1336	695
Wash 2 with distilled water	>1000	670	>1000	>1000	>1000	112	184
0.05% Na Hypochlorite	530	430	60	3335	3415	14.5	102
Saline 1	130	360	20	2975	2145	4.8	34
Saline 2	110	80	10	2620	1445	14.5	29
Saline 3	80	60	10	1795	675	0	4.8
Bioburden (3 replicates)	5.3	5.3	5.0	5.9	NA	68.9 (gauze)	NA

AD – Air-dried amnion
FD – Freeze-dried amnion

Earlier, we conducted the same validation for bone processing as published in Yusof *et al.* (1994) [5]. The findings showed that the pasteurisation treatment meant for inactivation of HIV could inactive or reduce significantly the bioburden of bones. Unfortunately our laboratory could not conduct inactivation test on HIV.

Monitoring of bioburden

1. Amnion

Bioburden of amnions processed by the MINT Tissue Bank and National Tissue Bank HUSM for routine clinical applications over several years were compiled as in Table 4.

Bioburden in Table 4 would be able to show that the processing was improved over years due to skill acquired by tissue bank operators. Undeniably the premises too were improved over years and processing procedures were validated to be more efficient. Since 1992, the average bioburden of amnion were maintained below 100 cfu/item. Variation in bioburden throughout a year may due to variation in product sizes ranging from 80 to 200 cm^2. The most commonly found contaminants on the amnions isolated over several years were identified as *Bacillus sp.*, *Micrococcus sp.*, *Corynebacterium sp.* and *Rhodotorula sp.* [8].

2. Bone

Bioburden of freeze-dried bone processed by the MINT Tissue Bank and National Tissue Bank HUSM over several years were compiled as in Table 5. Unlike amnion production, only a small number of bones were processed per year therefore only a few samples could be taken for bioburden estimation.

The bioburden level was maintained low mainly around 10 cfu/item. The high count in 1993 suggests the procedure was not yet in place and operators were being trained.

Table 4. Bioburden estimation of amnion products using filtration method (average bioburden of samples taken randomly from at least 3 processing batches per year).

Year	Bioburden (cfu/product unit or item)	
	MINT Tissue Bank	National Tissue Bank, HUSM
1989	32.8 ± 24.7	NA
1990	140.8 ± 141.7	74.0 ± 34.3
1991	R&D	133.0 ± 120.9
1992	R&D	81.4 ± 127.8
1993	6.8 ± 2.7	9.3 ± 12.9
1994	21.7 ± 34.7	6.5 ± 7.1
1995	2.9 ± 3.8	7.7 ± 1.2
1996	2.7 ± 1.6	6.9 ± 4.2
1997	11.6 ± 18.6	NA
1998	NP	NA
1999	NP	NA
2002	NP	14.5 ± 12.7
2003	NP	15.6 ±7.5

R&D – Processed amnion mainly used for R&D, therefore no products to test.
NP – Amnion was no longer processed for routine production.
NA - No data available.

Table 5. Bioburden estimation of bones using filtration method (average bioburden of samples taken randomly from processing batches per year).

Year	Bioburden (cfu/product unit or item)	
	MINT Tissue Bank	National Tissue Bank, HUSM
1993	88.4 ± 110	NA
1994	13.6 ± 26.6	Human 7.0 ± 5.0 Bovine 9.8 ± 9.1
1995	21.4 ± 22.0	Human 1.5 ± 1.5 Bovine 6.7 ± 8.9
1996	11.5 ± 11.6	Human 10.8 ± 4.0
1997	4.0± 1.9	NA
1998	NP	NA
1999	NP	Bovine 16.0±10.0

NP – Bone was not processed for routine production.
NA - No samples were taken for bioburden estimation.

Radiation dose validation exercise

Amnion samples used in the radiation dose validation exercise were of several sizes. Samples were taken from a single mother of a processing batch. However, the number of samples was limited for bigger sample size of 10 cm x 10 cm, therefore samples were taken from multiple mothers in a processing batch. Table 6 describes the sampling. Samples of 10 cm x 10 cm were mainly used for treating burn wounds while smaller sizes were for ophthalmology and other skin disorders.

Table 6. Sampling of air-dried amnion pieces for dose validation exercise.

Size	No. of pieces produced from one amniotic membrane	Sampling for 20 pieces [10 pieces for bioburden and 10 pieces for dose validation exercise]
10 cm x 10 cm	5-6 pieces	From several mothers of a processing batch
5 cm x 5 cm	20-25 pieces	From single mother of a processing batch
3 cm x 3 cm	30-45 pieces	From single mother of a processing batch
2.5 cm x 2.5 cm	60-90 pieces	From single mother of a processing batch

Bioburden levels of amnion, as shown in Table 7, were reduced with size. The smaller sizes with very low bioburden may give rise to bigger error.

Table 7. Bioburden levels (cfu/item) of different sizes of amnion using filtration method.

Sample no.	Bioburden (cfu/item)			
	10 cm x 10 cm	5 cm x 5 cm	3 cm x 3 cm	2.5 cm x 2.5 cm
1.	23	0	0	1
2.	11	0	0	0
3.	2	0	0	0
4.	14	3	1	5
5.	13	1	0	1
6.	35	1	2	0
7.	9	1	1	0
8.	3	0	0	0
9.	0	0	0	1
10.	35	0	0	2
Average bioburden	14.5 ± 12.7	0.6 ± 0.97	0.4 ± 0.69	1.0 ± 1.6
Average per 100 cm^2	14.5	2.4	4.4	16.0

Dose validation exercise for 10 cm x 10 cm pieces is summarised in Table 8. Both exercises were invalid as the sterility test yielded more than 1 positive. The delivered dose in Sample A was initially meant for verification dose of Method (a) but it did not fulfil either Method (a) or Method (b) and lower than Method (c). In sample B, the delivered dose was lower than any of the required verification doses (±10%).

Table 8. Radiation dose validation exercise for 10 cm x 10 cm air-dried amnion.

Sample	Bioburden	Verification dose kGy ±10%			Delivered dose	Sterility test	Remarks
		Method (a)	Method (b)	Method (c)			
A	14.5	3.3 (bioburden 15)	3.2	7.5	Min 3.75 Max 4.36	3 + Not valid	Repeat, Consider Method (c)
B	15.6	3.4 (bioburden 16)	3.3	7.6	Min 2.5 Max 2.8	2 + Not valid	Repeat, Consider Method (c)

We improved our microbiology laboratory and improved the skill of conducting sterility test before we proceeded for the next exercise.

Our experience in dealing with small size amnions is summarized in Table 9. In all cases, the sterility test was accepted (< 1 positive) however the dose delivered still did not fulfil any of the verification doses. Since the products were challenged at doses lower than the verification dose of Method (c) and since it passed the sterility, the exercise was valid and the dose 25 kGy was substantiated.

Dose validation exercise was repeated for 3 cm x 3 cm of air-dried amnion as summarized in Table 10. For product AM12, the sterility test was accepted (< 1 positive), the delivered dose was lower than the verification dose of Method (c) thus the dose 25 kGy was substantiated. For AM17, sample was not exposed to any radiation dose since the bioburden was very low and the exercise was valid as no positive growth was observed. Product AM22 was challenged at two radiation doses, it failed according to Method (b) but was validated at the verification dose of Method (c), thus the dose 25 kGy could still be substantiated.

There are a few suggestions to improve the exercise. The recovery factor for low bioburden has to be reviewed and apply the recovery factor to the average bioburden estimation. Any average bioburden lower than 0 should be rounded up to 1, and perhaps instead of using average bioburden to set the verification dose; we might consider using the highest bioburden out of the ten samples, taking into account the worst scenario. From these exercises, we realised that the skill in conducting sterility test in a well-maintained laboratory is essential.

Another challenge is to deliver the intended dose within the range of ± 10%, which requires a thorough dose mapping study of the radiation chamber of the gamma cell. Problems encountered in the dosimetry were due to many factors. Variation in doses at different points in the irradiation chamber needs further study even though the uniformity is reasonably good at Uniformity = 1.1.

The use of Amber Perspex and Radiochrome dosimeters for low dose measurement was not reliable as the dosimeters were found to be not stable under tropical ambient conditions of high temperature and high humidity, which led to inaccurate dose rate determination. For future work we will use a ceric/cerous dosimeter for low doses to establish a dose rate calibration curve and conduct a dose mapping together with personnel from the Standard Dosimetry Laboratory before any validation exercises could be carried out.

Table 9. Radiation dose validation for different sizes of air-dried amnion.

Sample	Bioburden	Verification dose kGy ± 10%			Delivered dose	Sterility test	Remarks
		Method (a)	Method (b)	Method (c)			
AM24 5 cm x 5cm	0.6	1.0 (Table 2a - bioburden 0.65)	0.88	3.36	Min 1.7 Max 1.8	0 +	Valid based on Method (c) 25 kGy
AM19 2.5 cm x 2.5cm	1.0	1.3 (Table 2a - bioburden 1.0)	1.25	4.2	Min 2.1 Max 2.2	0 +	Valid based on Method (c) 25 kGy
AM19 5 cm x 5cm	4.0	2.2 (Table 2a - bioburden 4.0)	2.24	6.06	Min 2.7 Max 2.9	0 +	Valid based on Method (c) 25 kGy

Table 10. Radiation dose validation exercise for 3 cm x 3 cm of air-dried amnion.

Sample	Bioburden	Verification dose kGy±10%			Delivered dose	Sterility test	Remarks
		Method (a)	Method (b)	Method (c)			
AM12	0.4	Not available Table 2a	0.59	2.7	Min 2.15 Max 2.46	1 +	Valid based on Method (c), 25 kGy
AM17	0.1	-do-	0	0	No radiation	0 +	Valid based on Method (c), 25 kGy
AM22	0.2	-do-	0.096	1.42	Min 1.4 Max 1.5	1 +	Valid based on Method (c), 25 kGy
AM22	0.2	-do-	0.096	1.42	Min 0.87 Max 0.97	2 +	Not valid for Method (b)

CONCLUSIONS

In our work, we have used microbiological analysis in many aspects. Microbial levels in different samples throughout the processing could assist us in validating the procedures. Bioburden estimation for finished product is the best tool to qualify our tissue products for proper handling and processing. Therefore, bioburden estimation can act as a quality control check before the products can be released for sterilisation.

Its role in determining the radiation dose can ensure the product is sterilised at a high sterility assurance level. The radiation validation exercise can only be done after the bioburden of routine production is consistent and preferably low after efficient and validated processing procedure. We suggest that the IAEA Code of Practice is to be simplified for easy understanding and application by tissue bankers.

ACKNOWLEDGEMENTS

The authors acknowledge financial support from the Malaysian Government IRPA grant (No. 2-06-05-012-03), Technical Assistance under the IAEA/TC Programme and laboratory technical support by Ms Salahbiah Abd Majid.

REFERENCES

1. ISO, Sterilization of medical devices – microbiological methods – part 1: estimation of population of microorganisms on products, *ISO 11737-1*, **1995**, Geneva.
2. ISO, Sterilization of health care products – requirement for validation and routine control – radiation sterilization, *ISO 11137*, **1995**, Geneva.
3. IAEA, *Code of Practice for the Radiation Sterilization of Tissue Allografts: Requirements for Validation and Routine Control. (INT/6/052)*, **2002**, Vienna.
4. H. Mohamad and N. Yusof, Tissue banking in Malaysia – amniotic membrane, *Malaysian J. Nucl. Sc.*, **1991**, 9, 127-131.
5. N. Yusof, M. A. Noor Azlan, S. N. Selamat and C. M. Lee, Radiation sterilised freeze dried bone allograft – process validation, In: *Proc. 7th Internal. Conf. on Biomedical Engineering,* National University Hospital, Singapore, **1994**, pp. 303-305.
6. ISO, Sterilization of medical devices – microbiological methods – part 2, *ISO 11737-2*, **1998**, Geneva.
7. ISO, Sterilization of health care products – radiation sterilisation – substantiation of 25 kGy as a sterilisation dose for small or infrequent product batches, *ISO/TR 13409*, **1996**, Geneva.
8. N. Yusof, The use of gamma irradiation for sterilization of bones and amnions, *Malaysian J. Nucl. Sc.*, **1994**, 12, 243-251.

PROTECTIVE EFFECTS ON MICROORGANISMS IN RADIATION STERILISED TISSUES

Jolyon H. Hendry

Applied Radiobiology and Radiotherapy Section, Division of Human Health, International Atomic Energy Agency, Wagramer Strasse 5, 1400 Vienna, Austria {E-mail: J.Hendry@iaea.org}

ABSTRACT

Sterilisation of tissues means the inactivation of harmful bacteria and possibly viruses in the tissue. The tissue consists of many cell populations that already are reproductively inert, as well as structural extracellular matrices. Inactivation of these organisms is caused by ionisations that break the DNA molecule irreparably.

About 40% of the DNA damage is caused by direct ionisation of the DNA, and the remaining 60% is caused by indirect damage from the ionised products of water molecules surrounding the DNA and of free molecular oxygen. This means that desiccated or anoxic tissues are more radioresistant, and anoxic tissues require about 3 times as much radiation dose for equivalent injury as in oxygenated tissues. Frozen tissues are also more radioresistant, because there is less indirect effect caused by immobilisation of the active radicals. Radical scavengers can also be used to reduce the indirect contribution to DNA damage.

Organisms differ in inherent radiosensitivity for two main reasons, first because of target size, and second because of repair capability. It is interesting that the smaller organisms have both a smaller target size and often a high repair capability, so these factors work in the same direction to make the spectrum of differential radiosensitivity among organisms fairly large. Some bacteria are very resistant because of high repair capacity or some other feature of their genetic organisation, and viruses have an even smaller target size.

The resistance may be increased further by irradiation in hypoxic conditions, using various radical scavengers, at very low temperatures, or using low dose-rate irradiation. The magnitude of these dose modifications can be quite large. Hence the conditions for radiation sterilisation of tissues need to be specified and well controlled. Increased killing of microorganisms relative to injury to the host tissues might be achievable using radical scavengers that act on the indirect mode of radiation action, or alternatively by using neutron irradiation in the case of bone where the energy absorption can be substantially lower than in soft tissues.

INTRODUCTION

The need to sterilise allografts has become an increasingly important issue following a few isolated cases of transmission of disease in such transplants. Ionising radiation is a very effective method of tissue sterilisation, and this has been practised successfully for over 40 years in some countries. The conventional sterilisation dose is 25 kGy (in some cases up to 35-40 kGy) delivered at high dose rate from cobalt-60 sources or a linear accelerator. This dose level originates from the same dose used for many years to sterilise medical devices, and it is sufficient to reduce the levels of normosensitive bacteria by a million fold, which is the conventionally accepted depletion level required. Problems can arise if the dose delivery pattern is changed, for example if the dose rate is low so that more contaminant cells repair and survive, and also if the irradiation conditions change, for example if the contaminant cells are hypoxic and hence more resistant.

In addition, certain strains of bacteria are very radioresistant, requiring much more dose to sterilise them, and viruses and prions are even more radioresistant. These problems cannot be overcome simply by markedly increasing the radiation dose, because doses above 50 kGy can cause detrimental changes to the tissues. The relevant features of microorganism radiosensitivity will be summarised with respect to tissue sterilisation conditions.

Radiosensitivity

There is a wide range of radiosensitivities among organisms. In general, organisms with a larger critical target size are more radiosensitive, because the probability of inactivation increases with a multiplicity of potential inactivation sites. Also, smaller cellular targets often have better repair or restitution capabilities. The target for replicating cells is the DNA, and double-strand breaks (dsb) are potentially lethal lesions. One unrepaired dsb in repair-deficient yeast is a lethal event. At the other extreme there are very radioresistant bacteria, and the archetypal example is *Deinococcus radiodurans*. In this case the radioresistance is currently considered to be due to the unusual tight packing of its genome, so that the radiation-generated free DNA ends are held tightly together to facilitate template-independent yet error-free joining of the DNA breaks [1].

A summary of the sterilising doses for different microorganisms is given in Table 1. Most survival curves for inactivation of cells and microorganisms are exponential in shape on a semilog plot of survival versus dose (Figure 1). Some have curvature in the lower range of doses before the exponential portion begins at higher doses, and this is due to repair, which declines in efficacy at higher doses. The exponential form means that additional equal decrements in viability are produced by equal increases in dose. The shape of the curves (the radiosensitivity) can be modified by changes in a variety of parameters, namely water content, radical scavengers, oxygen levels, growth state, temperature, dose-rate, and type of radiation used.

Water content and radical scavengers

DNA damage is caused by both ionisations in the DNA helix itself (direct action), as well as indirect damage from ionisations in the surrounding water causing hydroxyl and other radicals to attack the DNA. Approximately 40% of DNA damage is cause by direct action and 60% by indirect action, and the latter percentage is higher for injury to membranes. This indicates that lyophilised material containing bacteria will be more radioresistant to sterilisation than hydrated material.

Exogenous radical scavengers may be used to reduce the contribution from indirect action to radiosensitivity. This latter aspect is under detailed investigation at present in order to try to identify scavenger protocols that will protect proteins, membranes and structural tissue elements, but not the DNA by as much. This would be a way of increasing the efficacy of radiation in sterilising microorganisms, in comparison to any structural damage to the irradiated tissue.

Table 1. Summary of the sterilising doses for different microorganisms [2].

Group	Organisms	Sterilising Dose (kGy)*
Sensitive	Vegetative bacterial (excl. some micrococci & streptococci) Animal viruses > 75 μm	0.5-10
Moderately resistant	Moulds and yeasts *Streptococcus faecium* (suspended in buffer) Animal viruses 20-75 μm	4-30 10-30
Resistant	Bacterial spores *Bacillus pumulus* *Clostridium botulinum* (some strains) Some viruses *Streptococcus faecium* (dried from serum broth) Animal viruses (except foot & mouth disease virus) < 20 μm	10-50 10-30 30 10-30 10-45 30-40
Highly resistant	Moraxella (some strains) *Micrococcus radiodurans* Bacillus spores (contrived mutants) Foot & mouth disease virus Bacterial viruses	~ 50 55-70 35-80 ~ 50 wide range

* Based on inactivation factor of 10^8 ($8D_{10}$).

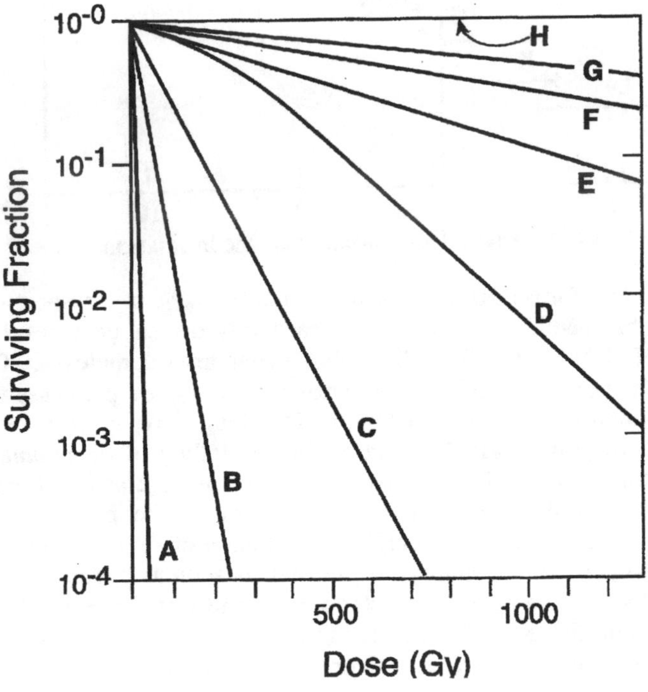

Figure 1. Cellular radio-sensitivities [4].

A – mammalian cells
B – *E. coli*
C – *E. coli* B/r
D – Yeast
E – *Phage Staph E*
F – *B. megaterium*
G – Potato virus
H – *Micrococcus radiodurans.*

Oxygen levels

Oxygen is used in the initial radiochemical reactions, and for cellular inactivation about 2-3 times as much radiation dose is needed in anoxic compared to in oxic conditions. This increase (the oxygen enhancement ratio, OER) is dependent on the particular endpoint. Although there are many exceptions, the general pattern is for the OER to be high (up to 10) for specific membrane damage such as enzyme inhibition or release, and lower (down to or even below 1) for specific nuclear damage in DNA or RNA (Figure 2). This formed the basis of the Type O (oxygen modifiable membrane damage) and Type N (not modifiable nuclear damage) hypothesis [3]. This indicates again the possibility of greater modification of non-nuclear targets.

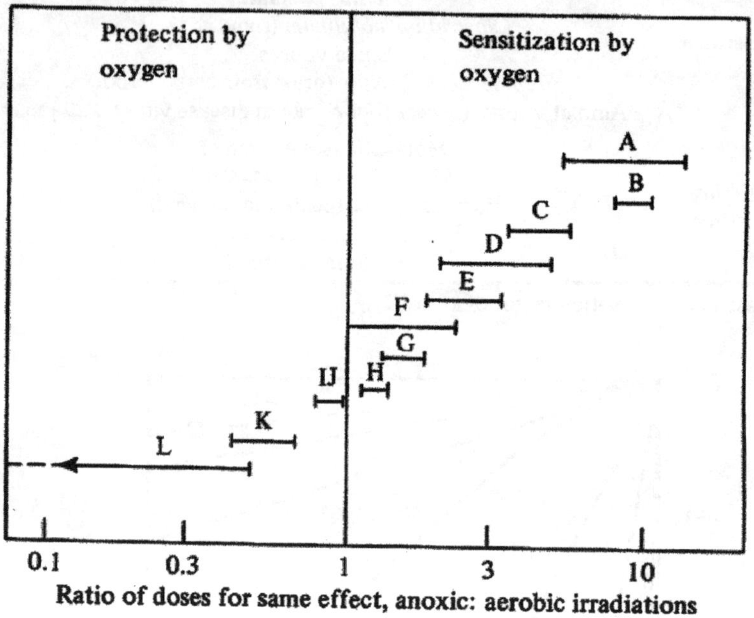

Figure 2. Ranges of factors for sensitisation or protection by oxygen in respect of biological or biochemical functions: A, release of lysosomal enzymes; B, inhibition of DNA synthesis by DNA-membrane complexes; C, inhibition of the induction of bacterial enzymes requiring a permease to transport the inducer across the membrane; D, killing of repair proficient bacteria, also of polymerase-deficient strains; E, killing of mammalian cells; F, killing of repair-deficient bacteria, yeast, an alga and a slime-mould; G, bacteriophage irradiated in suspension in the presence of radical-scavenging material and a high concentration of –SH compound; H, bacteriophage irradiated within the host cell as soon as possible after injection of the nucleic acid; I,J, bacteriophage and transforming DNA irradiated in broth; K, loss of biological activities of RNA; L, bacteriophage irradiated in buffer, with or without various solutes [3].

Growth state

The growth state of bacteria also influences radiosensitivity. There are marked fluctuations through the growth cycle, by more than two-fold in dose required for the same depletion level, e.g. Figure 3.

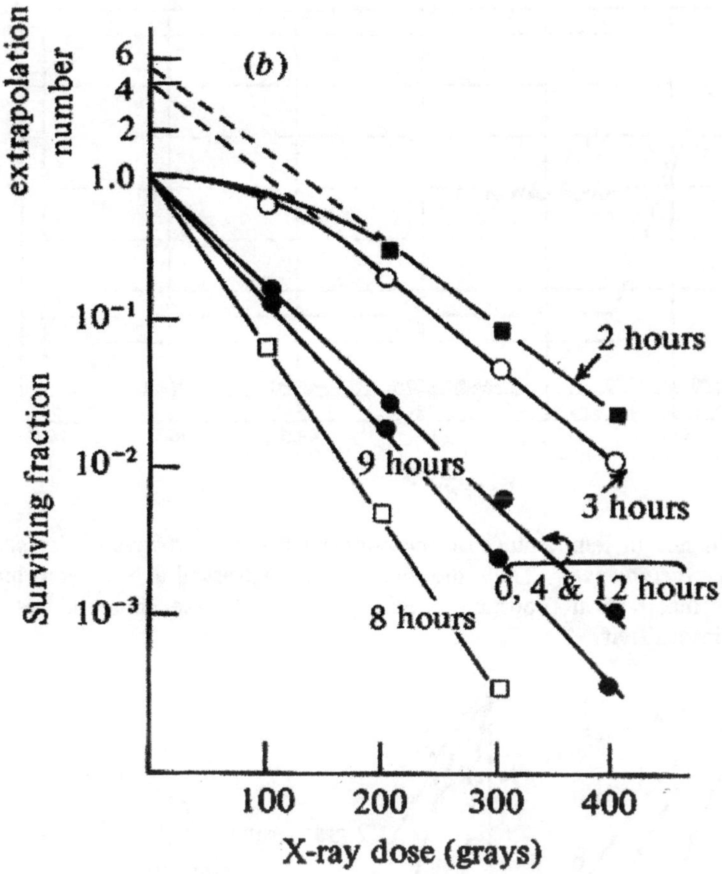

Figure 3. E. coli B/r, various hours after seeding – stationary to growth to stationary [3,5].

Temperature

At lower temperatures approaching 0°C repair is inhibited, but any effects on cellular radiosensitivity can be counteracted by prevention of growth and cell cycle related changes in radiosensitivity. At even lower temperatures there is increasing radioresistance caused by immobilisation of the radicals participating in the indirect action of radiation damage (Figure 4). Radioresistance almost plateaus at -120°C.

Dose rate

Lower dose rates are less efficient in killing microorganisms at ambient temperature and oxygenation because of repair during the irradiation. This can have a marked effect on radiosensitivity. There can be as much as a 4-fold increase in resistance when increasing the dose-rate from 1 to 22 Gy per minute (Figure 5), and even a 20% increase when going from 140 to 650 Gy per minute (Figure 6). At much higher dose rates achievable using linear accelerators there may be further resistance induced in tissues by radiochemical depletion of oxygen. In slightly hypoxic tissues this can occur when the instantaneous dose rate in the pulses is above 100 kGy per minute, even though the average dose rate may be much lower [6].

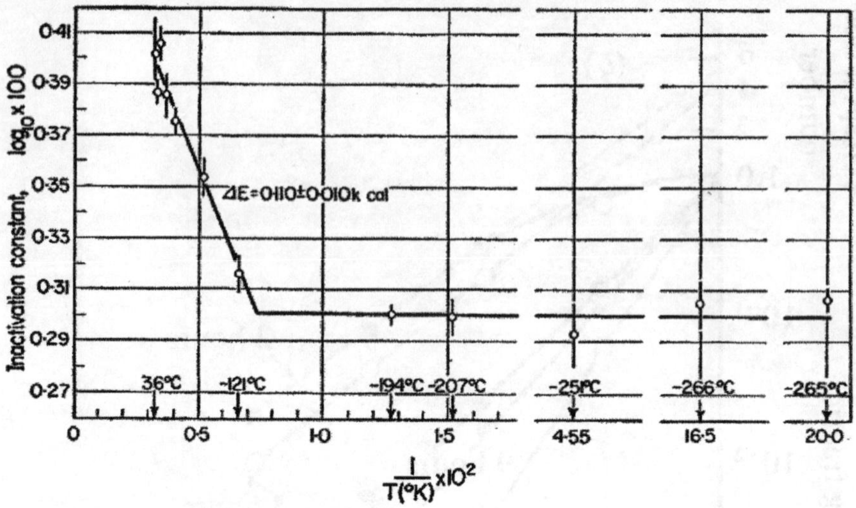

Figure 4. Influence of temperature on the radiosensitivity to X-rays of dry spores of *B. megaterium.* The radiosensitivity is expressed as the logarithm of the inactivation constant, which is an inverse function of the radiosensitivity [7].

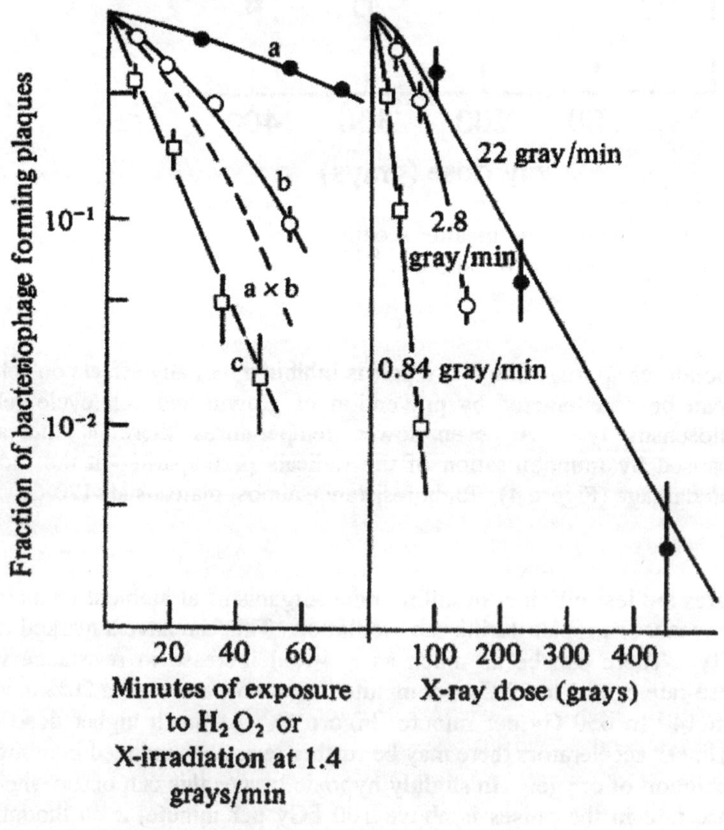

Figure 5. Phage S13 dose-rate effect [3].

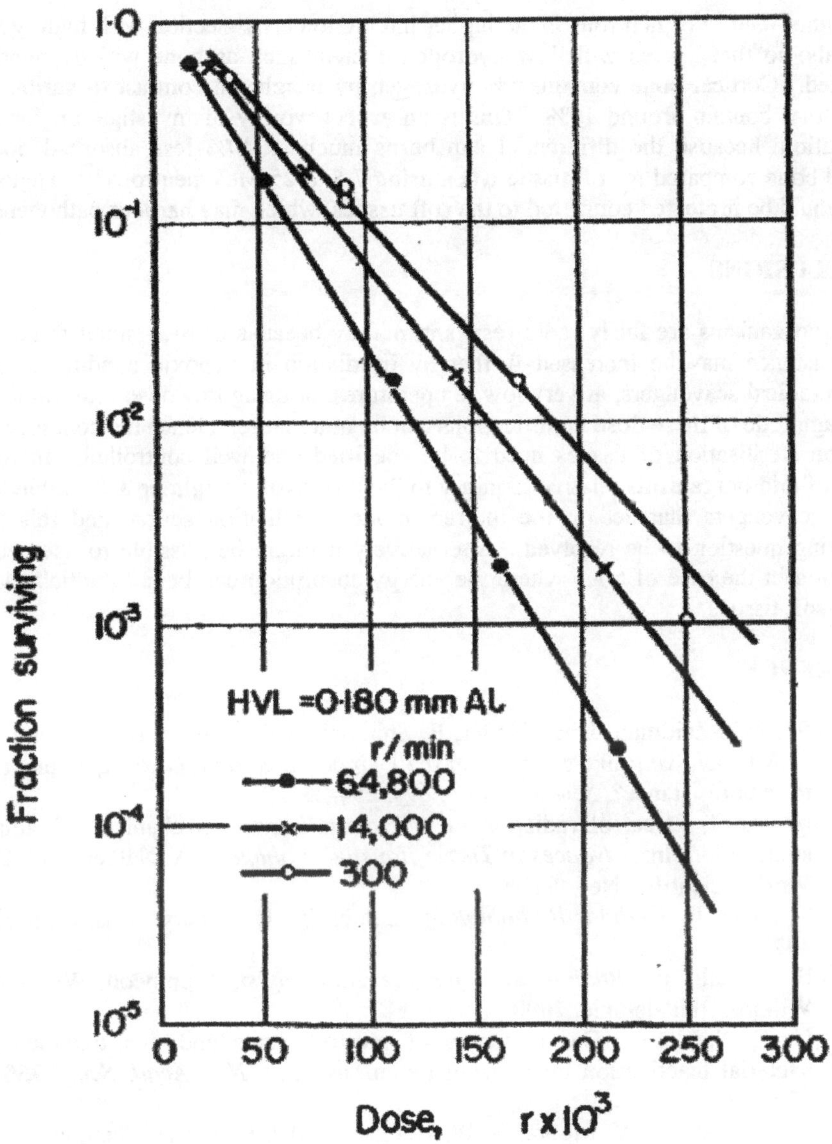

Figure 6. Effect of dose-rate [3]; closed circles, 650 Gy/minute; crosses, 140 Gy/minute; open circles, 3 Gy/minute [7].

Radiation quality

Densely ionising radiation (high linear energy transfer, LET) is more efficient in killing cells, by a factor of 3 or more in terms of lower sterilisation doses in the case of neutrons or hadrons. There is less indirect action, and hence the effects of changes in oxygen tension and temperature are less. There is also less repair, because of increasing complexity of the primary lesions as the density of the ionisations increases. Hence procedures designed to achieve differential modifications to radio sensitivity through the indirect action pathway will be less applicable in the case of high LET radiations.

Another feature of neutrons is the higher interaction cross-section with hydrogenous materials, so that tissues with low hydrogen content such as bone will be relatively protected. Cortical bone contains 5% hydrogen by weight, in contrast to various soft tissues that contain around 10%. This is an aspect worthy of investigation for bone sterilisation, because the differential can be as much as 40% less absorbed dose in cortical bone compared to soft tissue when using 7.5 to 25 MeV neutrons [8]. Hence the bone would be protected compared to the soft tissues, which may harbour pathogens.

CONCLUSIONS

Microorganisms are fairly radio resistant, mainly because of their small target size. The resistance may be increased further by irradiation in hypoxic conditions, using various radical scavengers, at very low temperatures, or using low dose-rate irradiation. The magnitude of these dose modifications can be quite large. Hence the conditions for radiation sterilisation of tissues need to be specified and well controlled. Increased killing of microorganisms relative to injury to the host tissues might be achievable using radical scavengers that act on the indirect mode of radiation action, and this is an interesting question to be resolved. Alternatively it might be possible to use neutron irradiation in the case of bone where the energy absorption can be substantially lower than in soft tissues

REFERENCES

[1] S. Levin-Zaidman, J. Englander, E. Shimoni, A.K. Sharma, K.W. Minton and A. Minsky, Ring-like structure of the *Deinococcus radiodurans* genome: a key to radioresistance?, *Science*, **2003**, 299, 254-256.

[2] N. Yusof, Effect of radiation on microorganisms – mechanism of radiation sterilisation, In: *Advances in Tissue Banking, Volume 5,* A. Nather (ed.), **2001**, World Scientific, New Jersey.

[3] T. Alper, In: *Cellular Radiobiology*, Cambridge University Press, Cambridge, **1979**.

[4] E. J. Hall, In: *Radiobiology for the Radiologist*, Lippincott Williams & Wilkins, Philadelphia, **2000**.

[5] G. E. Stapleton, The influence of pre-treatments and post-treatments on bacterial inactivation by ionising radiations, *Ann. N.Y. Acad. Sci.*, **1955**, 59, 604-618.

[6] J. H. Hendry, J. V. Moore, B. W. Hodgson and J. P. Keene, The constant low oxygen concentration in all the target cells for mouse tail radionecrosis, *Radiat. Res.*, **1982**, 92, 172-181.

[7] E. L. Powers, R. B. Webb and C. F. Ehret, Progress in nuclear energy, In: *Biological* Sciences, Volume 2, **1959**, Pergamon Press, London, p.178.

[8] M. Catterall and D. K. Bewley, In: *Fast Neutrons in the Treatment of Cancer*, Academic Press, London, **1979**.

INDEX

accelerators 81
alanine dosimeters 88
alloderm 118
allogenic tissue 235-254
allograft infections 105
allograft tissue 117-121, 303-309, 311-318
allograft, arterial 123
allografts 39-64, 65-73, 141-149, 176, 211, 248
allografts, musculoskeletal 133-139
allografts, structural 15-162
amnion anatomy 198
amnion grafts 197-220
amnion samples 320, 324
anitibiotic soaking 283
aorta, rabbit 123-129
arterial allograft 123
backscatter factor 97
bacteria, growth rate of 334
bacterial inactivation 281-286
bacterial reduction 283
basement membrane 204
bioburden 281-286, 305, 311-318, 319-329
bioburden levels 303-309
bioburden reduction 117,118
bioburden, determination 50
biological factors 102
biomechanical testing 289, 291
blood donors 168
blood testing of donors 267
blood tests 18
bone 180
bone allografts 65-73, 260
bone grafts 183
bone matrix 151-156
bone samples 320, 324
bone tissue transplants 235-254
bone tumours 157-162
bone, living 269
bone, radiation on 133-139, 141-149, 151-156
calorimeters 85
cardiovascular allografts 263
cartilage 163-172
cartilage allografts 263
cells 197-220
chemical dosimeters 85
chorion 199
clinical use of grafts 212
cobalt-60 gamma ray source output 89-90
code of practice 3, 39-64, 65-76
cold gamma radiation 151-156
cold gamma sterilisation 117-121
collagen 184
collagen bundles 118,120,137
collagen solubility 184, 188
colony counts 314
contamination of tissues 311-318
corneal allograft 256
cross-contamination 311-318
cryopreservation 123 -129
cytotoxicity 191
deep freezing 183
demineralised bone 151-156
dendritic cells 207
dermis 205
DNA damage 332
DNA fragments 125
donor selection 16, 265,175
donor tissue, irradiation of 105-108
donors 66,69
dose delivery 79-104
dose homogeneity 95
dose mapping 53
dose rate 335
dose validation 319 - 329
dosimetry 79 - 104
dosimetry for radiation processing 106 - 108
dosimetry, at low temperatures 87

electron beam irradiation 221-232
electron paramagnetic resonance spectrometry 179
endothelium 125-129
epidermis 207
ethical rules 9
fibroblast 206
fungal isolates 316
gamma irradiation 221-232, 235-254, 287-302
gamma irradiation, of donor tissue 105-108
gamma radiation, on bone 133-139, 141-149, 151-156
gamma sterilisation 117-121
grafts, integrity and functionality 197-220
Gram negative isolates 316
Gram positive isolates 315
HCV 238
health care products 39
heart valves 163-172
hemodilution 268
hepatitis B virus 237, 259
hepatitis C virus 238
human cortical allograft 141-149
human dermis 118,120
human hepatitis B virus 237
human hepatitis C virus 238
human immunodeficicency virus 236, 260
human T-lymphotrophic virus 262
Hyaluronan 221-232
IAEA code of practice 65-76
IAEA international standards 3-37
interaction coefficiency 82
international standards, for tissue banks 3 - 37
ionisation chambers 86
irradiated allografts 157 –162
irradiation of collagen 190
irradiation of donor tissue 105-108
irradiation process 73
legal and regulatory control 8
legal rules 9, 31
life cell 117-121
lyophilisation 183, 283
lysosomal enzymes 334
macrophages 206
mast cells 206
mechanical properties, of bone 141-149
microbial identification 306, 313
microbial standard distribution of resistance (SDR) 46-50, 79,80
microbiological testing 53
micro-organisms 331-338
micro-organisms survival curve 80
microvesicles 126
molecular weight 221-232
mono-tissue banks 176
MTT reduction test 191
multi-tissue banks 174
musculoskeletal tissue 133, 143
organ issue donation 256
osteochondral allografts 263
osteoinduction 154
oxygen levels 334
pathogen inactivation 287-302
pepsin digestion 189
preservation 173-195
procedure for preservation 173-195
processing of tissues 319-329
processing procedures 6, 23
protease sensitivity assay 290, 294
quality assurance 13
quality control policies 13
rabbit aorta 123-129
rabies 259
radiation dectors 85
radiation dose 319- 329
radiation of soft tissue 163- 172
radiation quality 337
radiation sources 79-104
radiation sterilisation 173-195, 214, 221-232, 234-254
radiation sterilised tissues 331-338
radiation sterilisation, of bone

radiation sterilisation, of bone allografts 65-73
radiation, effects on grafts 197-220
radical scavengers 332
radioactivity 84
radiographic dosimeters 87
radiometry 82
radionuclides 81
radiosensitivity 332
skin allografts 264
skin anatomy 200
skin grafts 197-220, 264
skin-function 207
soft tissues 163-172
solid-state dosimeters 87
sterilisation 173-195, 281 286, 303-309
sterilisation dose 45, 52, 303-309, 319-329
sterilisation of bone 133-139
sterilisation, cold gamma 117-121
sterilisation, irridiation 151
sterilisation, of tissue allografts 40
sterilisation, tissues for 79-104
sterilised tissues 331-338
sterility assurance 281-286
sterility assurance level 303-309
structural allografts 157-162
structure of hyaluronan 221-232
tendon allografts 263
tendons 163-172
tensile stress of tendons 292
terminal sterilsation 283, 303-309
tissue allografts 39-64, 66-76, 117-121, 287-302
tissue bank facilities 69
tissue banking 173
tissue banks, standards for 3-37
tissue donors 42
tissue processing 270, 282, 323
tissue quality 319-329
tissue retrieval 6, 19-24
tissue selection 282
tissue, soft 163-172
tissues, for sterilisation 79-104
tissues, sterilised 331-338
T-lymphocytes 207
transmission electron microscopy 290, 293
transmission of viruses 245
transplantation 79
transplants 235-254, 255-278
TUNEL method 123-129
validation of tissue processing 323
vascularisation 212
viral diseases 263
viral reduction 285
virus detection 240
virus inactivation 243
viruses 235-254, 255-278
xenografts 177, 212
X-ray tube 89-91